A CLINICAL COMPANION

to accompany

Biochemistry, Fifth Edition

Kirstie Saltsman

Jeremy Berg

Johns Hopkins University School of Medicine

Gordon Tomaselli

Johns Hopkins University School of Medicine

W. H. Freeman and Company
New York

ISBN: 0-7167-4738-3

Printed in the United States of America

First printing 2002

CONTENTS

PART IV: RESPONDING TO ENVIRONMENTAL CHANGES

Advances in biochemistry dramatically impact teaching for the health care professions. In a text such as *Biochemistry* 5e, students in premedical programs, in medical school, and in related pre-professional programs are confronted with the rich and complex fabric of biochemistry presented from a general perspective. To highlight aspects of the subject with particular relevance to medicine, the authors of *Biochemistry* 5e, have identified some sections of the text with a specific clinical icon. However, space limitations restricted the number and length of these sections. In this Clinical Companion, we have the opportunity to amplify this clinical perspective, more thoroughly revealing the deep interconnections between biochemistry and medicine. The approach parallels methods that we have used in teaching medical students at Johns Hopkins where basic scientists work with clinicians to present clinical cases and journal articles to enrich the core curriculum.

For each of the 34 chapters in *Biochemistry* 5e, a clinical case is presented, chosen to illustrate a particular concept, enzymatic reaction, or pathway discussed in the text. The cases are presented in a comprehensive manner, with much of the complexity present in real clinical situations. Interspersed in the case presentation are questions intended to focus the readers' attention on important points or to stimulate further investigation of the clinical practices used. A discussion follows that links the case to the principles discussed in *Biochemistry* 5e, and elaborates on the particular disease or treatment. Selected references and web sites as well as problems with solutions conclude each chapter. These cases truly complement the text; the cases make clear how the material in each chapter in *Biochemistry* 5e is relevant to medicine while the fundamental principles presented in the text make the cases more understandable. We have selected a wide range of topics covering metabolic disorders, infectious diseases, and genetic diseases with applications in internal medicine, pediatrics, oncology, cardiology, endocrinology, ophthalmology, immunology, radiology, and nutrition among other specialties.

A book such as this could never be written without help from our colleagues. We thank our editor Mark Santee for coordinating the project and for his constant support and encouragement. We thank Professor Jon Lorsch for advice and comments on the entire text and for assistance with a number of figures. We thank Gregory J. Gatto, Jr. for help with Chapter 4, Professor Wendie Berg for help with Chapter 6, and Dr. Angela Guarda for help with Chapter 30. Finally, we thank numerous other colleagues who provided clinical data and figures that are used throughout the text.

Prelude
A Newborn in the ICU:
A Candidate for Gene Therapy

CHAPTER 1

CASE HISTORY

A newborn boy was admitted to the intensive care unit because of premature birth and abdominal distension. The infant was the first born to a 30-year-old woman after a gestation of 31 weeks. The pregnancy was uneventful until the onset of lower abdominal pain the day prior to admission. An ultrasonogram on admission revealed no fetal abnormalities.

Both parents were of Danish descent. The mother's blood group was A, Rh-negative. Tests for antirubella antibodies, hepatitis B surface antigen, and syphilis were negative. There was no history of sexually transmitted disease. Two female maternal cousins were said to have cystic fibrosis. One male paternal cousin was known to have infertility due to azoospermia (absent or abnormal sperm).

Why were tests done for antirubella antibodies, hepatitis B surface antigen, and syphilis?

Is azoospermia associated with cystic fibrosis?

The Infant Is Delivered and Examined

On the morning of admission, contractions began to occur every five minutes, and the patient was given terbutaline (beta-adrenergic agonist) subcutaneously. The cervix was soft, closed, and slightly effaced (the neck of the cervix was shortened). Two hours later, it was 50 percent effaced and dilated to 2–3 cm, and the membranes ruptured with the passage of copious amounts of clear amniotic fluid. A normal spontaneous vaginal delivery occurred three hours later, and the Apgar scores were 6 at 1 minute, 8 at 5 minutes, and 9 at 10 minutes. The temperature was 38.4°C. The baby's head was normal size, and there were no nasal polyps noted. His respiratory rate was normal but scattered wheezes were audible on examination of the lungs. His heart rate was 160 bpm and the heart examination was unremarkable. The abdomen was protuberant, with decreased bowel sounds but without a fluid wave. The spleen was not palpable. The liver edge was palpated 2 cm below the right costal margin, and a three-vessel umbilical cord was present.

What are the Apgar scores and how are they arrived at? Are they within the normal range for this newborn?

What are nasal polyps?

What does the lack of fluid wave indicate?

Radiograms Indicate an Abnormality in the Lung and in the Intestine

The infant was taken to the neonatal intensive care unit. In the first day of life the infant developed bile-stained emesis and passed small amounts of inspissated (thick) meconium; however, the abdominal distension did not change. He had intermittent cough. Chest radiogram revealed evidence for right upper lobe atelectasis (small areas of collapse of the lung) but no evidence for aspiration. A radiogram of the abdomen revealed haziness in the right lower quadrant without air fluid levels, peritoneal calcification or free air under the diaphragm (the latter two are signs of meconium peritonitis). The hematocrit was 41%, the platelet count was 240,000 per cubic millimeter, and the white blood cell count was 3700 per cubic millimeter.

What is meconium?

What does bile in the emesis indicate about the location of the obstruction?

The Intestinal Obstruction Is Resolved by Enema

A presumptive diagnosis of simple meconium ileus (obstruction of the intestinal tract at the ileum) was made. A high Gastrografin (radiographic contrast) enema containing polysorbate and Tween 80 (a detergent) was administered. The enema demonstrated obstruction of the terminal ileum and was associated with passage of a small volume loose stool. The enema was repeated eight hours later with diarrhea and resolution of the intestinal obstruction. Subsequent stools were large in volume and steatorrheic (containing large amounts of fat).

What might substantial and steatorrheic stools indicate?

Diagnosis and Treatment

A pilocarpine stimulated sweat test was performed and the sweat chloride concentration was 48 mmol/l (a value greater than 40 mmol/l in an infant is highly suggestive of cystic fibrosis). DNA testing using a *cystic fibrosis (CF)* mutation panel revealed two copies of the DF508 mutation of the *cystic fibrosis transmembrane conductance regulator (CFTR)* gene in the infant and single copies of this mutation in both parents. The parents were referred for genetics counseling.

On the seventh hospital day the infant was discharged to his home to receive chest physiotherapy consisting of postural drainage and manual percussion. His pancreatic enzyme deficiency was treated with pancrelipase 3000U added to 120 ml of formula. On this regimen he was growing and developing normally over the first six months of life. He continued to experience cough and sinus congestion requiring intermittent administration of antibiotics.

Is the $\Delta F508$ mutation that causes cystic fibrosis a recessive or dominant mutation?

Do you think this infant will continue to develop normally and achieve a normal lifespan?

Do you think cystic fibrosis is a good candidate for gene therapy?

DISCUSSION

Gene Therapy

A discussion of gene therapy in this first chapter is apt as many believe it has the potential to revolutionize medicine. Gene therapy, which is the treatment or prevention of disease by gene transfer, aims to reverse the actual *cause* of disease. As will be seen in the following chapters, this is in contrast to most other therapies, which target various aspects of the pathophysiology of a disease. The first clinical gene therapy trial occurred in 1990 on a four-year-old girl with a severe, inherited immunodeficiency. Although she continues to do well today, it is unclear to what extent the improvement in her condition is attributable to gene therapy, as she had concurrently begun a new drug treatment.

Many subsequent gene therapy trials proved disappointing; however, some progress has recently been made, with the first reports of clinical benefit in humans from gene therapy. For example, the gene encoding *vascular endothelial growth factor (VEGF)*, an angiogenic factor, induced blood-vessel growth in those with severe limb ischemia or angina (ischemia of heart muscle), resulting in apparent clinical benefit.

The vector used for gene-transfer can be either viral or nonviral

The gene-delivery vehicle, or vector, used in gene therapy is either viral or nonviral, and each particular vector has its own set of advantages and disadvantages. It is unlikely that a single, generic vector will emerge for the treatment of multiple diseases, but rather that each vector will have to be chosen and engineered to fit its particular purpose. The earliest vectors were either *retroviral (Rv)* or *adenoviral (Ad)*, and although both types initially proved problematic in one respect or another, "second generation" versions are being used successfully today. *Adeno-associated virus (Aav)*, a nonpathogenic virus isolated from the respiratory and GI tracts of humans, is also among the most promising viral vectors currently used. Among the nonviral methods, the use of naked DNA has been surprisingly successful. Although transfection of naked DNA in vitro into tissue culture cells is ineffective, for reasons not yet clear, DNA appears to be taken up with reasonable efficiency when administered in vivo, directly into the patient. Complexing DNA with either cationic lipids or polymers seems to somewhat improve the efficiency of gene transfer into the target cells, and is also used in gene therapy.

Technical difficulties

The four major technical difficulties with gene therapy that have emerged in the decade since its inception are: (1) efficiency of transfection; (2) transience of expression; (3) immunogenicity (the transgene product or viral antigens produced by the vector can elicit an immune response in the patient); and (4) manufacture (viral vectors, especially, can be difficult to manufacture in quantities and qualities adequate for use in gene therapy). The severity of each of these problems depends upon the particular vector, and in fact, in some applications the "problem" does not actually interfere with the goal of treatment. For example, transient expression of a gene that induces cell death does not pose a problem when treating a malignant tumor. However, transient expression is unsatisfactory when treating chronic diseases such as hemophilia or cystic fibrosis. Once a suitable vector for the treatment of a given disease is identified, another major hurdle is determining the most effective way to deliver it to the target cells. Local administration, such as injection at the target site, is most common as it favors delivery to the target cells. Systemic administration, with the potential for transgene (the gene borne by the vector) expression in any cell of the body, is rarely used because of safety considerations.

Cystic fibrosis is a good candidate for gene therapy

Cystic fibrosis is a monogenic disease, caused by a mutation in a single gene, an important factor in its suitability for gene therapy. A variety of the available vectors have been employed, with use of Ad, Aav, and cationic liposome vectors yielding preliminary successes. Conventional therapies for treating cystic fibrosis simply target the symptoms, and do not significantly alter the course of the disease. Although progress has been made, with the median survival age now reaching nearly 40 years, it remains a debilitating and fatal disease. Thus, although cystic fibrosis is well understood at the genetic and physiological levels, it remains a serious disease with few good treatment options. This has made it an especially attractive target for gene therapy, and has facilitated approval from regulatory agencies for clinical trials.

Etiology

Cystic fibrosis is an autosomal recessive disease caused by mutation of the CFTR gene (Figure 1.1). The most common mutation, accounting for 70–80% of CF cases, is the ΔF508 (deletion of the phenylalanine at position 508 in the amino acid chain) mutation carried by this infant, however there are hundreds of other mutations in the gene that cause the disease as well. The CFTR gene product is found throughout the body in the epithelial cells that line the lungs, pancreas, genitourinary tract, sweat glands and colon. It is a chloride channel, but also modulates the transport of other electrolytes, as well as water, across epithelial cell membranes, and maintains normal mucus viscosity.

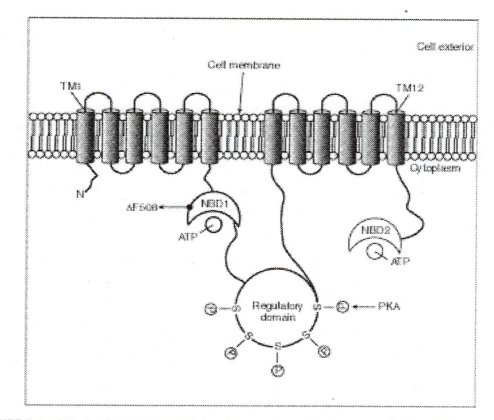

FIGURE 1.1 Schematic of CFTR structure. The structure consists of 12 transmembrane (TM1-12) domains, two nucleotide binding domains (NBD1 and NBD2), and a regulatory domain. The channel is regulated by phosphorylation of serine residues by protein kinase A (PKA). The ΔF508 mutation lies in NBD1. See Section 13.3 of *Biochemistry* 5e for a more detailed discussion of CFTR structures.

Interestingly, the ΔF508 mutation seems to have little effect on the function of CFTR, but instead prevents the protein from reaching the cell membrane. Therefore, patients with this mutation have dramatically reduced functional expression of CFTR due to a *protein-trafficking* defect. When ΔF508 CFTR is expressed in mammalian tissue culture cells, the trafficking defect can be corrected by lowering the temperature. Several other genetic diseases have been shown to be caused by such a protein-trafficking defect.

Epidemiology

Cystic fibrosis is the most common inherited disease amongst Caucasians. Approximately 1 in 20 Americans carries an abnormal CFTR gene, and 1 in 2500 American babies is born with the disease each year. Those of northern European extraction, like the infant discussed here, who is of Danish descent, are most commonly affected.

Pathophysiology

Lung disease is the most common cause of death in those with cystic fibrosis

Cystic fibrosis is characterized by the production of abnormally thick mucus in the intestinal and respiratory tracts, and in exocrine glands such as the pancreas. This can cause obstructions such as the intestinal obstruction that occurred in this infant (meconium ileus), and the respiratory tract obstruction that lead to atelectasis in the upper right lobe of his lung. Thick mucus can prevent alveoli from filling with air, with resultant contraction (or collapse) of the affected areas of the lung.

Still more troubling, the thick mucus results in repeated infections of the airways, leading to severe lung disease and respiratory failure, the most common cause of death among those with cystic fibrosis. The thick mucus causes impaired mucociliary clearance, the body's natural mechanism for clearing particulate matter and infectious organisms from the airways, resulting in susceptibility to lung infections. Infection with bacterial pathogens such as *Pseudomonas aeruginosa* is especially common, necessitating the frequent administration of antibiotics. Repeated infections eventually cause damage to lung tissue and severely compromise pulmonary function. Efforts to treat CF with gene therapy have focused on the lung because of the mortality associated with damage to this organ in those with the disease.

Meconium ileus in a newborn is a manifestation of the intestinal disturbances common in those with cystic fibrosis

Meconium ilius, obstruction of the ileum by unusually thick meconium, is one of the first signs of cystic fibrosis in newborns. The protuberant abdomen and decreased bowel sounds observed in this infant at birth, as well as the bile-stained emesis and inspissated meconium that he later discharged were consistent with an obstruction in the intestine. The haziness in the lower right quadrant of the abdomen (where the ileum is found) observed in the radiogram is typical of the condition, and is due to small air bubbles mixing with the meconium. The lack of air fluid levels is also typical and has also been attributed to the sticky, intralumenal meconium.

Intestinal disturbances are common in those with cystic fibrosis. The hyperviscous mucus in the intestinal tract hinders absorption of nutrients, which leads to malnutrition and slow growth. Obstruction of the pancreas, which normally delivers digestive enzymes to the duodenum, results in impaired digestion, especially of fats, and contributes to malnutrition. The substantial, steatorrheic stools observed in this infant are typical of those with the disease.

Diagnosis

The most common test for diagnosing cystic fibrosis is the pilocarpine-stimulated sweat test, in which the concentration of sodium chloride in the patient's sweat is measured. The sodium chloride concentration in sweat of those with cystic fibrosis is unusually high, with the concentration measured in this infant, 48 mmol/l, above the normal range and highly suggestive of CF. Meconium ileus is a very *sensitive* indicator of disease, with virtually all cases of the condition occurring in those with cystic fibrosis. However, meconium ileus is not a *specific* indicator of disease, as only 10–20% of infants with CF are born with the condition. In newborns that do not have meconium ilius, the first sign of disease is usually poor weight gain or frequent coughing, wheezing, and respiratory tract infections.

Treatment

Conventional treatment of lung complications

Management of lung problems is the major focus of CF treatment. To improve breathing the airways are cleared of thick mucus by postural drainage and manual percussion. An infant is placed in a position to facilitate drainage while the chest and back are percussed (clapped) to loosen mucus and allow it to be coughed out. Bronchodilators may be used to widen the airways and mucolytics to thin the mucus. Antibiotics are frequently administered to control the lung infections that plague those with the disease.

Treatment of lung complications using gene therapy

Lung infections usually begin early in the life of a newborn with cystic fibrosis, and repeated inflammation caused by these infections progressively damages the lungs as the child grows to young adulthood, becoming life threatening. Thus, gene therapy early in life has the greatest potential for clinical benefit.

Several gene therapy clinical trials have been undertaken in cystic fibrosis patients. All involve the transmission of the CFTR gene to cells of the airways. The goal is to introduce the wild type gene into cells, rather than to replace the defective version, which is technically far more difficult. The earliest vectors used were adenoviral, which are exceptionally efficient at infecting cells (i.e., a large proportion of cells exposed to the vector take up the transgene). Unfortunately, transgene expression using adenoviral vectors is transient, necessitating repeat administration if the patient is to experience long-lasting benefit from the treatment. Repeat administration is ineffective in this case because adenoviral vectors typically elicit strong immune responses. Neutralizing antibodies directed against the viral particle abrogate gene transfer with repeated administrations, and at high doses inflammation can be a clinically significant problem as well.

Although improved adenoviral vectors with lowered immunogenicity are being developed, Aav vectors are showing perhaps still more promise. Trials have shown these vectors to be safe, to have low immunogenicity, and they provide sustained expression in a variety of cell types, including the epithelial cells of the airways targeted in those with cystic fibrosis. Additionally, the use of cationic liposomes has yielded encouraging results. This type of vector suffers from low transfection efficiency and transience of expression; however, its low immunogenicity allows for repeat administrations. Both Aav and cationic lipid vectors have been shown to at least partially correct the electrophysiological defects in cells of the airways.

While no gene therapy protocol has yet been approved by the FDA for the treatment of cystic fibrosis, clinical trials using the above-mentioned vectors are well under way. In addition, other vectors such as those based on lentiviruses, herpesviruses, and others, as well as some DNA–protein conjugates, are currently under development.

Treatment of gastrointestinal problems

Once meconium ileus in a newborn has been treated (if it has occurred), the gastrointestinal problems are more easily managed than pulmonary manifestations of CF. Patients are prescribed a high-calorie diet supplemented with vitamins to counter malnutrition, and pancreatic enzymes to improve digestion. The pancrelipase this infant was prescribed contains pancreatic lipase, amylase, and protease to assist in the digestion of fat, starch, and protein, respectively.

QUESTIONS

1. What is the difference between ex vivo and in vivo gene transfer?

2. How might one target a specific cell population (e.g., B lymphocytes) by systemically administering a transgene?

3. Genetic testing for defects in the CFTR gene is now available and is used to inform couples of the risk of having a child with cystic fibrosis. Why are these tests only 80–85% effective at detecting a disease-causing gene?

4. How does the ΔF508 mutation impair the function of CFTR?

5. What are air-fluid levels in the gastrointestinal tract and what do they indicate?

6. Are the current trials involving CFTR gene transfer into the lungs of cystic fibrosis patients an example of *somatic cell* gene therapy or *germ-line* gene therapy?

7. What are the chances that a subsequent child of this couple will have cystic fibrosis?

8. Briefly describe the structure and regulation of CFTR.

FURTHER READING

1. Mountain, A. Gene therapy: The first decade. *Trends in Biotechnology* (2000) 18:119–128.

2. Baumgartner, I., Pieczek, A., Manor, O., Blair, R., Kearney, M., Walsh, K., and Isner, J. M. Constitutive expression of phVEGF165 after intramuscular gene transfer promotes collateral vessel development in patients with critical limb ischemia. *Circulation* (1998) 97(12):1114–1123. (This is the first published report of clinical benefit from gene therapy.)

3. Jaffe, A., and Bush, A. Cystic fibrosis: Review of the decade. *Monaldi Arch. Chest Dis.* (2001) 56(3):240–247.

4. Flotte, T. R., and Laube, B. L. Gene therapy in cystic fibrosis. *Chest* (2001) 120:124S–131S.

For further information, see the following web sites:

The National Lung, Blood and Heart Institute:
www.nhlbi.nih.gov/health/public/lung/other/cf.htm

The Cystic Fibrosis Foundation: www.cff.org

NIH Office of Biotechnology Activities: www4.od.nih.gov/oba/

CHAPTER 2

Biochemical Evolution
The Emergence of Multidrug-Resistant *M. tuberculosis*, or, How to Survive a Multipronged Assault

CASE HISTORY

Patient History

A 48-year-old man was admitted to the hospital because of fever, cough, and sweats. He had been in excellent health until one month earlier, when he began to experience intermittent drenching night sweats in association with a dry cough. Three weeks prior to admission he developed an unrelenting fever.

The patient was a physician who had immigrated to the United States two years earlier from Romania. He is currently employed as a manager of a clinical chemistry laboratory. A tuberculin skin test (purified protein derivative [PPD], 5 Tuberculin Units), performed at the time of immigration, was positive. The patient had no symptoms, and the physical examination and a radiograph of the chest were normal. Isoniazid prophylaxis was not prescribed in view of his age.

He has no history of chest pain, shortness of breath, hemoptysis, abdominal pain, nausea, vomiting, or diarrhea. He has no risk factors for human immunodeficiency virus (HIV) infection, other than his previous occupation as a physician. He has not traveled outside the United States since his immigration from Romania. He has no known history of tuberculous infection or exposure. He lives with his wife and two children, ages 16 and 14, who are well, without significant medical problems. He does not smoke or drink alcohol.

Physical Examination and Laboratory Findings

On physical examination, the patient was thin but not wasted and in no acute distress. He had a low-grade fever of 38.3°C, a normal pulse, blood pressure, and respiratory rate. No rash was evident. Dullness and diminished breath sounds were noted at the base of the left lung; the remaining lung regions were clear. The heart and abdominal examination were normal.

The laboratory examination—including serum chemistries, blood count, urinalysis, and clotting times—was normal. A specimen of arterial blood, drawn while the patient was breathing room air, showed that the partial pressure of oxygen was reduced at 73 mmHg, the partial pressure of carbon dioxide was 37 mmHg, and the pH was 7.45. A radiograph of the chest revealed a left pleural effusion without definite air-space disease.

Specimens of blood were obtained for culture. A test for anti-HIV antibodies was negative. Induced sputum specimens revealed no acid-fast bacilli or *Pneumocystis carinii,* and a culture yielded normal respiratory tract flora.

The Patient Is Treated for Pneumonia

Erythromycin (500 mg every six hours) and cefotaxime (1 gm every eight hours) were administered intravenously for a presumed community-acquired pneumonia. The fever was treated with oral acetaminophen as required. There was no significant improvement in his condition over the first three hospital days. A left-sided thoracentesis (removal of fluid from around the lung by insertion of a needle into the pleural space) was performed. A total of 1000 ml of turbid yellow fluid was removed. Microscopic examination of stained specimens of the fluid revealed abundant red cells and moderate numbers of neutrophils, with no acid-fast bacilli or other microorganisms; cytologic examination showed no atypical cells.

A radiograph of the chest after the thoracentesis revealed a decrease in the left pleural effusion and an infiltrate in the left lower lobe of the lung. The right lung was clear, and the size of the heart was normal. The patient's shortness of breath improved; however, he remained febrile without a change in his cough.

What is a tuberculin skin test (PPD) and what constitutes a positive result?

What is air-space disease and how is it distinguished from pleural effusion on a chest radiograph?

What might cause a reduced arterial partial pressure of oxygen?

Why was he treated for pneumonia when he tested negative for Pneumocystis carinii? What are the most common causes of community-acquired pneumonia?

What atypical cells might be expected to be present in the fluid removed by thoracentesis?

Does the patient show any signs of being immunocompromised?

Diagnosis and Treatment

A needle biopsy of the parietal pleura was performed. Examination of the specimen revealed a few caseating granulomata (clusters of mononuclear cells surrounding a central area of "cheese-like" necrosis) within a dense lymphocytic infiltrate. Acid-fast stains showed one- or two-beaded mycobacteria within small areas of necrosis. *M. tuberculosis* was cultured from the specimen. Subsequently, *M. tuberculosis* grew out of the sputum cultures as well.

In view of the patient's recent immigration from an endemic area of drug-resistant tuberculosis, he was treated with a four-drug regimen and direct observation of therapy (DOT). The drugs prescribed were isoniazid (10 mg/kg/d), rifampicin (600 mg/d), pyrazinamide (25 mg/kg/d), and ethambutol (25/mg/kg/d) administered as a single oral dose. His symptoms improved over the next two weeks, and he was discharged from the hospital to continue the antimycobacterial regimen as an outpatient. Four-drug therapy was continued for two months, followed by nine months of treatment with isoniazid (5mg/kg/d), rifampicin (600 mg/d), and pyrazinamide (25 mg/kg/d). Sensitivity testing revealed that the strain of mycobacteria was not drug resistant.

What are mononuclear cells and what is their function?

What is the parietal pleura?

How is drug-sensitivity testing determined?

What is the basis for the acid-fast staining procedure and what does it reveal?

What is DOT?

DISCUSSION

Microbiology of *M. tuberculosis*

Tuberculosis (TB) long confounded physicians and was widely believed to be hereditary until the discovery of the bacillus in 1882. The disease is caused by *Mycobacterium tuberculosis,* a small, rod-shaped aerobic bacillus. It is exceptional for the presence of mycolic acids, long-chain fatty acids, in its cell wall, which is the basis for its acid-fast staining properties (Figure 2.1). Mycolic acids are common to all mycobacteria (Figure 2.2), and hence acid-fast staining merely indicates the presence of a member of the genus and not necessarily *M. tuberculosis*. The cell wall mycolic acids contribute to the imperviousness of *M. tuberculosis* to the host's immune defenses, but they are also their Achilles heel in that their synthesis and deposition in the cell wall are the targets of two of the most effective drugs used against the disease, isoniazid and ethambutol. *M. tuberculosis* is also distinguished by its unusually slow growth rate for a bacterium, with a doubling time of 12–24 hours.

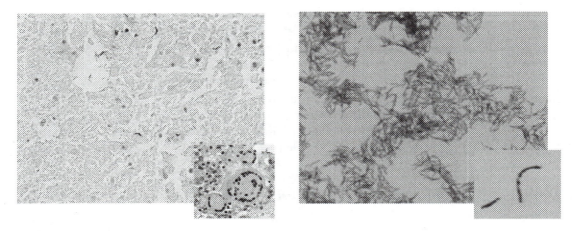

FIGURE 2.1 (Right) Mycobacteria tuberculosis in lung tissue M.TB are indicated by the arrow head . The inset shows a multinucleated giant cell in a caseating granuloma. (Left) Pure mycobacteria in culture stained by the Ziehl-Neelsen (cold acid fast) stain method. The inset shows a higher power view of the organism. Images courtesy of Drs. Raghunand Tirumalai, Mark Yoder and William Bishai.

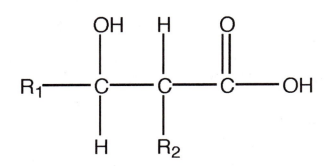

FIGURE 2.2 Structure of mycolic acid. Mycolic acids are complex, hydroxylated fatty acids with 60–90 carbon atoms. They may contain a variety of functional groups, such as methoxy, keto, and epoxy ester groups.

The Resurgence of Tuberculosis

Tuberculosis has been viewed as a disease of the past, long since brought under control; indeed, since the mid-1940s the use of antibiotics and improved sanitation have contributed to the steady decline in incidence of the disease in industrialized countries. In the mid-1980s, however, the number of cases began to increase, and this has largely been attributed to the emergence of multidrug-resistant strains and to the AIDS epidemic. In most individuals infection with *M. tuberculosis* is limited by the immune response and does not progress to the disease state. In contrast, in immunocompromised individuals such as those afflicted with AIDS, infection can progress rapidly to active disease. In addition, the tuberculosis bacillus can survive in a latent form for years after infection, and those with AIDS are much more susceptible to reactivation with progression to full-blown disease than the general population. This explains why the patient was tested for the presence of HIV antibodies upon admission.

The emergence of multidrug-resistant *M. tuberculosis* (MDR-TB) strains has also been a major contributing factor in the resurgence of the disease. The principles used to describe molecular evolution in the generation of life forms (see Section 2.2 of *Biochemistry* 5e) can also be applied to the emergence of drug-resistance in bacteria. When exposed to an antibiotic, bacteria possess all the basic requirements for evolution: they can *reproduce* (rapidly, compared to most organisms); they *vary*, thanks to genetic mutation and the ready acquisition of mobile genetic elements; and they are under *selective pressure*, applied by the antibiotic. The bacteria within the population will compete with one another for survival, with those resistant to the drug persisting at the expense of those that are not. Multidrug-resistant tuberculosis can be treated with second-line drugs, but these are less effective, more expensive, and more toxic, and disease caused by these strains is associated with higher fatality rates.

There are now more than eight million new cases of tuberculosis yearly, with sub-Saharan Africa and Southeast Asia bearing the greatest burden. In some areas, up to 20% of cases are caused by MDR-TB. Russia has the worst MDR-TB problem, concentrated in its prisons, but other areas, such as Mozambique, Iran, India, and China, are also among the most seriously affected. Eastern Europe also has a substantial problem, including Romania, from where this patient emigrated.

Mechanisms of Drug Resistance in *M. tuberculosis*

Resistance to multiple drugs often occurs simultaneously, with the acquisition of a mobile genetic element that encodes multiple resistance genes, or an efflux pump that has the ability to extrude multiple drugs (see this section in *Biochemistry* 5e). Remarkably, multidrug-resistance in *M. tuberculosis* arose cumulatively, through successive genomic mutations in separate genes. Monotherapy can give rise to a resistant strain, which can eventually take over the population. Use of a second drug can then give rise to a second resistance mutation, and so on. Even the simultaneous use of multiple drugs can lead to multidrug-resistance, if use is irregular, as repeated cycles of killing and regrowth of the bacterial population favors the selection of resistant strains. Resistance to all of the first-line drugs used to treat *M. tuberculosis* (rifampicin, streptomycin, isoniazid, pyrazinamide, and ethambutol, Figure. 2.3) has been reported, and in many cases resistant strains are immune to more than one drug. Multidrug-resistant *M. tuberculosis* is defined as resistance to isoniazid and rifampicin, with or without resistance to other drugs.

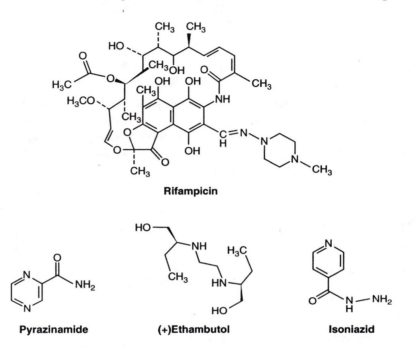

Rifampicin

Pyrazinamide   **(+)Ethambutol**   **Isoniazid**

FIGURE 2.3 Structures of first-line drugs against *M. tuberculosis*.

Resistance to rifampicin, streptomycin, and ethambutol result from genomic mutations that modify the drug's target. For example, rifampicin, one of the most potent drugs against *M. tuberculosis*, inhibits bacterial transcription by targeting the bacterial RNA polymerase. Rifampicin binds to the β subunit of the polymerase, encoded by the *rpoB* gene, which functions in the formation of phosphodiester bonds between the ribonucleotides of the growing RNA chain. Ninety-five percent of strains resistant to rifampicin have been found to have mutations in an 81 bp core region of the *rpoB* gene (Figure 2.4).

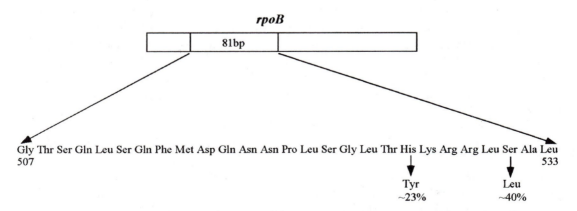

FIGURE 2.4 Over 95% of rifampicin-resistant clinical isolates lie within an 81-bp region of *rpoB* that encodes amino acids 507–533. The histidine → tyrosine and serine → leucine changes at positions 526 and 531, respectively, are the most common, together accounting for over 60% of resistant isolates.

In contrast, isoniazid and pyrazinamide resistance result from mutations in enzymes that convert the drugs into their active forms within the cell, or from mutations that cause over-expression of the drug target (Figure 2.5).

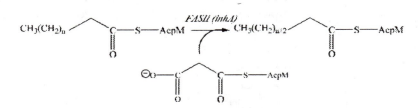

FIGURE 2.5 Elongation of fatty acids in the biosynthesis of mycolic acids. Elongation occurs by repeated addition of two-carbon subunits in an AcpM (acyl carrier protein) and FASII (fatty acid synthase II) dependent fashion. Enoyl-ACP-reductase (encoded by *inhA*) is a component of the FASII enzyme complex. Following elongation, further enzymatic steps complete the biosynthesis of mycolic acids. Isoniazid inhibits Enoyl-ACP-reductase by binding the NADH binding site, thus inhibiting enzyme activity.

Pathophysiology

Tuberculosis is primarily a disease of the respiratory tract. It is contagious and is transmitted through the air, and is usually acquired by inhalation. It does not release toxins as do many other bacterial pathogens, but rather, damage to the host is a result of the host's immune response to the organism.

Primary infection usually does not lead to disease

Infection begins when inhaled bacilli reach the alveoli in the lower lobes of the lung. There they are ingested by alveolar macrophages, in which they continue to divide, and a local immune response ensues. A granuloma, which is an encapsulated lesion formed by macrophages drawn to the site of infection, forms around the bacilli, walling them off from the rest of the body. Two to four weeks later, the interior of the lesion softens, forming a caseous (cheese-like) center, and bacilli and macrophages harboring them drain into the regional lymph nodes. In greater than 90% of infected individuals, the immune response limits the infection at this point, and the disease does not proceed to clinical illness. However, deposits of calcium and collagen can encapsulate the granuloma, forming a rigid coat in which the bacteria can survive, dormant, for decades. These are sometimes visible on a chest radiograph, although they were not picked up on the radiograph of this particular patient when he entered the United States. A radiograph can also reveal signs of active disease, and because it did not in this case, and the patient had no symptoms, he was not recommended for treatment upon entry into the United States.

At this stage, bacilli may escape from the lymph nodes and enter the bloodstream, implanting at other sites within the body and setting the stage for possible future extrapulmonary disease.

Progression to the disease state

Only approximately 1% of infected individuals rapidly progress to the disease state upon infection, and these are usually immunocompromised individuals or those at the extremes of age. A further 5% will develop the disease within a year of infection, and in another 5% the infection will resurface to cause disease anywhere from one year to decades later. Thus, the majority of individuals become clinically ill from tuberculosis from reactivation of a latent infection. Dormant bacteria will begin to multiply, usually in the lung, but at any site where it may have implanted during the initial infection. Pleural TB usually occurs when a subpleural pulmonary lesion ruptures, releasing bacilli into the pleural space, which can then progress to active pulmonary disease, as it did in this case.

Symptoms

Common symptoms of tuberculosis disease include fatigue, night sweats, fever, weight loss, and dry cough, which progresses to a productive cough and can be blood-tinged.

Vaccination

The Bacillus Calmette-Guérin (BCG) vaccine consists of an attenuated form of *Mycobacterium Bovis*, a close relative of *M. tuberculosis*. Although it is not commonly used in the United States, BCG vaccination of newborns is routine in many countries where tuberculosis is prevalent, including Romania. The degree of protection conferred by the vaccine is variable, and it does not appear to protect into adulthood.

Diagnosis

Diagnosis of tuberculosis infection

The tuberculin skin test (sometimes called the Mantoux test), in which PPD, a mixture of tuberculosis antigens, is administered intradermally, is the preferred method of detecting tuberculosis infection. A PPD test will be positive 2–10 weeks following infection, even in the large majority of individuals where infection does not progress to clinical illness. In addition, previous vaccination with the BCG vaccine can give a positive result, which can confound treatment decisions. A chest X ray may also be useful in identifying those previously infected, as any remaining calcified granulomas may be visible.

Diagnosis of tuberculosis disease

A chest X ray may reveal signs of tuberculosis disease such as lung cavities or fluid build-up, but the results are often not definitive. For example, in this case the patient was initially misdiagnosed as having a community-acquired pneumonia. His lack of response to erythromycin and cefotaxime were the first clues that the cause for illness was other than bacterial pneumonia.

Although rapid, DNA-based assays such as the polymerase chain reaction (PCR) are increasingly being used, evidence of mycobacteria in samples of bodily fluids, either by microscopic examination of the sample or following a culture, remains the most definitive way of diagnosing tuberculosis. The acid-fast staining procedure can rapidly detect the presence of mycobacteria, and although the strain may be other than *M. tuberculosis* (other mycobacteria, although less virulent, cause disease in humans), treatment for TB is usually initiated upon obtaining a positive result. Because the physician treating this patient observed a pleural effusion in the radiograph, pleural TB was suspected and the fluid removed by thoracentesis was examined microscopically for the presence of mycobacteria. Although the sample was negative, a sample of parietal pleura tissue, which is typically more heavily infected than the fluid, revealed evidence of the presence of mycobacteria, and subsequently led to the diagnosis.

Treatment of tuberculosis disease

Recommended tuberculosis treatment lasts at least six months and involves the use of four antimicrobials: rifampicin, isoniazid, pyrazinamide, and ethambutol or streptomycin. The initial intensive two-month treatment involves the use of four drugs simultaneously, followed by at least four months of treatment with rifampicin and isoniazid alone, the most potent drugs against *M. tuberculosis*. This intensive, multidrug regimen is used to rapidly quell the infection and prevent the emergence of drug resistance. DOT is a strategy initiated by the World Health Organization (WHO) for curing TB and controlling its spread. Its main feature is the direct observation by health care workers of diagnosed patients taking their medication, thus ensuring adherence to drug treatment. It has been remarkably effective, curing upward of 90% of cases in areas where it has been implemented.

Treatment of latent tuberculosis infection

Treatment for 6–9 months with isoniazid of those with latent tuberculosis infection, as determined by the tuberculin skin test, has been shown to be greater than 98% effective in preventing reactivation. However, it is contraindicated in those over 35 due to increased risk of hepatotoxicity (toxic effects in the liver).

Treatment of drug-resistant strains

When the organism is found to be resistant to first-line drugs, second-line drugs must be resorted to. Among the commonly used second-line drugs are para-amino salacylic acid, ethionamide, kanamycin, and capreomycin. Fortunately, the patient here was not found to be ill with a resistant strain, and the regimen he was prescribed should succeed in curing the infection.

QUESTIONS

1. Why did the evolution of life take place over the course of billions of years, whereas the evolution of drug resistance in bacteria occurs very rapidly, with it usually being reported within the first year of use?

2. What are some contributing factors in the rise in prevalence of drug resistance in bacteria?

3. Why did the culture of the sputum sample initially come up negative for acid-fast microorganisms, and later come up positive?

4. BCG vaccination of newborns is routine in Romania. Given this fact, do you think the patient tested positive for the tuberculin skin test when he entered the United States because he had been vaccinated, or do you think he had been infected prior to arrival in the United States?

5. In some states, legal measures can be taken against those who repeatedly fail to complete drug treatment for tuberculosis. Why do you think this might be?

6. The "second-line" drugs used to treat MDR-TB are prohibitively expensive in many areas where tuberculosis is endemic. If you were a microbiologist working for an international public health organization, what points might you stress to the manufacturer of a second-line drug as incentive to reduce its cost?

7. The development of resistance to a drug can come at a cost to an *M. tuberculosis* strain. What might this might be?

8. What mechanisms might a bacterium use to evade a toxic drug?

9. Why was the patient's sputum sample tested for the presence of *Pneumocystis carinii*?

10. Nucleic acid-based techniques are being developed to rapidly assess the drug resistance status of bacterial pathogens. Do you think these types of assays will be useful in detecting drug-resistance in clinical isolates of *M. tuberculosis*?

11. Although the DOT strategy initiated by the WHO has been very successful, can you think of some of its limitations?

12. Does resistance to first-line drugs in *M. tuberculosis* occur by genomic mutation, acquisition of a mobile genetic element, or both?

13. Ethambutol has been shown to act synergistically with both rifampicin and streptomycin, but not with isoniazid. Why might be the reason for this?

FURTHER READING

1. Wallis R. S., and Johnson J. L. Adult Tuberculosis in the 21st century: pathogenesis, clinical features, and management. *Current Opinion in Pulmonary Medicine* (2001) 7(3):124–132.

2. Rattan, A., Kalia, A., and Ahmad, N. Multidrug-resistant mycobacterium tuberculosis: molecular perspectives. *Emerging Infectious Diseases* (1998) 4(2):195–209.

3. Willcox, P. A. Drug-resistant tuberculosis. *Current Opinion in Pulmonary Medicine* (2000) 6(3):198–202.

4. Piatek, A. S., Telenti, A., Murray, M. R., et al. Genotypic analysis of mycobacterium tuberculosis in two distinct populations using molecular beacons: implications for rapid susceptibility testing. *Antimicrobial Agents and Chemotherapy* (2000) 44(1):103–110.

For further information, see the following web sites:

World Health Organization tuberculosis site: www.who.int/gtb/

Centers for Disease Control, Division of Tuberculosis Elimination: www.cdc.gov/nchstp/tb/

American Lung Association: www.lungusa.org/

Johns Hopkins Center for Tuberculosis: www.hopkins-tb.org

National Institute of Allergy and Infectious Diseases: www.niaid.nih.gov/factsheets/tb.htm

CHAPTER 3

Protein Structure and Function
A Fatal Outcome When Protein Folding Goes Awry

CASE HISTORY

A 58-year-old woman who had been admitted for terminal care with rapidly progressive dementia had died and was undergoing a limited autopsy of the central nervous system. The patient worked as a CPA in an accounting firm and had been well until several months ago when she began to experience memory loss and transient episodes of "confusion" associated with occipital headaches, neck stiffness, and blurring of vision. She was easily distracted and had difficulty finishing even trivial mental tasks. Neurological examination at that time revealed orientation to time, place, and person; deficits in short-term memory; no cranial nerve abnormalities; a broad-based gait; and equivocal plantar response. Within two months she began to experience difficulties with both fine and gross motor skills. She could no longer write and began using a walker at home and a wheelchair when going outside. Her cognitive ability further deteriorated, her speech was slow and dysarthric, and her memory loss became worse. She developed increased muscle tone and abrupt vigorous involuntary muscle contractions (myoclonus) in both upper extremities. The patient died seven months after the onset of her illness.

What is a plantar response and its significance?

What is the differential diagnosis of such a rapidly progressive neurological disorder?

The patient had no history of receiving a corneal transplant or dura mater grafts or human pituitary growth hormone, nor had she received any fertility therapy. There were no family pets. There was no history of travel to Europe or outside the continental United States since her youth. There was no family history of significant neurological disease. There was no known history of consumption of beef raised outside the country, and in general she did not consume wild game or have other food eccentricities.

What is the importance of the absence of receiving a neurological transplant or human growth hormone?

What is the significance of the travel history and absence of food eccentricities?

Brain Scans, CSF Analysis, and Diagnosis

Head magnetic resonance imaging (MRI), computerized tomographic scanning (CT), and routine cerebrospinal fluid (CSF) analyses did not reveal any abnormality that would explain the patient's neurological symptoms. An electroencephalogram (EEG) was remarkable for semiperiodic slow-wave patterns consisting of one-to-two-cycles-per-second triphasic sharp-wave activity. On the basis of the presentation and EEG a diagnosis of Creutzfeldt-Jakob disease (CJD) was made.

What is CSF and how is it obtained?

What is the significance of the results of the EEG?

What are the diagnostic features of CJD?

Pathological Examination and Immunoblot Analysis

Gross pathological examination of the brain showed no abnormalities. The histopathological examination indicated a widespread fine spongiosis with astrogliosis with no visible amyloid plaques and minimal inflammatory infiltrate. Tissue was prepared from fresh unfixed brain and treated with proteinase K followed by separation on a polyacrylamide gel, and transfer to polyvinylidene fluoride (PVDF) matrix. Blotting with anti-prion protein (PrP) antibodies revealed a band in the 27–30 kd range. This is consistent with protease resistance of prion protein (i.e., PrP^{sc}) and the diagnosis of CJD.

What is spongiosis? What is amyloid?

What are prions?

Retrospective analysis of CSF revealed the presence of a normal brain protein known as 14-3-3. Elevated levels of this protein are found in patients with sporadic CJD disease, viral encephalitis, and during the first month after cerebrovascular accidents (CVA) or strokes.

Genetic Analysis

The patient was part of a larger study of the molecular genetics of prior diseases. Analysis of her prion protein gene *(PRNP)* revealed methionine homozygosity at the polymorphic codon 129 and absence of genetic mutations.

What is significant about the polymorphism at codon 129 and the absence of genetic mutations in PRNP?

DISCUSSION

Prion Diseases

Creutzfeldt-Jakob disease (CJD) is one of a handful of remarkable neurodegenerative diseases known as *prion diseases* or *transmissible spongiform encephalopathies* (TSEs), which also include kuru, fatal familial insomnia (FFI), and Gerstmann-Sträussler-Scheinker disease (GSS) in humans, scrapie in sheep, and bovine spongiform encephalopathy (BSE) in cows. These diseases mystified scientists for decades, and lay at the heart of a heated controversy within the medical research community.

Prion diseases were eventually found to occur by a completely novel mechanism. Although they are transmissible, prion diseases are not caused by microscopic pathogens such as viruses or bacteria, but rather by misfolding of a normal, cellular protein known as *prion protein (PrP)*. All previously known pathogens contained nucleic acids (DNA or RNA), and the idea of an infectious protein was revolutionary. Only recently, with the accumulation of a large body of evidence, has the idea of an infectious protein as the source of prion diseases become widely accepted.

Prions Diseases Are a Result of a Change in the Tertiary Structure of PrP

Prion diseases occur upon conversion of PrP^c (normal cellular PrP) into PrP^{sc} (the disease version). While the amino acid sequences of PrP^c and PrP^{sc} may be identical, the tertiary structures—and hence the physicochemical properties of the proteins—are dramatically different. Although it remains unclear how the conformational change from normal to disease versions occurs, it results in a reduction in the α-helical content of PrP, with a concomitant increase in β-sheet content (see Section 3.3 of *Biochemistry* 5e). The change in tertiary structure increases the propensity of the protein to form insoluble aggregates, both inside the cell and in the form of extracellular *amyloid plaques*. These aggregates are extremely resistant to treatments that normally destroy nucleic acids and proteins, which led to much confusion as to the chemical nature of the infectious agent in the early years of research. The accumulation of PrP^{sc} eventually kills neuronal cells (via mechanisms still unclear), which results in the rapidly progressive dementia and neuromuscular abnormalities associated with prion diseases.

Structure of PrP

PrP^c is attached to the extracellular side of the cell membrane via a glycolipid called a *glycosylphosphatidylinosital (GPI) anchor* (see Section 12.5.3 of *Biochemistry* 5e). Nuclear magnetic resonance (NMR) studies of PrP^c in solution have shown that the C-terminal half of the protein consists of three α-helices, and a short antiparallel β-sheet. A disulfide bond between helices 2 and 3 stabilizes the structure, and two asparagine residues are glycosylated (Figure 3.1A). PrP^c thus exhibits two features typical of extracellular proteins: disulfide linkage and carbohydrate groups. NMR studies suggest that the N-terminal half of the protein is unstructured, although it contains two copper binding sites that may induce folding of this region in the presence of copper.

PrP^{sc} is isolated from diseased tissue in aggregates that are not amenable to high-resolution structural studies. However, lower resolution optical spectroscopy methods have indicated that 45% of PrP^{sc} is β-sheet, whereas PrP^c has very little β-sheet content. The mechanism by which PrP^c is transformed into PrP^{sc} is unknown; however, the transition is

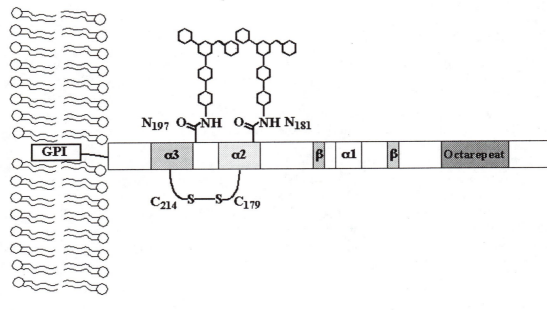

A

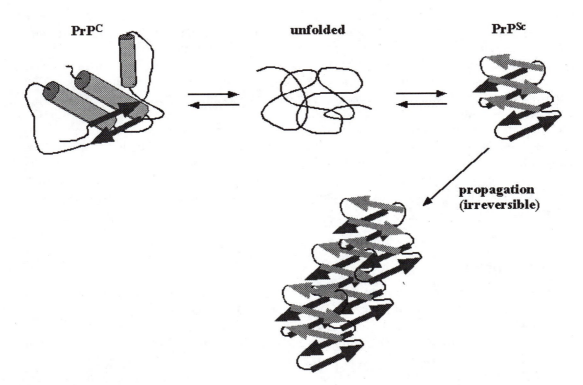

B

FIGURE 3.1 (A) Basic structure of human PrP. The α-helical regions and β-strands are indicated by the "α" and "β" symbols, respectively. The carbohydrate groups at asparagines 181 and 197, and the disulfide bridge between cysteine residues 179 and 214 are indicated. The glycine- and proline-rich octapeptide repeats, which confer copper-binding, are located in the N-terminus. (B) Conversion of PrP^c to PrP^{sc} involves passage through an unfolded intermediate. Once formed, PrP^{sc} has a high propensity to form aggregates, which leads to disease.

thought to involve passage through a virtually completely unfolded intermediate. Although rare, once the transition has occurred, the chain reaction leading to the irreversible formation of PrP^{sc} aggregates may take place (Figure 3.1B).

The existence of two stable forms of PrP has challenged the well-accepted notion that a given amino acid sequence dictates a *single* three-dimensional structure. It has been proposed that the cycling of PrP through an acidic cellular compartment called an *endosome* may set the stage for disease. The acidic and reducing environment of this compartment favors protein unfolding, which may lead to conversion of PrP^c to PrP^{sc}, and the chain reaction that eventually leads to disease.

Prion Diseases Are Transmissible

The etiology of prion diseases can be infectious or genetic, and in fact, in many cases it is simply sporadic and cannot be traced to either a genetic mutation or to contact with diseased tissue. Prion diseases are not contagious by casual contact, but they are nevertheless considered infectious.

Although transmission can occur by the oral route, by consumption of diseased tissue, it is inefficient and requires large doses of inoculum. Peripheral inoculation (e.g., subcutaneous), and especially intracerebral inoculation, are far more effective means of conveying disease. Individuals with seizure disorders treated with neurosurgical electrodes used previously on a CJD patient contracted the disease in the 1970s (although the electrodes had been sterilized) in what became one of the first instances of *iatrogenic* (that is, caused by a medical procedure) CJD. Other iatrogenic cases occurred from corneal transplants, dura mater grafts, human pituitary growth hormone treatments, and fertility treatment with human gonadotropins.

Laboratory studies have shown that spongiform encephalopathies such as CJD can cross species barriers, which has contributed to the crisis in Britain that began in the late 1980s when a BSE ("mad cow disease") epidemic was cited as a possible cause for a rash of cases of what has become known as *new-variant Creutzfeldt-Jakob disease* (v-CJD). The mechanism of transmission of prion diseases appears to involve the refolding of PrP^c into PrP^{sc} by using the latter as a *template* (Figure 3.2).

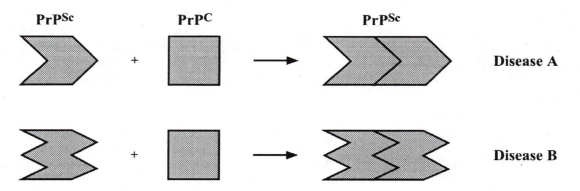

FIGURE 3.2 Schematic of proposed mechanism whereby PrP^c is converted to PrP^{sc}. PrP^c is thought to be converted into PrP^{sc} by using the latter as template, and the various prion diseases are thought to arise via different conformers of PrP^{sc}.

Different PrP^{sc} Conformers Give Rise to the Various Prion Diseases

All prion diseases are caused by aberrantly folded PrP, yet each disease has distinctive characteristics. This appears to be due to the existence of various *conformers* of PrP^{sc}, each with a slightly different structure, which ultimately gives rise to a different prion disease (see Figure 3.2).

The structural differences between the various conformers are not yet well understood, nor is it known how these differences translate into individual disease characteristics. The various prion diseases are differentiated based on incubation times, as well as neuropathological, epidemiological, and symptomatic differences.

Etiology of Prion Diseases

Prion diseases can have genetic, infectious (see above), or sporadic etiologies. GSS and FFI are examples of inherited human prion diseases, and there are inherited versions of CJD as well, although they are responsible for only a minority of cases. The inherited diseases are each associated with particular mutations in the *PRNP* gene; for example, mutation of codon 102, changing proline to leucine, causes GSS. Each mutation is thought to favor the conversion of PrP^c into a particular conformer of PrP^{sc}, thus predisposing an individual to acquiring the disease. Thus, a *single* amino acid change can dramatically alter protein structure and lead to a devastating disease.

How the disease arises in sporadic cases is not known, but possibilities include spontaneous conversion of PrP^c into PrP^{sc}, or a somatic mutation that favors the conversion.

CREUTZFELDT-JAKOB DISEASE

Epidemiology

The large majority of CJD cases—at least 85%—are sporadic, with most of the remaining cases being hereditary. Only a very small percentage (less than 1%) of cases are acquired from contact with neuronal tissue. The patient described here appears to have had sporadic CJD; there is no familial history of spongiform encephalopathies, nor did she appear to be the victim of iatrogenic CJD. She had not traveled to Europe since the BSE outbreak, hence she was unlikely to have contracted v-CJD from consumption of BSE-infected meat.

The disease is extremely rare, occurring in approximately one in one million people worldwide per year. There is no seasonal distribution or evidence of geographic clustering, except in areas with large numbers of familial cases. There has been no evidence of changing incidence over the past few decades. The disease typically strikes those between the ages of 50 and 70 years and has a rapidly fatal course, with 90% of patients succumbing within one year of the onset of symptoms. This patient, who began experiencing symptoms at the age of 58 and who died seven months after the onset, is thus typical.

Genetic susceptibility

Aside from the clearly inherited forms of CJD caused by specific mutations in PRNP, there appears to be a genetic susceptibility to sporadic and iatrogenic CJD and v-CJD determined by polymorphisms (genetic characters that occur in more than one form) at codon 129. This codon either encodes methionine or valine, but whichever it may be, homozygosity (both chromosomes carry the same form of a genetic character) is disproportionally associated with disease. Close to 90% of individuals who develop sporadic CJD are homozygous for this codon, compared with approximately 50% of the general population. The large majority of those with sporadic CJD are homozygous for methionine, as was the patient described here, which is associated with a more rapid disease progression than valine homozygosity.

Diagnosis

Typical disease progression

Among the common initial symptoms of CJD are fatigue, disturbed sleep, headaches, failing memory, depression, uncharacteristic behavior, episodes of confusion, and visual impairment. Ataxia (shaky movements and unsteady gate) and aphasia (impaired expression or comprehension of language) often develop, as they did in this patient, with her muscular coordination deteriorating to the point of requiring a walker and a wheel chair. As the disease progresses cognitive abilities rapidly deteriorate and myoclonic jerking (involuntary muscle jerking) develops, as they did here. In the late stages of the disease patients become unable to move and speak, and enter a coma. The mean survival time from onset of symptoms is five months.

Differential diagnoses

CJD is extremely rare and is thus frequently misdiagnosed. Some patients, especially those that present with aphasia or hemiparesis (paralysis of one side of the body), are initially misdiagnosed as having experienced a cerebrovascular accident (stroke), and symptoms are also often confused with those of other neurodegenerative disorders such as Alzheimer disease or Huntington's disease. CJD is distinguished by the very rapid progression of dementia (in a matter of weeks, or even days) and the development of myoclonic jerking.

Brain scans and laboratory findings

MRI and CT scans are usually normal in the early stages of disease, but may reveal abnormalities later on. The EEG, which measures electrical activity in the brain, is one of the most useful diagnostic tools for CJD. A specific pattern, characteristic of CJD, is observed in at least 60% of patients: periodic (biphasic or triphasic), synchronous (regular), sharp wave complexes (bursts of electrical activity) at a frequency of 1–2 cycles/second superimposed upon a slow background rhythm (Figure 3.3). The presence of protein 14-3-3 in the CSF is also a very useful diagnostic tool. When there is no sign of pleocytosis (elevated levels of lymphocytes in CSF), an elevated level of protein 14-3-3 in patients with progressive dementia is 96% effective at detecting sporadic CJD.

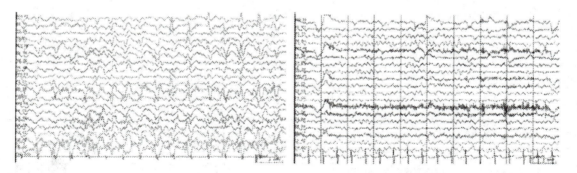

FIGURE 3.3 EEG from a patient with CJD (left) and a normal control (right). Periodic biphasic and triphasic waves are seen on a slow background rhythm. Images courtesy of Amy Abdallah and Dr. Gregory Krauss.

The only certain means of diagnosing CJD is by postmortem examination of the brain

A brain biopsy in a patient with characteristics of CJD can be used to confirm the diagnosis or exclude other treatable diseases; however, in practice the invasiveness of the procedure combined with the possibility of a false-negative result (the biopsy may miss the affected tissue) limits its utility. Thus, the disease is not usually definitively diagnosed until brain tissue is examined postmortem.

Although the gross anatomy of the brain is typically normal, histophathological examination of tissue sections reveals the neuronal loss and astrogliosis (growth of astroglial cells) that are hallmarks of the disease. Astroglial cells carry out a supportive function in the brain and are activated in response to injury. Unusually abundant and large vacuoles in neuronal cells are typical, giving the tissue section the sponge-like appearance characteristic of the disease (spongiosis). Amyloid plaques are sometimes observed, but are far more common in v-CJD cases. Areas of the brain affected are the cerebral cortex, the basal ganglia, the thalamus, and the cerebellar cortex. The autopsy of this patient's brain thus exhibited all the hallmarks of sporadic CJD: a normal gross anatomy, with widespread spongiosis and astrogliosis, and no evidence of amyloid plaques. The lack of inflammatory infiltrate is typical as prion diseases do not evoke an immune response.

A further diagnostic test takes advantage of the relative protease resistance of PrP^{sc} compared to PrP^{c}. Limited proteolysis of PrP^{Sc} using the proteinase K protease yields a truncated "protease resistant core" represented by a 27–30 Kd band on an immunoblot (full length protein is ~35 Kd). When treated in the same way, PrP^{c} is completely degraded and no band is observed. In this case, the presence of the 27–30 Kd band on the immunoblot confirmed the diagnosis of CJD.

Treatment

There is currently no effective treatment for controlling CJD.

QUESTIONS

1. How are *incidence* and *prevalence* of a disease defined? They are roughly equal in the case of CJD. Why?

2. What is the most unusual feature of prion diseases compared to other infectious diseases?

3. PrP is attached to the exterior side of the cell membrane via the posttranslational addition of a GPI anchor. Given that it is a cell-surface protein, what other type of covalent modification might you expect PrP to undergo?

4. What is an amyloid plaque?

5. The *PRNP* gene in humans encodes a 253 amino acid protein, yet the protein found in cells is only 208 amino acids in length. How might this be explained?

6. How do PrP^{c} and PrP^{Sc} differ in their biochemical properties?

7. Chaperones are a class of proteins involved in mediating protein folding during protein synthesis, and in preventing aggregation of proteins under conditions that cause protein unfolding, such as heat shock. Do you envision a therapeutic role for this class of proteins in treating prion diseases?

8. FFI is a hereditary prion disease characterized by several months of almost total insomnia followed by death. What part of the brain is likely to be affected in this disease?

FURTHER READING

1. Prusiner, S. B. Prions. *Proc. Natl. Acad. Sci. USA* (1998) 95:13363–13383. Nobel Lecture.

2. Johnson R. T., and Gibbs C. J. Creutzfeldt-Jakob disease and related transmissible spongiform encephalopathies. *NEJM* (1998) 339:1994.

3. Soto, C., Saborio, G. P. Prions: disease propagation and disease therapy by conformational transmission. *Trends in Molecular Medicine* (2001) 7(3):109–114.

For further information, see the following web sites:

Creutzfeldt-Jakob Disease Foundation: http://www.cjdfoundation.org/

National Institute of Neurological Disorders and Stroke:
www.ninds.nih.gov/health_and_medical/disorders/cjd.htm

CHAPTER 4

Exploring Proteins
Fragile: Handle with Care

Patient with History of Fractures Presents to the ER

A 13-year-old girl, accompanied by her mother, presented to the emergency room with acute trauma to her left forearm. According to the patient, she sustained the injury when she hit her arm on the banister while running down a flight of stairs. She reported that she did not hit her arm particularly hard, and was surprised by the intensity of the pain. According to the patient and her mother, she had sustained multiple fractures in the past, which had healed without complication. Upon further questioning, both the patient and her mother denied any history of abuse, and the pattern of injuries was not consistent with abuse. She has two sisters who are normal, there was no history of still births, and no family history of multiple fractures. She was the product of a normal gestation with normal birth weight, head size, and apgar scores.

Physical Examination

On physical exam, the patient was a well-developed adolescent female in moderate distress, and clearly was favoring her right arm. She was of normal height with no obvious skeletal abnormalities. There were no bruises or other evidence of trauma to the rest of her body. Head, eyes, ears, nose, and throat examination revealed that her sclerae had a pale blue hue, her teeth were normal, and there were no other abnormalities. Her chest cage was normal configuration, and the lung and heart examination was normal. There was no bowing of the long bones of the legs. Neurological examination was unremarkable, including normal hearing.

Radiographs of the patient's left forearm showed a simple transverse fracture of the left radius. Radiographs of the long bones of the arms and legs revealed slight osteopenia (loss of bone mass), and radiographs of the skull showed mottled areas, a reflection of irregular ossification.

What concerns should you have about a young patient with a history of multiple trauma?

What is the significance of the finding of blue sclerae?

Diagnosis and Treatment

From the clinical findings the patient was diagnosed with type I osteogenesis imperfecta. A skin biopsy was taken with the appropriate consent, and a sample was sent for DNA sequencing. In addition, skin fibroblasts from the sample were cultured in medium containing [^{3}H]proline and [^{35}S]cysteine, and the labeled collagen secreted from the cells was analyzed by SDS-polyacrylamide gel electrophoresis (SDS-PAGE) under both reducing and nonreducing conditions (Figure 4.1). The nonreducing gel revealed an abnormal, low-mobility band, indicative of a type I collagen defect. This was confirmed by the DNA test, which revealed a point mutation in the *COL1A1* gene (which encodes α1(I)), resulting in the replacement of glycine 94 with cysteine.

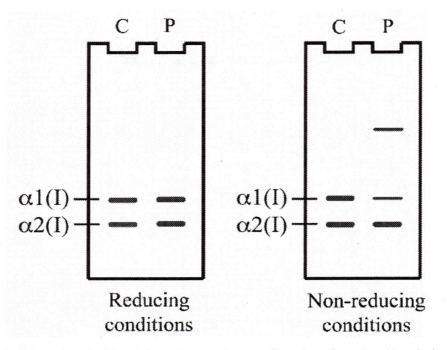

FIGURE 4.1 SDS-polyacrylamide gel electrophoresis of type I collagen from the patient (P) and a healthy control (C) under reducing (+ dithiothreitol) and and nonreducing conditions. Type I collagen is comprised of two α1(I) chains and one α2(I) chains. A high-molecular-weight species is present in the sample from the patient under non-reducing conditions.

What types of mutations can be detected via the reducing gel?

What could be responsible for the higher molecular-weight band in the non-reducing gel?

The patient was referred to the orthopedic clinic and was fitted with a light cast. She was also given a referral for physical therapy, which was to be initiated once the pain in her arm subsided. A mild exercise program was recommended to maintain bone mass.

DISCUSSION

Osteogenesis Imperfecta Has a Highly Variable Clinical Course

Osteogenesis imperfecta (OI) is an inherited disorder that results from mutations within the genes that comprise *type I collagen*. As type I collagen plays an important role in the structural integrity of bone, patients with OI is predisposed to easy bone fractures and skeletal abnormalities. Another feature associated with OI is a bluish hue of the sclera, which is probably the result of low levels of collagen, allowing the choroid behind it to be visualized. Thin skin, joint laxity, hearing loss, dental abnormalities, and cardiovascular problems may also occur.

The clinical spectrum of OI can vary quite widely, from patients who survive into adolescence and adulthood with brittle bones (as with the patient described here) to infants who are born with multiple fractures and skeletal abnormalities and die within several months of birth. The most serious forms are lethal in utero. In order to understand how these different phenotypes can arise, one must first be familiar with the structure, folding, and biosynthesis of the type I collagen molecule.

Structure and Folding of Type I Collagen

Over 19 types of collagen have been identified; however, the most abundant, and the primary constituent of bone, is type 1 collagen. Type I collagen forms a triple helical structure involving two $\alpha 1(I)$ chains and one $\alpha 2(I)$ chain.

The helical region of each chain consists of over 300 repeats of the tripeptide sequence Gly-X-Y, where X is typically proline and Y is usually 4-hydroxyproline, a posttranslationally modified form of proline. The glycine at every third residue is critical for the proper assembly of the trimer, as larger residues will form unfavorable steric interactions with the center of the helix.

Each chain is synthesized as a longer *procollagen* molecule, with peptide extensions at both the N- and C-termini (propeptides). The procollagen molecules, *pro$\alpha 1$(I)* (1464 amino acids, 138 kDa) and *pro$\alpha 2$(I)* (1366 amino acids, 129 kDa), are products of the *COL1A1* and *COL1A2* genes, respectively. Formation of the triple helix occurs in the endoplasmic reticulum (ER), and starts at the C-terminus, extending progressively toward the N-terminus in a zipper-like fashion. Interactions between C-terminal propeptides are instrumental in initiating the assembly of the triple helix. Once the collagen molecules have been secreted from the cell, the N- and C-propeptides are cleaved, reducing the solubility of the molecule, which favors the formation of higher order structures called *fibrils*. Collagen helices are cross-linked to one another through lysine and hydroxylysine residues that are first converted to their aldehyde derivatives by the enzyme lysyl oxidase.

Molecular Genetics of OI

Over 300 mutations in the *COL1A1* and *COL1A2* genes have been identified in patients with OI. Nearly all mutations are dominant, and although in most cases they are inherited from a parent, approximately 25% of cases are the result of a spontaneous mutation. The mutations are spread throughout the *COL1A1* and *COL1A2* genes, and no mutational "hot spots" have emerged. There are large variations in the clinical phenotypes between mutations, and even among similar mutations in the same region of the protein. This renders the assessment of prognosis based on DNA testing difficult. As a result, the *type* of OI (see below) is determined based on the clinical presentation, rather than on the nature of the mutation.

Type I OI is usually a result of decreased expression

Clinically, OI has been classified into four different types. Type I has a mild phenotype, consisting of blue sclerae and brittle bones without skeletal deformities. Mutations that yield type I OI usually result from decreased expression of either the *COL1A1* or *COL1A2* genes. For example, a null allele of *COL1A1* would result in half of the normal quantity of $\alpha 1(I)$ chains and, consequently, a reduced number of type I collagen fibrils would be formed; however, these would be normal in structure, which accounts for the fairly mild phenotypes associated with these types of mutations.

Mutations that disrupt helix formation cause more severe OI

Types II through IV vary in severity, with type II being the most severe and types III and IV having intermediate phenotypes. Patients affected with type II OI often have severe skeletal abnormalities, including both fractures and deformities, as well as dark sclerae. These patients often die at a very young age, often within one month of birth. Mutations that cause OI of these types are usually point mutations that replace one of the glycine residues in the helix region with an amino acid with a larger side chain. As a large residue at this position interferes with the core of the triple helix (Figure 4.2), these substitutions can severely disrupt the normal collagen structure. In general, the variability seen in the clinical phenotype is based on the location of the mutation along the chain. Mutations closer to the C-terminus of the protein are more disruptive because this is where folding of the triple helix initiates. Hence, these mutations lead to a more severe disease phenotype. The most severe mutations usually result from spontaneous mutations that occur in the germline or during development of the embryo. Although this patient has a mild version of the disease, it appears that it arose spontaneously because there is no evidence of OI in any of her family members.

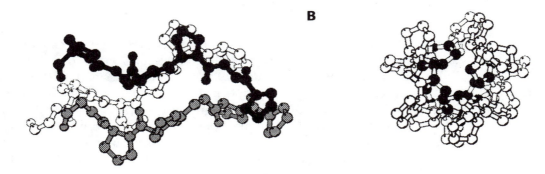

FIGURE 4.2 The structure of the collagen triple helix. (A) The three strands of the helix are shown in black, gray, and white. (B) A view looking down on the helix from the end, showing the glycine residues in black and the proline residues in white. Note the confined space for the glycine residues at the center of the structure.

Severity of these mutations may be explained by a dominant negative effect

Mutations that interfere with helix formation are examples of dominant negative mutations. Type I collagen is an oligomer of three chains, and a mutation in only one chain is sufficient to disrupt the entire triple helix. Suppose that a mutation is present in one of the two $\alpha 1(I)$ chain alleles and that each allele produces the normal amount of protein. Assembly into the triple-stranded structure would result in 25% of the collagen molecules having two normal $\alpha 1(I)$ chains, while 50% would have one normal and one mutant chain, and the final 25% would have two mutated chains. Thus, 75% of the collagen molecules would be expected to be defective, and these mutations may be still more disruptive if they affect the formation of higher order structures.

Incidence

Type I OI has an incidence of approximately 1 in 30,000, whereas the incidence of type II OI has been estimated to be 1 in 60,000. The incidence of types II, III, and IV combined has been estimated to be as high as 1 in 20,000.

Understanding This Patient

Mild disease belies apparent severity of mutation

Based on clinical manifestations, the patient described here appears to have a milder form of OI. She has suffered from multiple fractures, but does not appear to have any skeletal deformities. However, analysis of her type I collagen genes revealed that she has a type of point mutation in the COL1A1 gene that is not typically seen in patients with type I OI. The replacement of glycine with cysteine in the helical region of the $\alpha 1(I)$ chain would be expected to disrupt the formation of the collagen molecule and to have a dominant negative effect, leading to a serious disease phenotype.

Polyacrylamide gel electrophoresis indicates a likely disulfide-bonded species

DNA sequencing revealed that one of the $\alpha 1(I)$ chains contained a glycine to cysteine substitution near the N-terminus of the triple helix. As expected, the reducing gel showed no difference between this patient and the control, as a single point mutation would change the molecular weight of the peptide by such a small amount that it would not be detectable by this technique. However, the larger molecular weight band on the nonreducing gel indicates that a new protein product has been formed. Given that this band appears only under nonreducing conditions, it seems likely that this band represents a disulfide bridged species, most likely a dimer of two $\alpha 1(I)$ chains formed by a bond between the two cysteine residues.

Several features of the amino acid substitution contribute to the mild phenotype

Why does this mutation lead to a mild phenotype? First, the mutation is located very close to the N-terminus of the helix, where the helix folding process is nearing termination. Second, cysteine has a relatively small side chain and is likely to be less disruptive than other amino acid substitutions. Finally, one might anticipate that, in a triple helix with two mutated $\alpha 1(I)$ chains, the cysteine residues might be able to form a disulfide bond, which would then stabilize the entire assembly. This is supported by the gel electrophoresis analysis.

Diagnosis and Treatment

Diagnosis of OI is made on the basis of clinical findings, in particular, skeletal abnormalities, blue sclerae, and a history of bone fractures with little or no trauma. Molecular analyses, such as the polyacrylamide gel electrophoresis and DNA sequencing carried out on this patient, may be used to confirm the diagnosis. Patients with type I OI may usually be managed conservatively with treatment of fractures and physical therapy; however, more serious cases may require aggressive surgical intervention. Experimental therapies that are showing some promise include bisphosphonates, which reduce bone resorption, and transplantation of bone marrow stromal cells (which may differentiate into osteoblasts) from an HLA-matched donor. The wild-type collagen expressed in the donor cells would contribute to the formation of normal bone, despite the presence of residual mutant peptides produced by host cells.

QUESTIONS

1. $\alpha1(I)$ and $\alpha2(I)$ contain few aromatic amino acids. What might explain this?

2. Scurvy is a disease caused by vitamin C deficiency that affects the biosynthesis of collagen. What is the role of vitamin C in collagen biosynthesis?

3. Can all types of mutations in the *COL1A1* and *COL1A2* genes be identified by either reducing or non-reducing polyacrylamide gel electrophoresis of the secreted polypeptides?

4. How might cyanogen bromide (CNBr) have been used to localize the region of the gene affected by the mutation in the patient described here?

5. In the patient described here, the disease causing mutation was found to be a glycine-to-cysteine substitution at amino acid 94 of $\alpha1(I)$. Do you think the severity of disease would be greater or less for the equivalent mutation in $\alpha2(I)$?

6. Mutations in the type I collagen genes that result in impaired formation of the triple helix often lead to *overhydroxylation,* which may be visualized by electrophoresis as broadening of the bands representing each chain. What accounts for overhydroxylation in these mutants?

7. What are some of the variables that influence disease severity among mutations that cause OI?

FURTHER READING

1. Kocher, M. S., and Shapiro, F. Osteogenesis imperfecta. *J. Am. Acad. Orthop. Surg.* (1998) 6(4):225-236.

2. Myllyharju, J., and Kivirikko, K. L. Collagens and collagen-related diseases. *Ann. Med.* (2001) 33(1):7-21.

3. Dalgleish, R. The human type I collagen mutation database. *NAR* (1997) 25(1):181-187.

4. Cole, W. G. The Nicholas Andry Award—1996. The molecular pathology of osteogenesis imperfecta. *Clin. Orthop.* (1997) 343:235–248.

5. Starman, B. J., Eyre, D., Charbonneau, H., et al. Osteogenesis imperfecta. The position of substitution for glycine by cysteine in the triple helical domain of the pro$\alpha1(I)$ chains of type I collagen determines the clinical phenotype. *J. Clin. Invest.* (1989) 84:1206–1214.

For further information, see the following web sites:

Osteogenesis Imperfecta Foundation: www.oif.org/site/PageServer

National Institutes of Health, Osteoporosis and Related Bone Diseases, Osteogenesis Imperfecta: www.osteo.org/oi.html

Online Mendelian Inheritance in Man (OMIM), Osteogenesis Imperfecta type I: www3.ncbi.nlm.nih.gov:80/htbin-post/Omim/dispmim?166200

The Flow of Genetic Information
Retroviruses Break the Rule

CASE HISTORY

Early Symptoms and Hematologic Findings

A 34-year-old Japanese man was admitted to the hospital with a presumptive diagnosis of acute leukemia. The patient had been well until four weeks earlier, when he began to experience malaise, fever, and lymphadenopathy. Epistaxis (nose bleed) occurred and required cauterization. Two weeks before admission, the fever resolved, but the lymphadenopathy and malaise increased; his gums became swollen and painful and bled intermittently. He began to experience headaches with intermittent blurred vision. Hematologic tests revealed a hematocrit of 24%, *white-blood-cell count (WBC)* of 44,000 per cubic millimeter with 5% neutrophils, 9% lymphocytes, 2% eosinophils, and 84% blast forms. The platelet count was 22,000 per cubic millimeter. He was referred to this hospital for further evaluation and therapy.

Are the results of the hematologic tests normal?

Besides the hematocrit, what other type of test is commonly used as a measure of the number of erythrocytes?

Why is this patient's ethnic origin of concern?

Patient History

The patient is a computer programmer. He has no significant occupational exposures and denies knowledge of exposure to bone marrow toxins or ionizing radiation. There was no history of the use of tobacco, alcohol, or illicit drugs. Prior to four weeks ago he denied any history of fever, chills, sweats, weight loss, easy bruising, hematuria or hematochezia. There is no significant family illness and both parents are alive and well in their early seventies. He has two older brothers, one with hypertension and the other without significant medical illnesses, although he is a smoker. He has lived in the United States for the past 15 years, and his only significant travel is yearly visits back to Japan. He takes vitamin C (500 mg daily) and vitamin E (400 international units daily), and uses Tylenol on an as-needed basis for headaches.

Physical Examination and Laboratory Results

His physical examination was remarkable for a temperature of 37.6°C, the pulse was 102 per minute, and the respirations were 22 per minute. The blood pressure was 110/60 mmHg. Examination of the integument revealed petechiae (red spots under the skin) on the legs and the buccal mucosa. There was enlargement of the palatal tonsils, cervical, submaxillary, and axillary lymph nodes. The chest was resonant to percussion and clear to auscultation. The heart examination was remarkable for tachycardia and a grade II/VI systolic ejection murmur. The abdomen was soft and nontender, and both the liver and spleen were enlarged (hepatosplenomegaly).

Hematological test results were essentially unchanged from those reported at the other hospital. Serum chemistries were remarkable for the following: blood urea nitrogen (BUN) 35 mg/dl; creatine 1.6 mg/dl; uric acid 9.0 mg/dl; calcium 11.0 mg/dl; magnesium 1.8 mmol/l; *lactate dehydrogenase (LDH)* 400 IU/l. The urine was positive for protein and blood. The urine sediment contained five red cells and four white cells per high-power field, with rare hyaline casts.

Why are the serum levels of uric acid and LDH elevated?

What are hyaline casts?

Tests for CNS Involvement of the Disease

A lumbar puncture was performed and revealed a normal *cerebrospinal (CSF)* pressure, and the spinal fluid was clear and colorless with 54 red blood cells and 14 white blood cells per cubic millimeter, with no evidence of leukemic cells. The glucose and protein concentrations of the CSF were 80 and 37 mg/dl, respectively. When cultured, the CSF did not grow any bacteria or fungus.

How is a lumbar puncture performed? Is it a risk-free procedure?

Radiographs of the chest were normal; however, *computed tomographic (CT)* scans of the head, before and after intravenous contrast material, showed thickening of the posterior nasopharyngeal soft tissues, suggesting neoplasia. *Magnetic resonance imaging (MRI)* of the head, performed before and after the administration of gadolinium (a paramagnetic contrast agent), also showed prominent posterior nasopharyngeal soft tissues, but there were no abnormalities of the eyes or orbits.

What is the evidence for CNS involvement? Against?

Immunophenotyping and DNA Analysis of Bone Marrow Cells Indicate Diagnosis of Acute Adult T-cell Leukemia (ATL)

A bone marrow aspirate was performed and revealed increased cellularity with a predominance of blasts. The cells were myeloperoxidase and CD13 negative (myeloid markers). Blasts from the bone marrow aspirate did not express surface or cytoplasmic immunoglobulin (B-cell marker), but were CD4 and CD25 (T-cell markers) positive, and exhibited a "flower cell" morphology. Southern blot analysis revealed the presence of the *human T-cell leukemia virus type 1 (HTLV-1)* provirus in cellular DNA. These cytochemical, morphological, and DNA analyses confirmed the diagnosis of acute lymphoblastic leukemia of a T-cell origin.

How was the subtype of leukemia arrived at? Why is it important to identify the particular subtype?

What is a Southern blot?

Treatment

The patient was treated with a standard multidrug regimen for induction therapy. He received four doses of vincristine (1.5 mg/m^2 of body surface area) over 21 days, two doses of cyclophosphamide (600 mg/m^2) separated by one week, and oral prednisone (60 mg/m^2). He received one dose of methotrexate, cytarabine, and hydrocortisone intrathecally (into the spinal fluid). Radiation was administered to the retinas, optic nerves, and chiasm. The temperature became normal on the fourth hospital day. On the sixth hospital day, the white-cell count declined to 700 per cubic millimeter. He continued the induction chemotherapeutic regimen with a decline in WBC but no evidence of infection. Repeat bone marrow aspiration after the induction regimen revealed less than 5% blast forms.

What are the phases of treatment of leukemia?

Why were intrathecal chemotherapy and cranial irradiation administered?

Has this patient gone into complete remission (CR)?

Why might this patient be susceptible to infection?

DISCUSSION

Retroviruses

The path of genetic information generally flows in the DNA-to-RNA-to-protein direction; however an exception to this rule is exhibited by the retroviruses. This class of virus is notable because it possesses an RNA genome that is transcribed into DNA upon entry into the cell (see Section 5.3.2 of *Biochemistry* 5e). The flow of genetic information thus travels in the *reverse* direction, and hence the class name, derived from the Latin *retro,* or backward. The RNA-to-DNA transcription is achieved by an enzyme unique to retroviruses called *reverse transcriptase,* which is an appealing target for antiretroviral therapy. Specific inhibition of reverse transcriptase offers the possibility of inhibiting viral replication without harming the host because of the absence of the enzyme in host cells. The reverse transcriptase inhibitor, *zidovudine (AZT)*, was the first medication used to combat HIV, the most notorious amongst the retroviruses, and is still a critical part of treatment regimens.

Once transcribed into DNA the viral genome can integrate into host cell chromosomes, and be replicated along with them when the cell divides. At this point the viral DNA can remain integrated for extended periods without causing disease, and is said to be *latent*. Many diseases caused by retroviruses, including the T-cell leukemia discussed here, have long incubation periods during which the virus remains in the inactive, latent form.

The virus becomes reactivated (through processes not completely understood) when its genes begin to be transcribed, generating copies of the viral RNA genome and the proteins necessary for formation of the viral particle. At this point the virus can spread to other cells and cause disease, in this case, *acute adult T-cell leukemia (ATL)*.

HTLV-1 and ATL

HTLV-1 specifically infects T-lymphocytes, a type of immune cell. The association between HTLV-1 infection and ATL is clear: nearly all patients with the disease are *seropositive* for the virus (possess antibodies specific for the virus), and the *provirus* (double-stranded DNA copy of viral genome) is found in malignant cells but not in normal cells from the same individual. The HTLV-1 genome (Figure 5.1) encodes a protein called Tax, a transcriptional activator, which is necessary for cellular *transformation* (conversion of a cell into the malignant state). Tax most likely transforms cells through its capacity as a transcriptional activator: not only does it potently stimulate the transcription of viral genes, but it also acts upon host cell genes that promote cell growth and division.

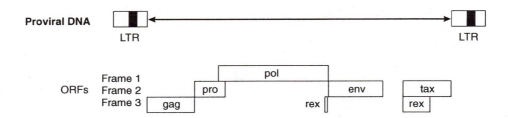

FIGURE 5.1 Genetic map of human T-cell leukemia virus-1 (HTLV-1). The genome is 9032 nucleotides long and contains the *gag*, *pro*, *pol*, and *env* genes common to all retroviruses, as well as the *rex* and *tax* genes. The various gene products are generated by frameshifting, alternative splicing, and proteolytic processing. LTR: long terminal repeat, ORF: open reading frame.

Epidemiology

ATL is especially prevalent in southern Japan, the Caribbean, central and southern Africa, and South America, especially Brazil. It is transmitted through sexual contact, contaminated needles, blood transfusions, and from mother to infant through breastfeeding. This patient may thus have become infected with HTLV-1 as a child in Japan before immigrating to the United States or during his yearly visits there. The lifetime risk of acquiring the disease is 1–4 % in those infected with the virus, and a long latency period (10–40 years) typically precedes the appearance of illness.

Pathophysiology

Impaired hematopoiesis is usually the first indication of leukemia

This patient presented with typical features of leukemia. The first manifestations of disease are usually the result of impairment of hematopoiesis (blood cell development) in the bone marrow. The accumulation of malignant cells there inhibits the development of other, normal blood cells. Thus, the red blood cell count is inversely related to the degree to which the bone marrow is infiltrated with leukemic cells (or the tumor burden). The general malaise and unusual bleeding experienced by this patient are hallmarks of reduced levels of erythrocytes and platelets, respectively, and the low hematocrit and platelet count confirmed this. The rapid onset of illness and elevated white blood cell count with a large percentage of blast forms were indications that the disease was an acute form of leukemia, rather than a less aggressive *chronic* form. At this point the patient was referred to the second hospital for further evaluation and therapy.

Physical examination and laboratory tests reveal further signs of leukemia

Physical examination of the patient revealed additional signs of leukemia. The petechia observed on the legs and buccal mucosa are further signs of abnormal bleeding, and the enlarged lymph nodes, tonsils and spleen are indicative of leukemic cell accumulation in the lymphatic system. The blasts also invaded the liver, causing hepatomegaly, a common symptom. The elevated LDH and uric acid levels are typical, and are a result of high leukemic cell turnover. Uric acid has limited solubility in the blood and precipitation in the joints will cause gout and in the kidneys can cause nephropathy, a common complication of leukemia that can lead to kidney failure and the requirement for dialysis. The protein found in the urine of this patient is indicative of kidney damage, as is the elevated BUN, serum creatinine, and magnesium levels. The elevated calcium level found in the patient is also a common feature of the disease, and is a result of bone resorption stimulated by the malignant cells.

Leukemia often affects the central nervous system

The headaches and blurred vision that this patient experienced as among the first symptoms are signs of CNS involvement of the disease. Although no evidence of leukemic cells was found in the CSF, the CT and MRI scans both revealed thickening of the posterior nasopharyngeal soft tissues, suggesting neoplasia. Although a minority of patients initially present with CNS involvement, as appears to be the case here, it usually eventually occurs unless CNS prophylaxis treatment is administered.

Diagnosis

There are over a dozen different types of leukemia, and identifying the specific subtype is important as it assists the physician in selecting the optimal treatment regimen. Diagnosis of ATL is based on several criteria, beginning with immunophenotypic analysis. The malignant T cells have a characteristic appearance, with a multilobuled nucleus, which has given rise to the term "flower cells" in describing them. They also almost invariably express the T-cell-specific markers, CD4 and CD25, and often express CD2, CD3, and CD5 as well. The myeloid specific markers, myeloperoxidase and CD13, are absent.

HTLV-1 serology is usually positive, but the diagnosis rests upon the detection of proviral DNA in malignant cells derived from the patient, as determined by Southern blot or (polymerase chain reaction) PCR.

Treatment

Acute lymphocytic leukemia (ALL) is generally uncommon in adults, with most cases occurring in children between the ages of 3 and 5. Although there has been much success in treating the childhood forms of the disease, the adult forms are more refractory to treatment. Although complete remission is often initially achieved, relapse usually occurs and the median survival time from diagnosis is approximately six months for those with acute ATL. Unfortunately, this patient, with a large tumor burden, has a poor prognosis.

Treatment of leukemia is generally divided into three phases: *induction, intensification,* and *maintenance.* The goal of induction is *complete remission (CR),* which is defined as the presence of fewer than 5% blast cells in a bone marrow aspirate. Even when undetectable microscopically, sensitive methodologies such as PCR usually uncover the presence of residual leukemia cells or cells with leukemic features in patients in CR (so-called *minimal residual* disease). Aggressive postinduction intensification has thus been a focus of therapeutic regimens and targets minimal residual disease. The goal of maintenance is the preservation of remission.

CNS involvement will occur in 50–75% of patients unless prophylactic treatment is administered. Hence introduction of chemotherapeutic agents intrathecally, into the CSF, is common practice. Cranial irradiation was additionally performed in this case due to evidence of leukemic cell invasion of the posterior nasopharyngial soft tissues, which lie in close proximity to CNS structures.

QUESTIONS

1. What is the difference between an *acute* leukemia and a *chronic* leukemia?

2. Although excessive numbers indicate a malignancy, why is it considered normal to find a limited number of blast forms (up to 30%) in a bone marrow aspirate?

3. Why are malignant cells more sensitive than normal cells to drugs that inhibit DNA replication, such as cyclophosphamide and methotrexate?

4. Upon entry of HTLV-1 into a host cell, reverse transcriptase generates a DNA copy of the single-stranded RNA genome. Initiation of this process requires a primer with a free 3'-OH group (Section 5.3.1 of *Biochemistry* 5e). What serves as the primer in this case?

5. The discovery of retroviruses and reverse transcriptase generated much excitement among molecular biologists because it was the first demonstration of the flow of genetic information traveling in the reverse direction. Does the flow of information travel in the forward direction at any point in the viral life cycle?

6. The error rate of reverse transcriptase is substantially greater than that of mammalian DNA and RNA polymerases. How does this unusually high error rate benefit the virus?

7. Retroviral genomes possess long terminal repeats (LTRs) several hundred nucleotides long at their 5'and 3' ends. What function(s) do these LTRs perform in the viral life cycle?

FURTHER READING

1. Bangham, C. R. HTLV-1 infections. *J. Clin. Pathol.* (2000) 53:581–586.

2. Bazarbachi, A., and Hermine, O. Treatment of adult T-cell leukaemia/lymphoma: Current strategy and future perspectives. *Virus Research* (2001) 78:79–92.

3. Coffin, J. M., Hughes, S. H., and Varmus, H. E. Retroviruses. *Cold Spring Harbor Press* (1997). [Can be viewed online at: www.ncbi.nlm.nih.gov:80/books/bv.fcgi?call=bv.View..ShowTOC&rid=rv.TOC]

4. Baltimore, D. RNA-dependent DNA polymerase in virions of RNA tumour viruses. *Nature* (1970) 226(252):1209–1211.

5. Temin, H. M., and Mizutani, S. RNA-dependent DNA polymerase in virions of Rous sarcoma virus. *Nature* (1970) 226(252):1211–1213. [These two *Nature* articles describe the discovery of reverse transcriptase.]

For further information, see the following web sites:

National Center for Biotechnology Information: www.ncbi.nlm.nih.gov/retroviruses/

National Institutes of Health, Medline Plus Health Information: www.nlm.nih.gov/medlineplus/leukemiaadultacute.html

National Cancer Institute, Cancernet: http://cancernet.nci.nih.gov/cgi-bin/srchcgi.exe?DBID=pdq&TYPE=search&SFMT=pdq_statement/1/0/0&Z208=208_01024H

The Leukemia and Lymphoma Society: www.l3.leukemia-lymphoma.org/hm_lls

CHAPTER 6

Exploring Genes
New Tools for Predicting the Course of Cancer

CASE HISTORY

Mammogram Reveals Abnormality

A 53-year-old, postmenopausal woman underwent an annual screening mammography. She had no known risk factors for breast cancer and was not on hormone replacement. A 5-mm mass with indistinct margins and pleomorphic calcifications, which was not seen on previous mammograms, was noted in the lower inner-right breast (Figure 6.1). The remainder of both breasts was fatty.

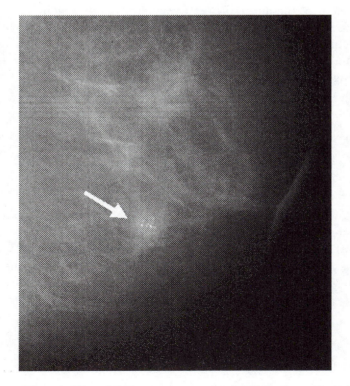

FIGURE 6.1 Mediolateral oblique (MLO) mammogram shows indistinctly marginated 5-mm mass (arrow) with associated pleomorphic microcalcifications.

What is the significance of the mammography findings?

What are calcifications?

Biopsy Revealed Invasive Ductal Cancer, but No Evidence of Metastasis

Stereotactic (using mammography image-guidance) percutaneous core biopsy was performed on this lesion, and revealed grade-III invasive ductal cancer. Needle-localized excision and sentinel lymph-node biopsy were subsequently performed. The cancer was 4 mm in size as determined microscopically, consistent with a T1a tumor. The estrogen receptor (ER) and progesterone receptor (PR) levels were determined through the use of specific antibody staining. No significant expression of these proteins was detected. High levels of the cell-surface marker Her-2 were detected using a similar test. The margins were clear and the sentinel lymph node showed no evidence of tumor, even with cytokeratin staining. The malignancy was assigned the stage I status.

What is the advantage of sentinel lymph-node biopsy compared with biopsy of a range of lymph nodes adjacent to the affected breast?

What is the significance of the immunohistological findings?

How is breast cancer staged?

What is cytokeratin staining?

Adjuvant Therapy

Radiation treatment of the affected breast was performed, which is standard treatment after breast-conservation surgery. Systemic adjuvant chemotherapy with the cytotoxic drugs adriamycin and cyclophosphamide (AC therapy) would not generally be given for a stage-I malignancy; however, her tumor was grade III, ER and PR negative, and Her-2 strongly positive. With these negative prognostic factors, the patient opted for chemotherapy.

What are the pluses and minuses of AC therapy?

Patient Enrolls in a Clinical Research Study

After the chemotherapeutic treatment was complete, she enrolled in a research study intended to test the use of gene microarray analysis in predicting the likely course of progression of breast tumors. Examination of RNA isolated from her tumor revealed that the gene expression pattern matched that associated with the occurrence of distant metastasis. While this finding was disturbing with regard to the potential aggressiveness of the excised tumor, it supported her decision to have adjuvant chemotherapy.

How can microarray analysis be used to predict the outcome of disease?

DISCUSSION

There Are Two Main Types of Breast Cancer

Breast cancer is a disease that originates in breast tissue, most often in the glands that produce milk following a pregnancy, or in the ducts that carry the milk to the nipple. When it originates in the milk-producing glands, or *lobules*, the tumor is said to be a *lobular carcinoma*; when it originates in cells of the *ducts*, it is referred to as a *ductal carcinoma* (Figure 6.2). Invasive ductal carcinomas are the most common type of breast cancer, accounting for 80% of cases. The term *carcinoma* indicates that the cancer is a malignancy of epithelial cells. Carcinomas are the most common human cancers, accounting for 90% of cases.

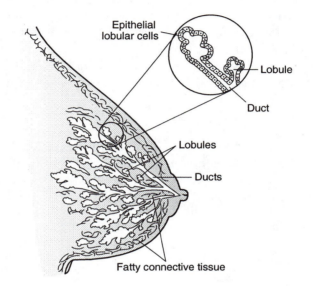

FIGURE 6.2 Breast tissue contains a network of ducts and lobules. Breast cancer usually originates in the epithelial cells lining these structures.

Epidemiology

Breast cancer is the second most common (after skin cancer) and second most deadly (after lung cancer) malignancy amongst women in the United States. Approximately 1 in 9 women who live to the age of 80 will develop breast cancer in North America and Western Europe; however, the risk is only 1/5 to 1/10 of this among Asian women. Men may also develop breast cancer; however, the risk is small, with men accounting for only 0.1% of cases. Mortality from breast cancer has been on the decline in the United States since 1995, largely as a result of routine screening and improved treatments.

Risk Factors

Genetic factors

Certain genetic factors are associated with breast cancer, with the most notorious, *BRCA-1*, encoding a protein involved in DNA repair. Individuals who inherit certain mutated alleles of this gene have between 60% and 80% lifetime risk of developing breast cancer. However, most cases of breast cancer are not related to *BRCA-1*: it has been estimated that only 5% of cases occur in women with *BRCA-1* mutations. A second gene, *BRCA-2*, has also been linked to breast cancer; however, mutations in this gene are less common. Mutations in these genes are also associated with increased risk of ovarian cancer. Mutations in the p53 tumor-suppressor gene are also associated with breast cancer, as well as a number of other malignancies. Mutations in p53 increase one's risk of developing breast cancer, and worsen the prognosis if the disease occurs.

Age

Age is a significant risk factor for the development of breast cancer, with risk increasing in an almost linear fashion with each year of life. A woman's chance of breast cancer increases from 1 in 2200 at age 30 to 1 in 9 by age 80, according to statistics compiled by the National Cancer Institute. In addition, earlier age of menarche, later age at the first pregnancy, and later age at menopause all increase the risk of developing the disease. Thus, the duration of the menstrual term of a woman, especially the length of the fraction prior to the first pregnancy, is a key determinant in breast cancer risk, likely due to prolonged exposure to estrogen.

Dietary and hormonal factors, and smoking

The contribution of dietary factors, such as fat and caloric intake, is controversial and no definitive links have yet been made with either of these. However, most physicians agree that moderate consumption of alcohol increases one's risk. Smoking has also been associated with slightly increased risk. The risks associated with oral contraceptives or hormone replacement therapy are low to none and are counterbalanced by the multiple protective effects conferred by these exogenous hormones.

Prognostic Factors

Staging

A number of features of the tumor have been found to influence the chance of tumor relapse and the prognosis of the patient. These prognostic factors are used in making treatment decisions as well as in predicting prognosis, and they are thus usually carefully assessed. The size of the tumor, spread to regional lymph nodes, and metastasis to a distant site are among the most important factors and are used in the *TNM staging* (0–IV) of the patient. Larger tumors, especially those with diameters greater than 5 cm, are associated with a poorer prognosis, as is spread to the regional lymph nodes.

The lymphatic system is a network of vessels that carries *lymph,* which includes tissue waste products. The lymphatic vessels of the breast drain into the lymph nodes found under the arm, the *axillary* lymph nodes, and thus these are sampled in breast cancer patients. The sentinel lymph-node sampling performed on this patient involves injection of a radioactive isotope of technetium (^{99m}Tc as a sulfur colloid), as well as isosulfan blue dye in the region of the breast adjacent to the lesion. The radiation emitted by the technetium and the location of the blue dye are then used to identify the first node through which the region of the tumor drains. Many physicians now more commonly use the sentinel lymph node sampling procedure rather than performing axillary surgery to remove and sample multiple nodes, as it is less invasive and evidence has shown it to be equally accurate in assessing the status of the axillary lymph nodes. Detailed examination of the sentinel node is performed, and if no tumor cells are found, then the chance of a "skip" metastasis is less than 5%.

Finally, metastasis to a distant site is a poor prognostic factor and corresponds to the most advanced stage of disease (IV), regardless of the size of the tumor or status of the lymph nodes. Common sites of distant metastases are bone, lung, and liver. This patient, with a small, invasive tumor and no lymph-node involvement or metastasis, was assigned the stage-I status. Her tumor was further subclasssified as TIa, indicating that it was ≤0.5 cm in size.

Tumors may also be *graded* (I-III) based on histologic features of the tumor, such as the number of mitotic cells and the size and degree of pleiomorphism among the cellular nuclei. Indistinct margins and pleomorphic calcifications are considered poor prognostic factors, and this patient's tumor was assigned a high grade of III.

Other factors

Other factors also contribute to the prognosis. Tumors often induce the formation of new blood vessels (angiogenesis) in their vicinity, which provide the nutrients and oxygen that promote tumor growth and progression. Once this process has begun the prognosis worsens. The tumor may also be evaluated at the cellular level through the use of molecular markers. The Her-2/neu (also called erb-B2) protein, a member of the epidermal-growth-factor (EGF) receptor family, is overexpressed in ~25% of breast cancers and is associated with a higher risk of relapse and death. Conversely, the presence of the ER is a favorable prognostic factor, indicating that the cells have maintained some degree of differentiation (neoplastic cells often fail to differentiate). In addition, the presence of ER indicates that the patient will likely respond to *tamoxifen* (see below), an estrogen receptor *antagonist* and one of the most effective drugs used to combat breast cancer. The presence of the progesterone receptor also contributes to a favorable prognosis. Thus, although this patient's tumor was small, the molecular markers of her tumor (no detectable ER or PR, and abundant expression of Her-2) suggested an aggressive malignancy.

Use of gene expression profiles may return us to the age of personalized medicine

The development of gene microarray analysis (see Section 6.3.2 of *Biochemistry* 5e) holds much promise for medicine. Whole genome analyses will allow for individualized assessment of disease risk, as well as assist in making therapeutic decisions should illness strike. Sensitivity to a number of drugs, such as codeine in the treatment of pain and tamoxifen in treating breast cancer, varies between individuals, and microarray analyses will assist physicians in choosing an effective therapeutic regimen for each individual patient. The more accurate assessment of disease risk that this technology will bring will also be clinically important as many of the most common illnesses can be effectively treated if identified early. Thus, although most physicians no longer make house calls, microarray analysis may nevertheless return us to an era of personalized medical treatment.

The use of gene expression profiles in predicting the outcome of breast cancer is still in the early stages, and its use in making treatment decisions is not yet approved by the FDA. The analysis is performed by isolating mRNA from a tumor sample and comparing it against an mRNA sample isolated from a *control* tissue sample. Analysis of the data identifies genes that are increased or decreased in expression in the breast cancer sample compared to the control tissue sample. Indeed, in one study comparison of samples from tumors that recurred with those that did not over a five-year period led to the identification of approximately 70 genes whose expression is associated with recurring tumors (see reference below, van't Veer et al., 2002). The set included genes encoding proteins involved in cell proliferation, blood-vessel growth, and the degradation of the extracellular matrix, consistent with a direct role in tumor growth and the potential for spread to distant locations in the body.

Treatment

Removal of the tumor

The first step in the treatment of breast cancer is the surgical removal of the tumor. This may be done either by *mastectomy*, which involves removal of the entire affected breast, or breast conservation surgery (BCS), which removes only the tumor along with a rim of healthy tissue surrounding it. Although mastectomy used to be standard treatment for breast cancer, more recently numerous studies have shown that BCS is equally effective as measured by 10-year survival rates for many types of tumors. *Lumpectomy*, the most common form of BCS, is thus becoming more common, and was the surgical procedure performed on this patient. She initially underwent a *stereotactic core biopsy*, which involves the use of several mammograms to localize the lesion, which is then sampled with a core (hollow) needle. Because in this case the lesion was small the tumor was removed using *needle-localized excision*, in which mammograms are used to guide the insertion of a needle (or fine wire) into the lesion, which is then surgically removed, together with some surrounding tissue. Larger tumors that may be felt may not necessitate the use of needle localization. Although mastectomy is used far less now than in the past, it is nevertheless recommended in some cases, such as when tumors are larger than 5 cm or when multiple tumors have arisen in different quadrants of the breast.

Adjuvant radiation

Adjuvant *radiation* treatment of the affected breast is standard practice for those who have undergone BCS. Radiation damages DNA and kills cells, especially those that are dividing, and it is thus used as a complement to BCS, killing any malignant cells that may not have been removed by surgery. Radiation therapy usually involves external beam radiation five days per week for a six-week period. There are few side effects associated with it and several studies have shown that it reduces the risk of recurrence. Radiation therapy is not routinely recommended when a mastectomy has been performed unless there are tumor cells at the deep margins of excision or the initial tumor was extensive.

Adjuvant chemotherapy

In contrast to radiation therapy, a number of factors must be considered before the decision to undergo adjuvant *chemotherapy* may be made. The decision is largely based on the risk of recurrence; however, other factors, such as the age of the patient and the presence of other medical conditions are also taken into account. With the exception of very advanced cases or those that are clearly non-invasive, risk of recurrence may be difficult to assess. In addition, the potential benefits of chemotherapy must be weighed against the discomfort associated with it, as well as the harmful effects. In contrast to radiation therapy, which is localized, chemotherapy is systemic, meaning that it enters the general circulation and affects cells of the entire body. It thus acts upon malignant cells that may have escaped from the primary breast tumor and lodged elsewhere in the body. AC therapy, the treatment undergone by this patient, is usually given cyclically, with alternating treatment and "rest" phases, for a period of 3–6 months. The most common side effects associated with it are hair loss, nausea, immunosuppression, as well as small chances of heart damage and leukemia.

Statistics have shown that women at all stages of breast cancer stand to benefit from chemotherapy; however, although it can reduce the risk of recurrence by as much as 15–20% among high-risk patients, the potential benefit shrinks to near insignificance in women with good prognostic factors. In addition, it is important to keep in mind that statistics measure *overall rates* within a population, which are not necessarily reflected in each individual. Invasive tumors have different characteristics, which may affect their response to chemotherapeutic agents. This patient, with no evidence of spread yet with a histologically aggressive tumor, was among the many in a "gray area" in which the decision to undergo chemotherapy is not clear-cut. Herein lies the extraordinary potential of microarray analysis: the identification of genetic factors associated with good responses to chemotherapeutic agents will allow for the straight-forward identification of patients who will benefit from treatment. This patient was not able to benefit from microarray analysis because of the preliminary nature of the research findings at the time of her diagnosis. However, the analysis performed subsequent to treatment validated her decision to undergo the chemotherapeutic regimen, as her tumor was found to have a gene expression pattern consistent with an aggressive tumor.

Adjuvant hormonal therapy and immunotherapy

Numerous lines of evidence implicate estrogen in the development of breast cancer, and hence the negative impact of a lengthy menstrual term (see Risk Factors, page 45). Thus, a number of hormonal therapies have been directed at inhibiting the effects of this hormone. The first adjuvant therapy to be used was the oophorectomy (removal of the ovaries), which dramat-ically reduces the level of circulating estrogens, and improves survival. Currently, the anti-estrogen, *tamoxifen,* is the most widely used hormonal adjuvant therapy. Tamoxifen binds the estrogen receptor, blocking estrogen binding, and thereby signaling through the receptor. Estrogen binding to its receptor can stimulate the growth of breast cells, hence the beneficial effect of tamoxifen. Tamoxifen is effective regardless of the levels of circulating estrogens; how-ever, the drug is ineffectual among ER-negative women. Other drugs used to limit estrogen levels are aromatase inhibitors or luteinizing hormone-releasing hormone (LHRH), both of which inhibit endogenous estrogen production.

Finally, trastuzumab (Herceptin), a monoclonal antibody to Her-2, is the first *immunotherapeutic* drug to show significant clinical benefit in breast cancer patients. Studies have shown that in patients overexpressing Her-2, trastuzumab combined with chemother-apy resulted in overall 25% increase in survival compared to patients undergoing chemother-apy alone. In the United States, trastuzumab is currently approved only for use in metastatic cancer, and thus it was not prescribed for this patient; however, clinical trials for its use in patients with less-advanced disease are underway.

Thus, as exemplified by tamoxifen and trastuzumab, analysis of the molecular determi-nants of a tumor have already assisted physicians in choosing treatment regimens for their patients. Microarray analysis will likely soon extend this to the many chemotherapeutic agents available for treating breast cancer, ultimately increasing survival and minimizing adverse ef-fects in patients.

QUESTIONS

1. An important assumption is made when using microarray analyses as a measure of cel-lular gene expression. What is this assumption, and do you think it valid?

2. Her-2 expression is increased in approximately 25% of breast tumors, and in most cases this was found to be the result of *gene amplification,* that is, the gene was present in mul-tiple copies in these tumors. Which of the following techniques could identify this ge-netic peculiarity: Southern blot, Northern blot, or Western blot?

3. Which of the following molecular biology tools is *not* used in microarray analysis: PCR, hybridization, generation of cDNA, probe-labeling, or Western blotting? Briefly describe the roles of the above-mentioned techniques that *are* used in microarray analysis.

4. What makes breast cancer particularly well-suited to microarray analysis, compared to diseases such as cystic fibrosis or Huntington's disease?

5. Why was this patient not treated with tamoxifen, one of the most effective drugs used in adjuvant therapy to combat breast cancer?

6. Both adjuvant chemotherapy and adjuvant hormonal therapy are systemic treatments; however, chemotherapy is generally associated with more side effects. What might account for this?

7. Genetic testing for mutations in *BRCA-1* and *BRCA-2* are recommended for some high-risk individuals because some alleles increase the chance of developing breast or ovarian cancer. The genetic test is usually carried out using *PCR*, followed by *allele-specific oligonucleotide hybridization*. Briefly describe how this test is carried out.

FURTHER READING

1. van't Veer, L. J., Dai, H., van de Vijver, M. J., et al. Gene expression profiling predicts clinical outcome of breast cancer. *Nature* (2002) 415(6871):530–536.

2. Perou, C. M., Sorlie, T., et al. Molecular portraits of human breast tumours. *Nature* (2000) 406(6797):747-752.

3. Alizadeh, A. A., Ross, D. T., Perou, C. M., and van de Rijn, M. Towards a novel classification of human malignancies based on gene expression patterns. *J. Pathol.* (2001) 195(1):41–52.

4. Eyster, K. M., and Lindahl, R. Molecular medicine: a primer for clinicians. Part XII: DNA microarrays and their application to clinical medicine. *SDJ Med.* (2001) 54(2):57–61.

5. Going, J. J., Mallon, E. A., Leake, R. E., Bartlett, J. M., and Gusterson, B. A. What the clinician needs from the pathologist: evidence-based reporting in breast cancer. *Eur. J. Cancer* (2001) 37(Suppl. 7):S5–S17.

For further information, see the followinng web sites:

American Cancer Society, Breast Cancer: www.cancer.org/eprise/main/docroot/CRI/CRI_2x?sitearea=LRN&dt=5

National Cancer Institute, Breast Cancer: www.nci.nih.gov/cancer_information/cancer_type/breast

Centers for Disease Control, Breast and Cervical Cancer: www.cdc.gov/cancer/nbccedp/

Nature Medicine, Special Focus on Breast Cancer: www.nature.com/nm/special_focus/bc/

Exploring Evolution
She may give you one, but can you inherit your mother's headache?

CASE HISTORY

Patient History

A 29-year-old woman was evaluated in the neurology clinic for headaches of increasing frequency and severity. The patient had been reasonably well until two years earlier, at which time she began experiencing episodic mild headaches that resolved spontaneously or with the administration of aspirin or other *nonsteroidal anti-inflammatory drugs (NSAIDs)*. She had recently relocated from her home in California to take a job as a congressional staffer in Washington, D.C., when she began to experience more frequent and severe headaches. The headaches are described as throbbing, bilateral head pain often heralded by flashing lights and "closing in" of her peripheral vision. The headaches are associated with photophobia and last for many hours (up to a day) in spite of aspirin, Tylenol, or other NSAIDs. Previously, she experienced a headache once every several months, but now headaches severe enough to cause her to miss work occur every two to three weeks. She denies dizziness, lightheadedness, blurred or double vision. She does not experience other neurological symptoms such as numbness, tingling, weakness, or clumsiness of movement. There has been no recent history of fever, rash, swollen lymph nodes, or scalp tenderness. When the headaches are particularly severe they are associated with nausea, but not vomiting. There has been no change in her cognitive abilities, however, while she has a headache her attention span is severely curtailed.

The symptoms this patient describes are consistent with migraine headache, of which there are two types: classic migraine (migraine with aura) and common migraine (migraine without aura). Which type do you think she has?

What is an aura?

There Are No Obvious Precipitants to the Headaches

She notes no obvious precipitants of the headaches. She denies any recent change in her sleep pattern, and has not significantly changed her diet. She drinks one or two cups of coffee per day and has for many years, and denies that this has an effect on her headaches. Her last menstrual period was two weeks ago and was normal in flow and duration, and the menstrual period prior to this was approximately six weeks ago, and again typical. She does not perceive any relationship between the headaches and menses. She currently is not sexually active. She drinks alcohol socially no more than once per week—typically white wine—and does not believe that alcohol consumption is associated with headache; however, she does not drink when experiencing a headache. She is a nonsmoker and does not use any illicit drugs. She takes a multivitamin daily but no other medicines.

What are some common precipitants to migraines?

A CT Scan Taken While She Was Experiencing a Headache Was Normal

One week prior to this clinic visit she was seen in the emergency room for a particularly severe headache. At that time she underwent a *computed tomographic (CT)* scan of the head with contrast which was normal. She was treated with an intramuscular injection of meperidine (Demerol) and hydroxyzine (Vistaril) with relief of her pain over several hours.

There Is a Family History of Migraine Headaches

Her mother experiences migraines that have gotten less severe as she has gotten older. She has two brothers and two sisters and both sisters and one brother have migraine headaches. None of the siblings has any other neurological diseases, nor do the siblings experience neurological symptoms when they have a headache. There is no family history of paralysis, ataxia, vertigo, or other significant neurological disease.

Is migraine headache a genetic disease?

Physical Examination and Laboratory Results

On examination in the clinic, the patient appeared well: her pulse was 84 per minute, and blood pressure was 108/66 mm Hg. The temperature was 36.6°C, and the respirations were 16 per minute. The head, eyes, ears, nose, and throat examinations were normal. The neck was supple with normal carotid pulsations and no thyroid enlargement. The chest and cardiac examinations were normal. The abdomen was soft and nontender with active bowel sounds, no masses were palpated, and there was no organomegaly (enlargement of the internal organs). The neurological examination revealed the patient to be completely oriented to time, place, and circumstances; she was fluent without significant dysarthria. Motor and sensory examinations were normal, and all reflexes were normal. There were no cranial nerve abnormalities, and cerebellar function was normal.

Laboratory evaluation included normal blood chemistries and a complete blood count remarkable for a mild anemia with a packed *red blood cell (RBC)* volume of 35% (normal 36–46%). Thyroid stimulating hormone level was normal at 1.5 μIU/ml. A magnetic resonance imaging (MRI) examination of the head and sinuses performed before and after the injection of gadolinium was normal.

Are migraines associated with any blood-chemistry abnormalities?

Diagnosis and Treatment

The patient was diagnosed with migraines with aura. She was treated with zolmitriptan (Zomig), 2.5 mg, with a repeat dose every two hours up to 10 mg to be taken at the onset of headache. She has been advised to keep a diary of her headaches, including notation of all potential triggers (e.g., food, drink, external stimuli).

Besides migraine, what are the other two main types of headache?

What are the principal treatment options for migraine?

DISCUSSION

Genetic Polymorphisms and Disease

As seen in Chapter 7 of *Biochemistry* 5e, sequence comparisons can identify related genes both within an organism and between distinct organisms. However, even comparison of the *same* gene between different members of a species often reveals slight genetic differences called *polymorphisms,* which result in slightly different gene products. In most cases the various *alleles* (versions of a gene) are considered normal and do not appreciatively alter the function of the gene product. However, under certain environmental conditions, or in combination with particular alleles of other genes, they may confer susceptibility to a disease or disorder. Thus, in many cases a particular allele or set of alleles simply *predisposes* one to a disease, rather than conferring it with near-certainty.

This is in contrast to the clearly genetic diseases, such as cystic fibrosis and Tay-Sachs disease (see Chapters 1 and 12, respectively), which adhere to classical Mendelian genetics, and for which predicting inheritance patterns is straightforward. It is far more difficult to identify genes that are but partial contributors to a disease, and as such the identification of such genes has lagged behind the identification of those involved in monogenic diseases.

Nevertheless, the search for these factors in identifying the underlying causes of diseases such as breast cancer, Alzheimer disease, alcoholism, asthma, type II diabetes, as well as migraine, which is discussed here, has been intense. This is because once identified, genetic screening tests may be developed, allowing for early diagnosis and preventive therapies. In addition, the tailoring of therapies to each individual patient to maximize effectiveness will become possible. The availability of the entire human genome sequence, as well as the "microarray" technology (see Section 6.3.2 of *Biochemistry* 5e) now commonly in use for analyzing whole genomes, will likely lead to an accelerated pace in determining the underlying causes of these multifactorial diseases.

Migraine Headaches

Epidemiology

Migraine headaches are a common affliction, affecting up to 12% of men and 24% of women. They typically begin between the ages of 10 and 40, and often subside after age 55. It has been estimated that 26 million Americans suffer from migraine headaches, which are often serious enough to prohibit daily activities. They are distinct from the two other main types of headache, tension headache and cluster headache. Tension headaches are very common and are characterized by mild to moderate pain, while cluster headaches are rare, but the pain associated with them is excrutiating.

Diagnosis

No laboratory tests are available for the diagnosis of migraine, and thus the diagnosis rests upon the pattern of symptoms. Various diagnostic tests are nevertheless often carried out, such as the neurological examination and CT and MRI scans performed on this patient, to rule out other causes.

Migraine is often preceded by a *prodromal* period (a period of premonitory signs) during which symptoms such as depression, euphoria, irritability, hyper- or hypoactivity, or food cravings are experienced. Subsequently, approximately 30% of patients experience transient neurological symptoms known as an *aura,* which occurs within an hour of the onset of headache, and which is often concurrent with it as well. The neurological symptoms can be visual, sensory, or motor, and can involve language deficits. Visual auras, like the flashing lights and restriction of peripheral vision experienced by this patient, are the most common.

The headache itself is characterized by its duration of 4 to 72 hours. The pain is moderate to severe, throbbing, and is often unilateral, although it can be bilateral, as it is here. It is often associated with sensitivity to light (photophobia) or sound (phonophobia), and with nausea and vomiting. Although having one or two migraine headaches in a lifetime is not unusual, recurrent attacks are abnormal and are recommended for treatment.

Pathophysiology

Although the pathophysiology of migraine is not fully understood, what is known is that it is associated with disturbances in intracranial blood flow: the pattern of intracranial *vasoconstriction* followed by *vasodilation* during a migraine attack has been recognized for decades. The aura is thought to occur when a depolarization wave spreads across the cortex of the brain (called a *cortical spreading depression [CSD]*). This results in the release of excitatory amino acids from nerve cells, and a reduction in intracranial blood flow.

A critical feature of the migraine headache is activation of the trigeminovascular system, and experiments have indeed shown that CSD can activate this system of intracranial nerves and blood vessels, providing a possible link between the aura and the headache. Activation of the trigeminal nerve (a mixed nerve that includes *nociceptive,* or pain-sensing, neurons; see Chapter 32 of *Biochemistry* 5e) causes pain as well as the release of vasoactive neuropeptides, which cause the vasodilation associated with this phase of the migraine.

Serotonin (5-HT) is released during a migraine attack and was once thought to be its cause. It is now believed to be a *homeostasis* (maintenance of equilibrium) mechanism, causing vasoconstriction and reducing neuronal excitability. Several of the most effective drugs used to treat migraine, including the zolmitriptan this patient was prescribed, are structurally related to serotonin (Figure 7.1).

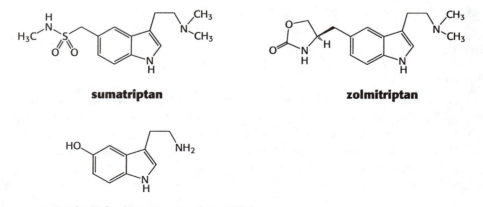

sumatriptan   **zolmitriptan**

serotonin (5-hydroxytryptamine, 5-HT)

FIGURE 7.1 Chemical structures of serotonin and some related drugs.

Etiology

Both genetic and environmental factors are involved in migraine. The etiology of migraine is unknown and the genetic factors involved are still under investigation; however, 70% of patients have a first-degree relative with a history of migraine. This has led to much interest in the search for "migraine genes." In addition, environmental factors appear to conspire with genetic determinants in producing a migraine: there exist several common environmental "triggers" associated with the onset of a migraine attack.

CACNA1A and familial hemiplegic migraine. The clearest genetic association with migraine is between the α1A subunit of a brain-specific, voltage-gated Ca^{2+} channel (CACNA1A) and a rare form of the disorder called familial hemiplegic migraine (FHM). This channel is thought to mediate neurotransmitter release, and hence its role in migraine pathophysiology has generated much interest. At least a dozen alleles of CACNA1A have been associated with FHM; however, only 50% of cases can be linked to CACNA1A, indicating that other genes are involved in the remaining cases.

The dopamine D2 receptor may be linked to some cases of migraine. The inheritance of FHM follows an autosomal dominant inheritance pattern, and thus adheres to the principles of Mendelian genetics. However, the inheritance of most cases of migraine is more complex, involving multiple interacting environmental and genetic factors. Distinguishing and identifying these factors in these cases has been a challenge and few persuasive genetic associations have been made to date. The link between CACNA1A and these more common types of migraine has been inconclusive. The clearest association has been made with the dopamine D2 receptor (DRD2). Two separate studies have identified an association between allelic variation of this neurotransmitter receptor and migraine, suggesting that hypersensitivity of the dopaminergic system could account for some cases. Despite the prominent role of serotonin in regulating cerebral vascular tone, genetic variation in the 5-HT receptor has not been linked to migraine.

Migraine triggers. Although the mechanisms through which they act are unknown, there are several generally recognized triggers for migraines. There is a body of evidence citing cycling estrogen as a trigger, which may explain why they are more common in women than men. In addition, lack of sleep, stress, exposure to loud noises or bright lights, certain foods, food additives (such as monosodium glutamate), and alcoholic beverages (especially red wine) are common triggers. This patient denies the existence of any triggers for her migraines, which may indicate that her threshold for initiation of an attack is so low that a trigger would go unnoticed.

Treatment

Treatment for migraine can be *prophylactic* or *abortive.* Prophylactic treatment is usually recommended when attacks occur more than two or three times per month, and medication is taken on a daily basis as a preventive measure. Commonly used drugs are β-adrenergic blockers and the anticonvulsant, sodium valproate; however, these have limited efficacy and can have severe side-effects, and are thus of limited utility.

Abortive therapy is initiated at the time the headache begins, or at the onset of the prodrome or aura, if they occur. NSAIDs may be used for mild attacks, but the most commonly used drugs are the *ergots* and the *triptans,* both of which act on the large family of *serotonin receptors.* They act through the serotonin 1B and 1D receptors to cause vasoconstriction and reduce pain. Ergots are often poorly tolerated due to their action on a wide variety of receptors and are thus not as frequently prescribed as the triptans.

The zolmitriptan prescribed to this patient is one of a handful of "second-generation" triptan drugs used to treat migraine. The prototype, sumatriptan (see Figure 7.1), is highly effective, reducing the severity of the headache in up to 80% of cases. The advantages of the second-generation triptans lie in their higher oral bioavailability and improved brain penetration, providing more effective and rapid relief. The triptans are generally well tolerated, with their main shortcoming being the high frequency of headache recurrence. Up to 40% of patients who initially respond well suffer recurrence within 24–48 hours.

QUESTIONS

1. Those who suffer from migraine headaches often have close family members with a history of migraines. Does this fact alone indicate a genetic component to the disorder? What else might account for this familial clustering?

2. What types of mechanisms might account for heritable variations in the response to a particular drug?

3. What is the difference between a genetic trait that is *completely penetrant* and one that is *incompletely penetrant?*

4. This patient has three siblings who suffer from migraine, one of which, a sister, was treated prophylactically with the β-adrenergic blocker, timolol. However, she soon developed a profound slowing of heart rate in response to this medication and stopped taking it. Could a genetic polymorphism account for this exaggerated response to timolol?

5. Why is caffeine sometimes used to treat mild migraine headaches?

6. You are a geneticist and through careful analysis of a large family with migraine headaches have determined that the disorder is linked to a previously cloned gene, *X,* which lies on chromosome 14. You set out to identify the polymorphism in gene *X* that confers susceptibility to migraine. You isolate DNA from a family member that suffers from migraine, amplify the DNA using gene specific primers and polymerase chain reaction (PCR), and clone it into a plasmid (see Sections 6.1 and 6.2 of *Biochemistry* 5e). You then sequence the gene and, indeed, find a *single nucleotide polymorphism (SNP)* that causes a missense mutation within one of the exons. Can you conclude that it is responsible for confering susceptibility to migraine?

FURTHER READING

1. Ophoff, R. A., van den Maagdenberg, A. M. J. M., Roon, K. I., Ferrari, M. D., and Frants R. R. The impact of pharmacogenetics for migraine. *European Journal of Pharmacology* (2001) 413:1–10.

2. Ferrari, M. D. Migraine. *The Lancet* (1998) 351:1043–1051.

3. Riley, J. H., Allan, C. J., Lai, E., and Roses, A. The use of single nucleotide polymorphisms in the isolation of common disease genes. *Pharmacogenomics* (2000) 1(1):39–47.

For further information, see the following web sites:

Journal of the American Medical Association, Migraine Information Center: www.ama-assn.org/special/migraine/

National Institutes of Health, Medlineplus Health Information, Headache and Migraine: www.nlm.nih.gov/medlineplus/headacheandmigraine.html

American Council for Headache Education: www.achenet.org/

CHAPTER 8

Enzyme Inhibition
The Basis of Human Immunodeficiency Virus (HIV) Therapy

CASE HISTORY

Patient Is Found to Be HIV Seropositive, but Declines Treatment

A 38-year-old white homosexual man was found to be HIV-seropositive 10 years ago. He denied history of intravenous drug use and had not received any transfusions or blood products, but had had sexual contact with an infected person. He had antibodies to hepatitis B surface antigen and did not have a history of hepatitis C. At the time of diagnosis he felt well and his CD4$^+$ T-lymphocyte count (CD4$^+$ count) was 450 cells/ml and his plasma HIV-1 RNA level was 500 copies/ml. He declined treatment, instead opting to follow his *white-blood-cell count (WBC)* and HIV-1 RNA levels. He had had no history of opportunistic infections and no other complications of HIV-1 infection. There was no history of other serious medical illness, and he takes no medicines other than a multivitamin.

What are the risk factors for HIV-infection?

What is the difference between being HIV-infected and having AIDS?

What is the normal level of CD4$^+$ T lymphocytes in the blood? How low must the level go before one is considered to have AIDS?

What is the length of time between exposure and development of clinical symptoms?

What are the problems with deferring antiretroviral therapy?

Signs of Acquired Immunodeficiency Syndrome (AIDS) Appear and Treatment Is Initiated

Eight years after the diagnosis was made, in the face of falling CD4$^+$ counts and a rise in plasma HIV-1 RNA levels, he was started on antiretroviral therapy. Upon initiation of therapy his CD4$^+$ count was 55 cells/ml and his plasma HIV-1 RNA level was 31,000 copies/mL. He was begun on antiretroviral triple therapy with the protease inhibitor *indinavir* (also known as *Crixivan;* see Section 9.1.7 of *Biochemistry* 5e), and the nucleoside analogs, *lamivudine (3TC)* and *zidovudine (AZT)*. Within two months of beginning therapy his CD4$^+$ counts had risen to >400 cells/ml and HIV-1 RNA levels were <50 copies/ml. Two months later he complained of mild gastrointestinal upset on the drug regimen but the CD4$^+$ count was 550 cells/mL and his plasma HIV-1 RNA level was undetectable.

What is highly active antiretroviral therapy (HAART)?

Why is multiple-drug therapy necessary?

The Patient Is Treated for Herpes Zoster, and Signs of HIV Replication Reappear

Six months after beginning therapy he developed herpes zoster in a lower thoracic distribution. He was treated with valacyclovir, 1 gm three times a day for a week. The rash subsided but he developed postherpetic neuralgia (intense pain along the course of a nerve) which was treated with amitriptyline, 25 mg daily, and acetaminophen with oxycodone. On his return to clinic two weeks later he continued to complain of disabling pain. His CD4$^+$ count remained above 500 cells/ml and plasma HIV-1 RNA level remained undetectable. At this time amitriptyline was discontinued and carbamazepine, 200 mg per day, was added for better control of the neuralgia. He returned to clinic two months later with improvement in pain and a reduction in the requirement for oxycodone. However, repeat testing revealed an increase in HIV-1 RNA to 1100 copies/ml and a CD4$^+$ count of 300 cells/ml.

What is herpes zoster?

Why did the HIV-1 RNA levels rise and the CD4$^+$ count fall?

The Drug Regimen Is Changed and the Infection Brought Under Control

In view of the rising HIV-1 RNA level the antiviral drug regimen was changed. He was begun on two other *nucleoside analog reverse transcriptase inhibitors (NRTI)*, *stavudine (d4T)* and *didanosine (ddI)*, and a *non-nucleoside reverse transcriptase inhibitor (NNRTI)*, *nevirapine*, with restoration of his CD4$^+$ count and suppression of the rise in HIV-1 RNA level.

DISCUSSION

The HIV Life Cycle

Virion structure

HIV is a member of a class of viruses called the *retroviruses* (see Section 5.3.2 of *Biochemistry* 5e). There are two variants, HIV-1 and HIV-2, but HIV-1 accounts for the large majority of infected individuals worldwide. The viral particle (virion) consists of two copies of the RNA genome and several viral enzymes enclosed within a protein shell known as a capsid (see Figure 8.1). A lipid bilayer envelope containing both viral and cellular proteins encases the capsid and its contents.

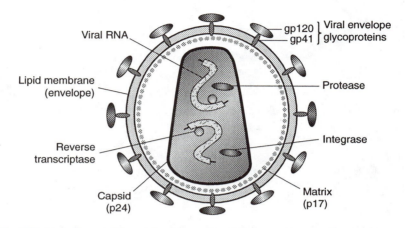

FIGURE 8.1 The HIV virion. The capsid contains two copies of the genome, as well as nucleocapsid (an RNA binding protein), and the viral enzymes, reverse transcriptase, protease, and integrase. The matrix protein is thought to interact with gp41, thus recruiting the viral glycoproteins to the particle during assembly and budding. The glycoprotien gp120 interacts with CD4 during viral attachment.

Attachment and formation of provirus

The main target of the virus is a type of immune cell called a *helper T lymphocyte,* or a *CD4+ T lymphocyte* (see Section 33.5 of *Biochemistry* 5e). The primary gateway through which the virus enters the cell is the CD4 molecule, a transmembrane receptor which lies in the plasma membrane (Figure 8.2). Once bound to the host cell via CD4, the viral envelope fuses with the cell membrane, releasing the viral RNA genome together with several enzymes into the cytoplasm. One of these, *reverse transcriptase,* generates a double-stranded DNA copy of the genome, which then passes into the nucleus and is integrated into a host cell chromosome via the action of a viral enzyme, *integrase.* At this stage the virus is said to be in the *proviral* form. Reverse transcriptase is a critical player in viral replication— there is no host cell counterpart to this enzyme and without it the viral genome can not replicate. Reverse transcriptase has thus been one of the major targets of HIV drug therapy. Although there are no currently approved drugs that target integrase, there are several now under investigation.

Completion of cycle and generation of new virions

Once integrated into the host cell chromosome, transcription, and subsequent translation of the viral genome lead to the accumulation of viral RNAs and proteins. These steps are carried out by host cell proteins, illustrating the dependence of viral replication on host cell processes. Subsequently, the viral proteins are processed into their active forms via the action of the viral *protease,* another important target of drug therapy. In the final step the various components of the viral particle assemble near the cell surface and "bud" away,

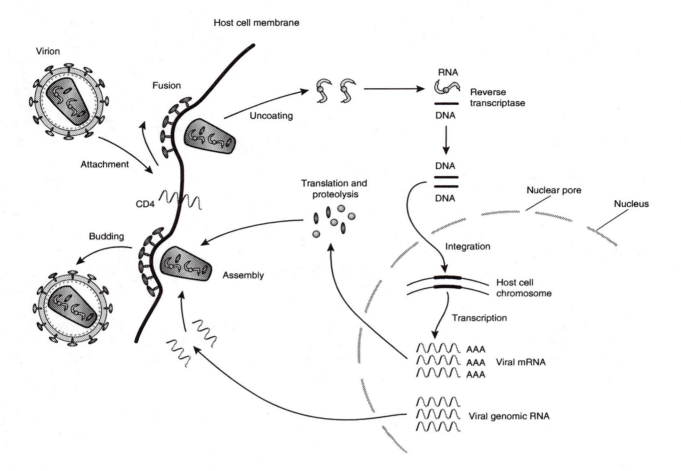

FIGURE 8.2 HIV life cycle. The cycle begins with attachment of the viral particle to the host cell membrane. Fusion subsequently occurs, followed by removal of the capsid protein, with the release of the viral genome into the cytoplasm. Reverse transcriptase then generates a double-stranded DNA copy of the single-stranded RNA genome, which then passes into the nucleus and is integrated into the host cell DNA via the viral integrase. Transcription generates viral mRNAs, the templates for viral protein synthesis, as well as copies of the viral genomic RNA. Translation of the viral mRNAs produces viral polypeptides, which are then cleaved by the viral protease into functional viral proteins. Assembly and budding at the cell membrane then produce new virions, which are then free to infect other cells.

forming a new particle which can then go on to infect other cells. This process weakens the host cell and can result in cell lysis.

Disease Progression

Acute phase

Transmission of HIV can occur through sexual contact, exposure to contaminated blood, or from an infected mother to her fetus or suckling infant. Approximately three weeks after infection, mononucleosis-like symptoms such as fever, muscle aches, and enlarged lymph nodes may appear; however, in a significant number of infected individuals these symptoms go unnoticed. During this initial *acute* phase viral replication and destruction of CD4$^+$ T cells is rapid, but the immune system is soon mobilized, reducing the rate of viral replication, with the resolution of symptoms within a week or two.

Chronic phase

Then begins the *chronic* phase, the duration of which may vary, but which lasts between 8 and 10 years on average. During this phase the virus is not completely eliminated, but rather a balance between the immune system and the virus is achieved in which the rate of viral replication is fairly low and steady. Although the replication rate is relatively low, it varies between patients and is an important determinant of how rapidly the disease progresses. The patient generally feels well and experiences few symptoms during this phase. Until this point the patient is said to be HIV-infected, but does not have AIDS.

Onset of AIDS

Eventually, the virus gains the upper hand and the CD4$^+$-T-cell count begins to fall and the patient becomes immunocompromised. The patient becomes increasingly unable to resist infections with common pathogens encountered in everyday life, and becomes more susceptible to certain forms of cancer. When the CD4$^+$-T-cell count falls below 200/ml (normal is 800/ml) or infections start to occur, the patient is diagnosed with AIDS. Illness due to opportunistic pathogens such as *Pneumocystis carinii* (which causes pneumonia) and *Toxoplasma gondii* (which causes toxoplasmosis) are common, as is herpes zoster, which occurred in this patient. Reactivation of the varicella-zoster virus, which has remained dormant in the dorsal root ganglion (a group of sensory nerve cell bodies leading from the periphery to the spinal cord) following an episode of the chicken pox (caused by the same virus), causes herpes zoster. Although herpes zoster can occur in those with competent immune systems, it is more likely to occur in immunocompromised individuals. Between 10% and 20% of those infected with HIV will suffer from herpes zoster.

Epidemiology

Since the beginning of the pandemic in the late 1970s it is estimated that 60 million people have been infected with HIV, and more than 20 million have died from AIDS. Forty million are currently living with HIV, with the largest burden borne by sub-Saharan Africa. AIDS is the biggest killer there, and life expectancy is expected to drop from 59 to 45 years of age in the next five years. In the most severely affected countries, Zimbabwe and Botswana, between 50% and 70% of 15-year-olds are expected to die of AIDS. Approximately one million individuals are estimated to be living with HIV in North America.

Treatment

Prevention is better than cure

Patients diagnosed with HIV infection no longer necessarily face the prospect of an invariably fatal disease. The effectiveness of drug regimens developed in the past few years have brought the disease within the realm of chronic conditions, much like diabetes or hypertension, which can be controlled for decades without significant loss of quality of life. However, the costly drugs are well out of reach of the large majority of patients in developing nations, where the greatest burden lies. The most promising approach towards bringing the pandemic under control lies in *prevention,* through the development of an effective vaccine, and therefore this is currently an area of intense research.

When to treat

Even without treatment one can remain well for years while HIV infected, and thus the question of when to begin treatment does not have a straightforward answer. The recommended drug regimens are complex, and the patient risks developing resistance to some of the most potent medications before there is a real need for them. However, viral replication is ongoing from the start of the infection, and viral burden is known to correlate with prognosis. Thus, physicians currently recommend that patients begin treatment even in the absence of symptoms if their CD4$^+$-T-cell count is low (<350–500 cells/ml) and/or the plasma HIV RNA level is high (>5000 copies/ml). Although upon diagnosis this patient had a fairly low CD4$^+$ T-cell count (450 cells/ml), his HIV RNA level was low (500 copies/ml), and he was thus somewhat on the borderline for initiation of treatment. He elected to postpone treatment until these values reached more dangerous levels.

Highly active antiretroviral therapy (HAART)

A large body of evidence indicates that the viral load carried by the patient correlates with prognosis, and thus the aim of therapy has been to inhibit viral replication. The principle drugs currently used to combat HIV are classified either as *reverse transcriptase inhibitors* or *protease inhibitors,* and the recommended treatment consists of a combination of these known as HAART. HAART consists of at least two nucleoside analogs (RT inhibitors) and a protease inhibitor, and is currently the most effective treatment plan, typically bringing the viral RNA level down to undetectable levels within six months, as it did for this patient.

Development of resistance

The high mutability of HIV results in the rapid development of drug-resistance (often within weeks) in patients undergoing monotherapy. A multidrug regimen is thus always recommended, as the simultaneous development of resistance to multiple drugs is far less likely (see Chapter 2 of *Biochemistry* 5e). Nevertheless, even resistance to a multidrug regimen may occur: the bioactivity of a drug may vary from patient to patient, and poor patient compliance with the complex regimen may result in subtherapeutic drug levels that allow resistance to develop.

In this case, the patient developed resistance to the antiretroviral drug regimen while undergoing treatment for neuralgia associated with herpes zoster. The anticonvulsant he was prescribed during this period, carbamazepine, is a strong inducer of a hepatic enzyme called cytochrome P450 3A4 (CYP3A4), one in a family of enzymes that metabolizes drugs in the process of their elimination. The protease inhibitors, such as indinavir, are substrates for CYP3A4, and thus a likely explanation for the development of resistance in this case is a reduction in the level of indinavir due to carbamazepine induction of CYP3A4.

Mechanisms of Antiretroviral Drug Action

Nucleoside analogs

The drugs initially prescribed to this patient were the nucleoside analogs, 3TC and AZT, and the protease inhibitor indinavir. The nucleoside analogs are reverse transcriptase inhibitors, and all drugs of this class inhibit the enzyme via the same basic mechanism. They are structurally similar to the nucleotides that are the natural substrates for reverse transcriptase

(Figure 8.3), and are recognized by the enzyme and incorporated into the growing DNA chain. It is important that, once the drug has been incorporated into the chain, no further nucleotides can be added, thus prematurely terminating strand synthesis. These drugs all lack the ribose 3'-OH group through which the subsequent nucleotide becomes incorporated into the chain (See Figure 5.22, *Biochemistry* 5e). In addition to the zidovudine and lamivudine initially prescribed, the didanosine and stavudine used later in the course of his disease are also in the class of nucleoside analogs.

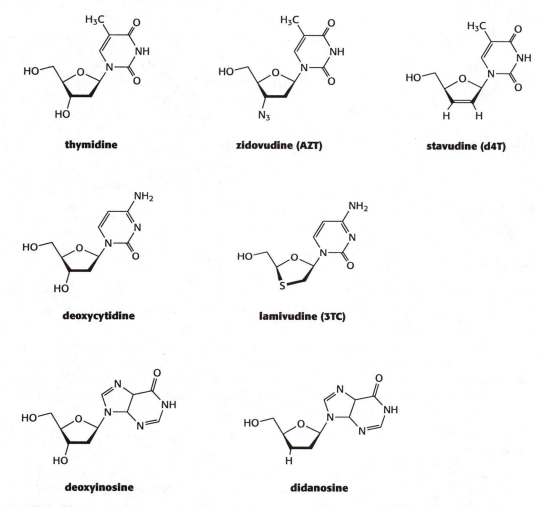

thymidine

zidovudine (AZT)

stavudine (d4T)

deoxycytidine

lamivudine (3TC)

deoxyinosine

didanosine

FIGURE 8.3 Structures of nucleoside analog drugs and their natural counterparts. Zidovudine and stavudine are thymidine analogs, whereas lamivudine is a deoxycytidine analog. Didanosine is a deoxyinosine analog, but the drug is converted into dideoxyadenosine in the cell and thus competes with deoxyadenylate for binding to reverse transcriptase.

Nevirapine

The mechanism by which nevirapine inhibits reverse transcriptase is distinct from the nucleoside analog drugs described above. Nevirapine is a noncompetitive inhibitor of reverse transcriptase, binding near, but not within, the active site and altering its structure. Nevirapine does not appear to affect binding of nucleotides to the active site, but rather slows their incorporation into the growing chain.

Indinavir

Indinavir is a competitive inhibitor of the viral protease. The HIV protease is an aspartyl protease (see Section 9.1.6), and cleaves viral polypeptides to form the various structural components of the viral particle and the viral enzymes. Although the protease acts at a late stage in the viral life cycle, it plays an essential role; without it, the viral particle cannot form, and the spread of the infection is halted. Indinavir, which resembles the phenylalanine-proline dipeptide (Figure 8.4), mimics the transition-state of the reaction, and is thus a very effective inhibitor of the viral protease (see Sections 8.5.3 and 9.1.7 of *Biochemistry* 5e).

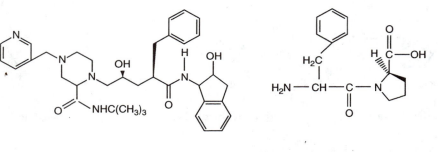

Indinavir   **Phenylalanine-proline dipeptide**

FIGURE 8.4 Structures of indinavir and the phenylalanine-proline dipeptide.

QUESTIONS

1. Upon initiation of therapy, the HIV-1 RNA level found in the patient was 31,000 copies/ml of plasma. What would you estimate the concentration of viral particles in the plasma to have been?

2. This patient's CD4$^+$ count began to fall upon administration of carbamazapine for the neuralgia associated with the herpes zoster infection. The CD4$^+$ count probably decreased on account of an increased rate of inactivation of the antiretroviral, indinavir. Why do you think the antiretroviral drug regimen was changed, rather than simply discontinuing carbamazapine or increasing the dose of indinavir?

3. Would you expect zidovudine to inhibit the cellular DNA polymerases? The cellular RNA polymerases?

4. Inside the cell zidovudine must be converted into the active form, zidovudine triphosphate, for it to serve as an inhibitor of reverse transcriptase. The rate-limiting step in this conversion appears to be the formation of zidovudine diphosphate from zidovudine monophosphate, which is carried out by the cellular enzyme, thymidylate kinase. This enzyme normally converts thymidine monophosphate into thymidine diphosphate, which is then converted into thymidine triphosphate for incorporation into DNA. Does the competition between zidovudine monophosphate and thymidine monophosphate for thymidylate kinase have any effect on the inhibition of reverse transcriptase?

5. Would you expect zidovudine and stavudine to be prescribed together?

6. What is the mechanism of action of valacyclovir in combating varicella-zoster virus?

7. The K_i of stavudine for reverse transcriptase is approximately 20 nM. However, the concentration of drug required to inhibit the replication of the virus by 50% (IC_{50}) in tissue culture cells is ≥9 μM. How might this apparent discrepancy be explained?

FURTHER READING

1. Bartlett, J. G, and Gallant, J. E. Medical Management of HIV Infection. Johns Hopkins University, Division of Infectious Diseases (2001). [Available online at: www.hopkins-aids.edu/publications/book/book_toc.html]

2. Tozser, J. HIV inhibitors: Problems and reality. *Ann. N.Y. Acad. of Science* (2001) 946:145–159.

3. Balint, G. A. Antiretroviral therapeutic possibilities for human immunodeficiency virus/acquired immunodeficiency syndrome. *Pharmacology and Therapeutics* (2001) 89(1):17–27.

4. Miller, M. D., and Hazuda, D. J. New antiretroviral agents: looking beyond protease and reverse transcriptase. *Current Opinion in Microbiology* (2001) 4(5):535–539.

For further information, see the following web sites:

Johns Hopkins AIDS Service: www.hopkins-aids.edu/

National Institute of Allergy and Infectious Diseases, Division of Acquired Immunodeficiency Syndrome: www.niaid.nih.gov/daids/

Centers for Disease Control, Divisions of HIV/AIDS Prevention: www.cdc.gov/hiv/dhap.htm

Joint United Nations Programme on HIV/AIDS: www.unaids.org/

CHAPTER 9

Catalytic Strategies
The Case of the Crushed Mailbox

CASE HISTORY

A 65-year-old African American man returns to clinic for routine follow-up of mild systemic hypertension that has been controlled with diet. He offers no specific complaints at this visit but his wife states that she has noticed that he frequently "bumps into things" such as doorways. He recently ran over the mailbox when pulling into his driveway, failing to notice that he had gotten so close to it. He has no other symptoms and does not complain of problems with his vision.

His physical examination was remarkable for a mild elevation of blood pressure (142/90 mmHg) and no focal neurological deficits. Detailed ophthalmological examination revealed no eye tenderness, no scleral or conjunctival redness, and normally reactive pupils. Slit lamp examination of the eye revealed a normal anterior chamber and no evidence for angle closure. Examination of the fundus of the eye demonstrated an increase in the size of the optic cup relative to the optic disk with no other abnormalities. His intraocular pressures were 32 mmHg OD (right eye) and 28 mmHg OS (left eye) (upper limits of normal: 20 mmHg). Visual field testing revealed bilateral visual field loss.

What part of the visual field appears to be affected in this patient?

What is a slit-lamp examination and what does it reveal?

How is the fundus of the eye visualized?

How is intraocular pressure measured?

Diagnosis and Treatment

The physician made the diagnosis of glaucoma and the patient was treated with a beta adrenergic blocker (Timolol maleate 0.25%) and an alpha adrenergic agonist (apracloni-dine 0.5%) administered topically in both eyes. He returned in one month with intraocular pressures of 29 mmHg OD and 26 mmHg OS and treatment with a topical carbonic anhydrase inhibitor (dorzolamide 2%) was initiated. The patient was seen two months later and intraocular pressures were down to within the normal range of 18mmHg OD and 17mmHG OS. Reexamination of his eyes and visual field testing indicated no further impairment of vision. The patient was advised to continue the course of treatment and to return for reexamination on a semiannual basis. His wife purchased a new mailbox and moved it well away from the driveway to avoid damage to it in the future.

What appears to be the goal of treatment?

Was the initial drug treatment effective?

Did the patient's vision improve?

DISCUSSION

Symptoms and Etiology

Primary open-angle glaucoma (POAG) accounts for 60–70% of all glaucomas. Its causes are largely unknown, but it is characterized by poor drainage of the aqueous humor from the anterior chamber of the eye (Figure 9.1), resulting in increased intraocular pressure (IOP). Typically, it has no symptoms besides the gradual loss of peripheral vision, which often goes

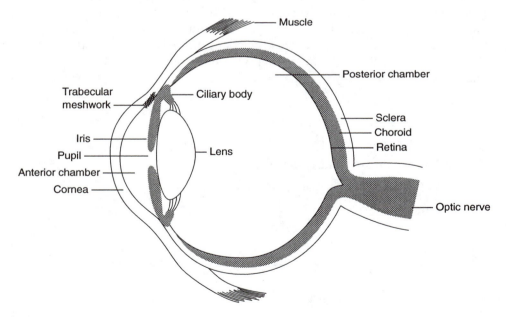

FIGURE 9.1 Anatomy of the eye.

unnoticed until significant damage has occurred. Unfortunately, the development of symptoms signals irreversible damage, and contemporary therapy is aimed at halting further progression of disease and preservation of the remaining vision.

Diagnosing Glaucoma

Three tests are commonly used to diagnose glaucoma. Examination of the optic nerve is done using an opthalmoscope, which allows the physician to view the fundus through the patient's pupil. Enlargement or notching of the optic cup are tell-tale signs of damage to the optic nerve, and glaucoma. Additionally, a tonometer is used to measure the pressure within the eye, and a visual field test will reveal the characteristic shrinking of the breadth of peripheral vision.

Management of Glaucoma

Although laser treatment and surgery are options in the treatment of POAG, drug treatment is usually first-line therapy for newly diagnosed patients. A number of types of drugs are used, among them beta-blockers, alpha-agonists, and carbonic anhydrase inhibitors. All of these agents act to decrease the IOP by reducing the flow of fluid into the anterior chamber and/or increasing the drainage out of it, and the choice of drug is based on effectiveness and severity of side effects for each individual patient. Often, a combination of drugs is used, as for the patient described here. When the combination of timolol maleate and apraclonidine proved ineffective for this individual, the carbonic anhydrase inhibitor, dorzolamide, was prescribed, this time with success.

Treatment of Glaucoma by Inhibition of Carbonic Anhydrase

Carbonic anhydrase (CA) is a zinc-containing enzyme that catalyzes the hydration of carbon dioxide to yield a bicarbonate ion and a proton (see Section 9.2 of *Biochemistry* 5e). Inhibition of the enzyme leads to decreased production of bicarbonate, a key step in aqueous humor production. Bicarbonate formed in the ciliary body associates with sodium ions and is secreted into the posterior chamber of the eye (see Figure 9.1). Due to osmosis, a passive flow of water ensues and continues on into the anterior chamber to form the aqueous humor. Thus, CA inhibitors reduce IOP by decreasing fluid flow into the eye.

Biochemical Basis for Dorzolamide's Action

Dorzolamide, introduced in 1995, was the first topical carbonic anhydrase inhibitor to appear on the market. The difficulty in developing topical drugs was that clinical effectiveness necessitated inhibition of CAII, the most active human isoform, by virtually 100%. This requires both a high degree of inhibitory activity and efficient penetration of the eye. Dorzolamide, with its high affinity for the enzyme ($Kd = 0.37$ nM) and efficient ocular penetration, was shown to reduce IOP to comparable extents as the previously used systemic CA inhibitors acetazolamide and methazolamide.

Dorzolamide is chemically classified as a sulfonamide (Figure 9.2) and acts by binding to the active site of CA and preventing the binding of substrates, water, and carbon dioxide. It thus acts as a competitive inhibitor. The nitrogen atom of dorzolamide binds the zinc atom and displaces the hydroxide group normally bound in this position (compare

FIGURE 9.2 Structure of a sulfonamide.

Figures 9.3a and 9.3b). The nitrogen also donates a hydrogen bond to the hydroxyl side chain of threonine-199, which in turn donates a hydrogen bond to the side chain of glutamate-106. In addition, one of the oxygen molecules of the sulfonamide group makes a hydrogen bond with the peptide backbone at threonine-199.

The similarities between binding of dorzolamide and the binding of the enzyme's substrates are clear. In addition, dorzolamide makes contacts with the walls of the cone-shaped active site via its bicyclic ring structure (−R).

Dorzolamide Is a Transition State Analog

A key concept in catalysis is that enzymes decrease the activation barrier of a reaction by lowering the free energy associated with the transition state (see Section 8.3 of *Biochemistry 5e*). The transition state is the most unfavorable, or highest energy state the reaction passes through, and enzymes act by lowering its energy, thereby increasing the rate of reaction. Thus, enzyme structures have evolved to specifically bind the transition state of the reactions they catalyze. Dorzolamide is an effective inhibitor of carbonic anhydrase because it mimics the transition state of the reaction, and hence binds tightly to the enzyme.

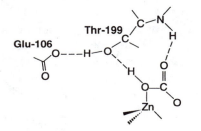

FIGURE 9.3A Schematic drawing of substrate binding to the active site of carbonic anhydrase.

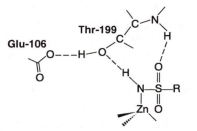

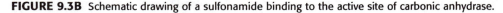

FIGURE 9.3B Schematic drawing of a sulfonamide binding to the active site of carbonic anhydrase.

QUESTIONS

1. Altitude sickness is caused by the reduced partial pressure of oxygen at high elevations, which results in hypoxemia, a lowered oxygen level in the blood. This stimulates hyperventilation, which can lead to alkalosis (increased alkalinity, or pH, of bodily fluids) and the symptoms of mountain sickness, which include fatigue, nausea, headache and a rapid and forceful heartbeat. The carbonic anhydrase inhibitor, acetazolamide, taken orally as of the first day of ascent, can be given as a prophylactic treatment to mountain climbers. How do you think the drug might work in this context?

2. If you were to design a new carbonic anhydrase inhibitor drug for the treatment of glaucoma, what physicochemical properties of the drug might you need to consider?

3. Sulfonamides are also used as antimicrobials (the "sulfa drugs") but those used to treat glaucoma do not have antibacterial activity. By comparing the structures of the following antimicrobial sulfonamides with the carbonic anhydrase inhibitors used to treat glaucoma (Figure 9.4) can you identify the structural component important for antibacterial activity?

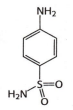
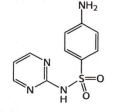
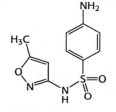

| **Sulanilamide** | **Sulfadiazine** | **Sulfamethoxazole** |

Examples of commonly used sulfa drugs

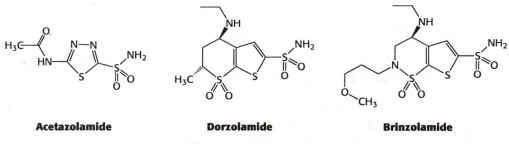

| **Acetazolamide** | **Dorzolamide** | **Brinzolamide** |

Some carbonic anhydrase inhibitor drugs

FIGURE 9.4

4. Prior to the development of dorzolamide, the only available carbonic anhydrase inhibitors were taken orally and were associated with a number of side-effects, including polyuria, fatigue, and gastrointestinal disturbances. Can you explain the physiological basis for these side-effects?

5. About 25% of patients experience a bitter or metallic taste upon administration of dorzolamide eye drops. Can you explain why this might occur? [Hint: Lacrimal fluid from the eye drains into the oropharynx.]

6. A patient who wears soft contact lenses would like to know if she can take dorzolamide to treat the glaucoma from which she suffers. Dorzolamide itself is not contraindicated in those with soft contact lenses, however, can you think of any other factors that should be considered before advising her?

7. A less common form of glaucoma called acute closed-angle glaucoma occurs when, for a variety of reasons, the root of the iris presses against the trabecular meshwork, from which fluid drains from the eye. With the blockage of the outflow channel, the IOP can rapidly rise to dangerously high levels, causing painful swelling of the eye, nausea, and dizziness. Acute closed-angle glaucoma can lead to blindness in as little as a day or two unless treated immediately. Do you think medication can be used to treat this form of glaucoma?

8. Latanoprost, a prostaglandin analog, was approved by the FDA in 1996 for the topical treatment of glaucoma. It functions by stimulating uveoscleral fluid outflow (flow of fluid through the front of the eye). Given their modes of action, would you expect latanoprost to have an IOP lowering effect on top of that produced by dorzolamide?

9. The use of systemic carbonic anhydrase inhibitors in post-menopausal women has been associated with inhibition of osteoporosis. Can you speculate as to how CAI might inhibit bone loss?

FURTHER READING

1. Infeld, D. A., and O'Shea, J. G. Glaucoma: diagnosis and treatment. *Postgraduate Medical Journal* (1998) 74:709–715.

2. Lindskog, S. Structure and mechanism of carbonic anhydrase. *Pharmacol. Ther.* (1997) 74(1):1–20.

3. Herkel, U., and Pfeiffer, N. Update on topical carbonic anhydrase inhibitors. *Current Opinion in Opthalmology* (2001) 12(2):88–93.

For further information, see the following web sites:

National Eye Institute: www.nei.nih.gov/

American Academy of Opthalmology: www.eyenet.org

The Glaucoma Foundation: www.glaucoma-foundation.org/info

The Glaucoma Research Foundation: www.glaucoma.org

National Library of Medicine-Medlineplus: www.nlm.nih.gov/medlineplus/glaucoma.html

CHAPTER 10

Regulatory Strategies
A Close Call for a Boy Scout

CASE HISTORY

The Patient Is Admitted with a Painful Crisis

A 14-year-old African American boy known to have sickle-cell disease was admitted to the hospital because of pain in the hip and lower back. The patient had relatively mild disease during childhood with three prior painful crises precipitated by upper respiratory infections requiring hospital admission and narcotic analgesics for pain control. He had received penicillin orally from two months of age until he was six years old and a multivalent streptococcal vaccine at two years of age with a booster at five.

Two days before admission, he was hiking with his scout troop but had no symptoms at that time. The day prior to admission he developed lower back pain and left hip pain. He had a low-grade fever of 38°C, tenderness over the lumbar spine, and pain with flexion of the left hip. Upon admission, his chest was clear to auscultation and abdominal exam was remarkable for enlargement of both the liver and spleen (hepatosplenomegaly). His hematocrit was 24% (normal >40%) on admission; he typically had a hematocrit of 25–30%. Blood chemistries revealed a mild elevation in the serum bilirubin. Urinalysis revealed microscopic hematuria (blood in the urine but not enough to be visible to the naked eye) with evidence for urinary tract infection. Radiographs of the hips and lower back revealed evidence for avascular necrosis of the left femoral head but no significant abnormalities in the lumbar spine. He was treated with oxygen delivered by a mask and intravenous hydromorphone (Dilaudid) 1 mg every four to six hours for pain control, and he began to improve.

What causes sickle-cell disease?

Why is hepatosplenomegaly common in those with sickle-cell disease?

Is the avascular necrosis of the femoral head observed in the radiograph likely to have arisen from sickle-cell disease?

What might be the cause of the anemia (low red blood cell count) and elevated bilirubin levels found in the serum?

His Condition Deteriorates

On the second hospital day he developed a cough, experienced a rigor, and became agitated, complaining of increasing back and hip pain and pain in his chest with breathing (pleuritic pain). His temperature was 38.8°C, and pulse was 150 bpm (normal range 60–100 bpm), and the respirations were 32 per minute (normal range 15–20 per minute). The blood pressure was slightly elevated at 135/85 mm Hg. The chest examination now suggested fluid in the left lung. The hematocrit had fallen from 24% to 18%. The arterial oxygen saturation was 91% while the patient was breathing oxygen through a mask at a rate of 4 liters per minute. Radiographs of the chest showed cardiomegaly (enlarged heart) and a small area of consolidation (opacification of the lung due to fluid collection) in the left lower lobe of the lung. An *electrocardiogram (ECG)* revealed a sinus tachycardia at a rate of 150 and findings suggestive of left ventricular hypertrophy (muscular enlargement of the wall of the left ventricle of the heart).

What is an ECG, how does it work, and what information does it yield?

Why might his heart rate and respiration rate be elevated?

Diagnosis and Treatment

In the course of a painful crisis, the patient has developed *acute chest syndrome (ACS)*, a life-threatening complication of sickle-cell disease.

The patient is treated with oxygen, an antibiotic, and a blood transfusion

The inspired oxygen was increased to 100% by a nonrebreathing mask, and the patient's oxygen saturation improved to 99%. The antibiotic, ceftriaxone (Rocephin) 50 mg/Kg was administered intravenously twice daily. An exchange transfusion was performed, with the removal of approximately 0.5 liters of the patient's red cells and then transfusion of 1.5 liters of packed red cells from a donor. After transfusion the percentage of total hemoglobin that was hemoglobin A (normal adult hemoglobin) was greater than 60% and the hematocrit was 34%.

Why was an antibiotic administered?

His condition improves and he is discharged

By the fourth hospital day chest, hip, and back pain had begun to subside. His blood oxygen saturation improved and was 99% on low-flow (2 liters/minute) oxygen delivered by a nasal cannula. Blood, sputum, and urine cultures were all negative and intravenous antibiotics were discontinued. The hematocrit was 29% and he underwent a transfusion of two units of packed red blood cells that raised the hematocrit to 33%. Intravenous hydromorphone was tapered and the patient was started on oxycodone 2 mg every 4–6 hours for pain control. He was discharged to home on the sixth hospital day with crutches to limit weight bearing on the left hip.

What is a nonrebreathing mask? Why was oxygen delivered by a nasal cannula once his symptoms improved?

The patient was treated with an exchange transfusion. What might be the rationale for such a procedure?

DISCUSSION

Sickle-Cell Disease Is a Genetic Disease

Sickle-cell disease is caused by an inherited mutation in hemoglobin, an oxygen transport molecule found in red blood cells (see Section 10.2 of *Biochemistry* 5e). Hemoglobin is a tetrameric molecule, consisting of two α-globin and two β-globin chains. Although there are several forms of the disease, the most common is caused by a point mutation in the β-globin chain of hemoglobin, which results in substitution of the glutamic acid residue at the sixth position of the amino acid chain with a valine residue. This mutant version is referred to as HbS (for *Sickle* hemoglobin), while the wild type counterpart is termed HbA (for *Adult* hemoglobin).

Heterozygotes are protected against malaria

Individuals that inherit a single mutant copy ("carriers" or "heterozygotes," also known as having "sickle-cell trait") are usually completely asymptomatic, and in fact, are somewhat protected against malaria, a disease caused by the *Plamodium falciparum* parasite. The relatively high prevalence of the HbS allele has been attributed to this protective effect. In support of this premise, the occurrence of the HbS allele is highest in the low-lying areas of west and central Africa, where malaria is endemic.

Homozygotes have the disease

In contrast to heterozygotes, homozygous individuals, who have had the misfortune of inheriting an HbS allele from both parents, are often severely affected by the disease. The substitution of a valine for glutamic acid increases hemoglobin's hydrophobicity, which favors self-association into long, rigid fibers that distort the red blood cell into the characteristic crescent, or sickle, shape. Sickled cells are rigid and irregularly shaped, and get trapped in the narrow capillaries that permeate the bodily organs. Once sickled red blood cells occlude a capillary, the obstruction can become still more impenetrable as more cells back up behind it. Herein lies the most problematic feature of the illness: as will be described below, almost all of the symptoms and pathophysiology of the disease have these episodes of vaso-occlusion at their source.

The anemia associated with the disease is a result of the increased fragility of red blood cells in affected individuals. Sickling damages the cell membrane and renders cells more susceptible to lysis, with the average half life of an erythrocyte being 10–20 days in a patient with sickle-cell disease rather than the usual 120. This shortened life span results in a chronic shortage of red blood cells (anemia) and an increased turnover of the hemoglobin within them. Bilirubin is a breakdown product of hemoglobin, explaining the elevated levels of the compound observed in the patient.

Epidemiology

Sickle-cell disease affects millions of people worldwide. It is most common in west, central, and east African countries, where the prevalence of sickle-cell trait in some regions reaches as high as 40%. Generally, the regions with the highest prevalence are low lying and wet regions where the Anopheles mosquito that transmits malaria is most often found. This supports the assumption that sickle-cell trait persisted because it provides a survival advantage in those suffering from malaria. In the United States, approximately 8% of African Americans have sickle-cell trait, and it is also found, but to a lesser extent, among individuals of Mediterranean, Middle Eastern, and East Indian origin.

What Causes Cells to Sickle?

Erythrocytes are packed with hemoglobin. In a healthy individual hemoglobin molecules are estimated to be separated by less than a molecule's width, and the concentration is typically even higher in those with sickle-cell disease due to dehydration of their red blood cells. In this context, the single valine substitution in HbS is sufficient to drive the protein to self-associate.

Formation of the double strand

The valine residue creates a hydrophobic "patch," which is well situated to interact with another tetramer via a second hydrophobic patch of the β-globin chain consisting of phenelalanine-85 and leucine-88. The patch made up of Phe-85 and Leu-88 is exposed only when hemoglobin is deoxygenated (the "T state" of the molecule), which explains why sickling usually occurs under conditions of low oxygen tension (such as in the organs or muscles). Both β chains of a tetramer participate in this type of contact, and the ensuing structure is a slightly twisted double strand, with the tetramers oriented in an energetically favorable staggered arrangement (Figure 10.1).

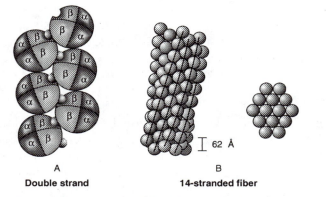

A	B
Double strand	14-stranded fiber

FIGURE 10.1. Polymerization of deoxyhemoglobin S. The left panel depicts the interactions between β subunits that result in the formation of the double strand. The right panel depicts the 14-stranded fiber.

Formation of higher order structures

The double strand of sickled hemoglobin then serves as a building block for higher-order structures. The fibers within sickled cells actually consist of seven pairs of strands, twisted into a rope-like helical arrangement, with two central pairs surrounded by five outer pairs (see Figure 10.1). These, then, can be ordered into larger domains in which parallel alignments of fibers form sheets, which transform the cell from its usual shape into the aberrant sickle or "holly leaf" forms typical of the disease (Figure 10.2).

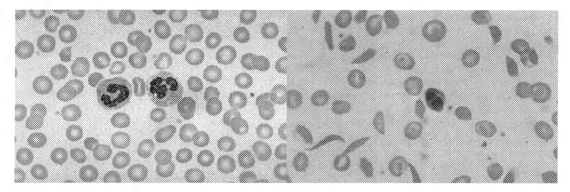

FIGURE 10.2. Smears of peripheral blood from a normal control (left) and a patient with sickle-cell disease (right). Photomicrographs courtesy of Dr. Gregory Kato, Johns Hopkins University.

Fiber formation is cooperative

Fiber formation has been found to be highly cooperative—although the initiation of a fiber is difficult, once a "nucleus" of 10 or so tetramers has formed the fiber will rapidly continue to grow. Thus, polymerization is exquisitely sensitive to HbS concentration, which explains why heterozygotes, with half as much HbS as homozygotes, are usually asymptomatic. Decreasing the concentration of HbS by half increases the time for fiber formation 1000-fold. By this time a red blood cell will have traversed the microcapillaries where sickling is most likely to occur and reached the relative safety of the larger veins leading to the lungs.

Episodes of Vaso-Occlusion Are the Most Troublesome Feature of Sickle-Cell Disease

The episodes are extremely painful

The vaso-occlusive events common in those with sickle-cell disease cause "painful crises," the most distressing symptom of the illness. Painful crises can be triggered by a minor illness, overexertion, minor trauma, or cold weather, but they often appear unpredictably. Note that the patient was admitted for a painful crisis following a hiking excursion, and thus the crisis was likely triggered by overexertion. During exercise the rapidly metabolizing muscle cells tend to deplete hemoglobin of its oxygen, and this together with dehydration can trigger a crisis. Crises vary greatly in severity and frequency. They can occur from less than once to 15 or 20 times per year, and can last for hours to weeks. Treatment for these crises is largely symptomatic but can also include some measures to minimize sickling. Note that upon admission the patient in the case was treated with an analgesic to control the pain and oxygen therapy to reduce sickling.

In addition to causing severe pain, vaso-occlusive crises, in blocking food flow and depriving bodily tissues of nutrients and oxygen, can cause damage to vital organs, which over time can compromise their function. Kidney damage is common, and the spleen is commonly affected as well, resulting in increased susceptibility to infections, especially those of bacterial origin. Prior to a landmark study in 1986 that showed marked improvement in prognosis with long-term treatment with penicillin, pneumococcal infections were a leading cause of death in children with sickle-cell disease. Hence the long-term treatment of the patient discussed here with penicillin throughout his early childhood period.

Sickling damages red blood cells and causes anemia

The anemia associated with the disease is a result of the increased fragility of red blood cells in affected individuals. Sickling damages the cell membrane and renders cells more susceptible to lysis, with the average half-life of an erythrocyte being 10–20 days in a patient with sickle-cell disease, rather than the usual 120. This shortened life span results in a chronic shortage of red blood cells (anemia) and an increased turnover of the hemoglobin within them. Bilirubin is a breakdown product of hemoglobin, explaining the elevated levels of the compound observed in the patient.

Acute Chest Syndrome

Acute chest syndrome (ACS) is a life-threatening complication of sickle-cell disease, which occurs in approximately 50% of patients and is fatal in 10–20% of cases. It is likely caused by an occlusion in the lung vasculature, and can lead to respiratory failure and death if not treated aggressively. It is often accompanied by an infection, and, in fact, infection may be the very cause for the episode. Infection can deplete a tissue of oxygen, thus triggering the sickling that leads to the pulmonary occlusion. Fortunately, the patient in the case was rapidly diagnosed and treated with antibiotics, oxygen, and blood transfusion. He was, in fact, treated with an exchange transfusion rather than a simple transfusion. In substituting his HbS erythrocytes with those from a healthy donor, the diseased cells were greatly diluted, reducing sickling and the risk of further vaso-occlusion.

Rapid diagnosis and treatment is one of the most important prognostic factors for ACS, yet diagnosis of the syndrome is notoriously difficult. The symptoms are variable and often mimic those of an infection such as pneumonia. Common symptoms are fever, cough, wheezing, chest pain, and a new pulmonary infiltrate often appears on a chest x-ray. The Boy Scout thus exhibited classic symptoms. ACS often occurs in the course of a painful crisis, and is more common though usually less severe in children than adults. Improved prognosis is associated with limited pulmonary infiltrates and mild hypoxemia (low blood oxygen). However, ACS tends to recur and repeated episodes are associated with poor long-term prognosis.

QUESTIONS

1. Hydroxyurea, a drug first used to treat cancer, was later found to be effective in the treatment of sickle-cell disease. In cancer patients the drug works by inhibiting DNA replication and cell division, whereas in the treatment of sickle-cell disease, it acts by stimulating the expression of fetal hemoglobin (HbF). HbF, like the adult form, is a tetramer consisting of the same two α chains as the adult form, but rather than two β chains, the HbF tetramer consists of two γ-globin chains. The non-copolymerization of HbS and HbF interferes with fiber formation and hence sickling. The boy described in the case study was not treated with hydroxyurea. Can you think of a reason why?

2. If you were asked to design a drug for the treatment of sickle-cell disease, which pathophysiological features of the disease might you target?

3. When a malaria parasite (plasmodium species) invades a red blood cell, its metabolic waste products result in acidification of the interior of the cell, which favors the T form of hemoglobin and can trigger sickling in carriers of sickle-cell disease (those with one disease gene copy and one normal copy). Sickling weakens the cell and makes it less hospitable to the parasite, rendering these individuals somewhat resistant to the disease. If infection with the malaria parasite had favored the R form of hemoglobin rather than the T form, do you think that the HbS mutation would have become as prevalent as it has?

4. Explain the therapeutic effect of supplemental oxygen in sickle-cell patients with painful crises.

5. Why do you think those with sickle-cell disease often have enlarged hearts?

6. Those homozygous for the HbS allele were previously referred to as having "sickle-cell anemia." The disease was renamed sickle-cell disease to include other illnesses caused by hemoglobin mutations, and because the term "sickle-cell anemia" was somewhat of a misnomer. Although HbS homozygotes are anemic due to the shortened life-spans of their erythrocytes, anemia is not the most troubling feature of the disease. In fact, it may even be somewhat of an advantage in terms of minimizing the severity of vaso-occlusive events. Can you explain why?

7. The patient in the case study was treated with an exchange transfusion when he developed ACS. A transfusion helps ameliorate the situation by diluting the patient's HbS erythrocytes with normal HbA erythrocytes. In addition, transfusion temporarily inhibits erythropoiesis (red blood cell production). Would you view this as an advantage or disadvantage in this situation?

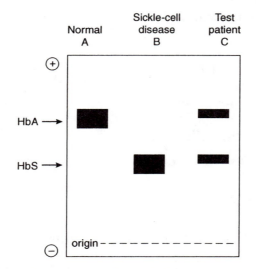

FIGURE 10.3. Gel electrophoresis pattern of hemoglobin from a normal individual (lane A), a person with sickle-cell anemia (lane B), and from a test patient (lane C).

8. The most widely used test for diagnosing sickle-cell disease is hemoglobin electrophoresis. The test is based on the different rates of migration of HbA and HbS in an electric field. Figure 10.3 illustrates a gel electrophoresis pattern in which lane A represents HbA and lane B represents HbS. Lane C represents the hemoglobin from a patient who may have sickle-cell disease. Does the patient have sickle-cell disease? How do you explain the difference in migration pattern between HbA and HbS?

9. A young couple, each known to carry a single copy of the HbS hemoglobin variant, comes to you, a genetic counselor, for advice regarding their wish to have children. What are their chances of having a child with sickle-cell disease?

10. ACS is often treated with blood transfusion, and once an individual has been struck with ACS, it typically recurs. Can you think of a problem that might occur with repeated transfusions?

11. 2,3-bisphosphoglycerate (2,3-BPG), also known as 2,3-diphosphoglycerate (2,3-DPG), is the predominant phosphorylated compound in red blood cells, accounting for about 2/3 of red-blood-cell phosphorus. 2,3-BPG stablizes the T form of hemoglobin and is increased in situations associated with hypoxemia (low levels of oxygen in the blood). Can you predict what effect 2,3-BPG has on the hemoglobin-oxygen dissociation curve and tissue delivery of oxygen?

12. Why have prophylactic antibiotics proven useful in extending the life of sickle-cell patients?

FURTHER READING

1. Bunn, H. Franklin Pathogenesis and treatment of sickle-cell disease. *New England Journal of Medicine* (1997) 337(11):762–769.

2. Edelstein, S. J. *The sickled cell: From myths to molecules.* Harvard University Press (1986).

3. Herrick, J. B. Peculiar elongated and sickle-shaped red blood corpuscles in a case of severe disease. *Archives of Internal Medicine* (1910) 6:517–521. (This is the first description of a case of sickle-cell disease.)

4. Steinberg M. H. Management of sickle-cell disease. *New England Journal of Medicine* (1999) 340(13):1021–1030.

For further information, see the following web sites:

The Sickle-Cell Disease Association of America: www.sicklecelldisease.org

National Institutes of Health–MedlinePlus: www.nlm.nih.gov/medlineplus/sicklecelldisease.html

National Center for Biotechnology Information. Online Mendelian Inheritance in Man (OMIM): www3.ncbi.nlm.nih.gov:80/htbin-post/Omim/dispmim?603903

CHAPTER 11

Carbohydrates
Nothing to Sneeze At

Patient Presents to Clinic with Signs of Respiratory Tract Infection

A 66-year-old man presented to the clinic with fever, chills, anorexia, and muscle aches (myalgias). He was well until two days ago when he experienced the sudden onset of fever, shaking chills, occipital headache, severe myalgias, malaise, and anorexia. The symptoms occurred in the afternoon after a round of golf with friends. He had a dry cough with sore throat and minimal clear nasal discharge. He also noticed hoarseness of his voice over the last day. The cough was accompanied by mild, nonpleuritic substernal chest pain, and mild shortness of breath. He denied dizziness, lightheadedness, palpitations, or chest pain when he was not coughing. He thought that he had the "flu" and was taking acetaminophen and pushing fluids. He denied rash and gastrointestinal or urinary tract symptoms (frequency, urgency, dysuria, cloudy, or foul-smelling urine).

What is influenza? What is its incubation period and typical duration of symptoms?

Patient History

He had no history of significant pulmonary illnesses such as asthma, emphysema, or chronic bronchitis. However, he is a cigar smoker and former cigarette smoker, having quit two years ago. He has prostatic hypertrophy but no other medical problems and takes no medicines regularly. Previously he had been hospitalized only once, for a fracture of the left tibia. He is a retired salesman with no significant history of occupational exposures (i.e., no exposures to noxious chemicals, dusts, or asbestos). He receives the influenza vaccine yearly but had not yet seen his doctor this year.

What is the significance of his smoking history in this setting?

What complications of influenza are smokers more likely to develop?

What strains of influenza does the trivalent vaccine target?

Physical Examination and Laboratory Evaluation

On physical examination the patient appeared acutely ill with flushing of the face, and the skin was warm and moist. His blood pressure was 144/78 mmHg, pulse was 96 beats per minute and regular, and the respiratory rate was 20 and unlabored. He was febrile with an oral temperature of 102.4°F. His eyes were watery and injected (bloodshot), and he had an erythematous pharynx without exudates. There was enlargement and tenderness of cervical lymph nodes. The chest examination revealed no utilization of accessory muscles, resonance to percussion, and scattered rhonchi throughout both lung fields, without rales or signs of consolidation (evidence for pneumonia).

Laboratory evaluation was remarkable for a normal blood count and normal serum chemistries. The chest radiogram revealed no infiltrates and was normal except for degenerative joint disease of the thoracic spine. The electrocardiogram was normal. He was discharged to home with instructions to stay on bed rest, continue vigorous intake of fluids, and to take acetaminophen or aspirin for fever.

Why would aspirin be avoided in children with febrile illnesses?

Why was the presumed influenza infection not specifically treated?

Why were viral cultures not sent?

Patient Improves, Then Worsens

He improved over the next day but then began to experience increasing cough, with fever and shaking chills despite antipyretics. The chest pain changed in character and became localized to the right chest with a significant pleuritic component (i.e., increased with respiration). The cough had now become productive of yellow-green sputum. He returned to the clinic where he again appeared acutely ill. He was flushed and his blood pressure was 96/60 mmHg, with a heart rate of 116 beats per minute. His temperature was 101.6°F, and the respiratory rate was 24 per minute. The chest examination now revealed rales at the base of the right lung with changes consistent with consolidation.

What complication of presumed influenza has he suffered?

Diagnosis and Treatment

The laboratory evaluation now revealed a white blood cell count of 16,600 per mm^3 with 78% polymorphonuclear cells, 10% band forms, 10% lymphocytes, and 2% eosinophils. Serum chemistries were normal. An arterial blood gas revealed a pH of 7.44, partial pressure of CO_2 (pCO_2) of 36 mmHg, and pO_2 of 88 mmHg on room air. A chest radiogram revealed consolidation of the right lower lobe without evidence for a pleural effusion. Sputum stain revealed mixed flora but with a predominance of gram-positive cocci in chains, consistent with pneumococcal pneumonia. He was started on supplemental oxygen by nasal cannulae and intravenous erythromycin and cefoxitin for presumed pneumococcal pneumonia. His sputum cultures grew *Streptococcus pneumoniae* and his antibiotics were changed to penicillin. He defervesced over the next two days and was well at the time of hospital discharge seven days later.

The blood gases were in the normal range in this patient, but what trends might you expect in a patient with a respiratory infection?

Why was the drug regimen changed?

DISCUSSION

Carbohydrates Are Important Mediators of Cellular Communication with Extracellular Molecules

Membranes serve as barriers, separating cells from the extracellular medium and allowing them to maintain intracellular conditions tailored to their particular functions. However, molecules embedded in membranes allow cells to communicate with their extracellular milieu, thus allowing the coordination between cells and extracellular molecules to effect various physiological processes. Carbohydrates, in the form of glycoproteins, proteoglycans, and glycolipids found in cellular membranes, are important players in many of these processes. Their structural diversity and hydrophilicity make them ideally suited for mediating specific interactions between cells and extracellular molecules.

Not surprisingly, a number of microorganisms have exploited cell-surface carbohydrates to their own pathological ends. For example, bacterial colonization of a host is often facilitated by interactions between bacterial *adhesins* and cell-surface carbohydrate molecules, and, as will be discussed here, influenza viruses also exploit cellular carbohydrates in order to gain entry to their target cells. A detailed understanding of the binding interactions that occur between host and pathogen is clinically important as it will allow for the rational design of drugs that interfere with this process, thus blocking the infection at a very early stage. Indeed, the elucidation of the binding interactions between the influenza hemagglutinin protein and its cellular receptor, sialic acid, has provided the basis for numerous research efforts in the development of antiviral drug therapies.

Influenza (the "Flu") Has Killed Millions Worldwide

Influenza is a viral disease of the respiratory tract that has been recognized for centuries, and has caused periodic *pandemics* (worldwide epidemics) since the 19th century (Table 11.1). The most deadly, in 1918–1919, is estimated to have killed over 20 million people, more than were killed in combat in World War I. This pandemic was also complicated by the appearance of disabling postencephalic Parkinsonism (von Economo encephalitis) in many patients that survived. The virus was isolated in the 1930s and was first observed by electron microscopy in 1943. The virion (viral particle) is typically spherical or oval, and is unusual in that it contains a *segmented* negative-stranded RNA genome. The approximately 13 Kb of genomic sequence is divided among eight distinct RNA segments. This allows for the *reassortment* of segments when more than one viral strain infects a single cell, which is thought to account for the *antigenic shifts* (major antigenic variations) that have been linked to the occurrence of pandemics.

TABLE 11.1 Influenza pandemics and their associated viral subtypes.

Year	Subtype	Designation
1889–1890	H2N8	—
1918–1919	H3N8	"Spanish Flu"
1957–1958	H2N2	"Asian Flu"
1968–1969	H3N2	"Hong Kong Flu"

Note: All the above strains are of type A.

More minor antigenic variations are referred to as *antigenic drifts,* and are probably the result of mutation. Influenza viruses are thought to have originated in wild birds (where the disease is usually asymptomatic), but depending upon the type, they can infect a number of other organisms besides humans, including horses and swine. These have been hypothesized to serve as "reservoirs" for the virus in years between human pandemics. Certain animal strains have been found to be capable of infecting humans, and in fact evidence indicates that a strain derived from swine was the cause of the 1918–1919 pandemic.

The Genome Encodes Ten Proteins

Ten proteins are encoded by the influenza virus genome (Figure 11.1). A helical *nucleoprotein* is closely associated with the RNA, as are three *polymerase* proteins that transcribe and replicate the genome. The *matrix proteins,* M1 and M2, line the inner surface of the envelope and may assist in viral assembly or stabilization of the envelope. *Hemagglutinin (HA)* and *neuraminidase (NA, also called sialidase)* are transmembrane glycoproteins embedded in the envelope. Hemagglutinin binds sialic acid residues found on the cell surface and is thus critical in the first step in the viral life cycle, cellular attachment. Conversely, neuraminidase functions at the end of the life cycle, in releasing newly formed viral particles from the cell, such that they can go on to infect other cells. The enzymatic activity of neuraminidase cleaves the glycosidic bond between sialic acid and the penultimate galactose residue, releasing the viral particles and allowing the infection to spread through the host. Two nonstructural proteins (NS) have been found in infected cells; however, their functions are unknown.

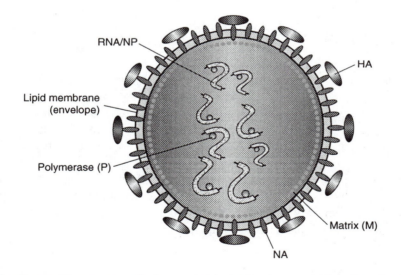

FIGURE 11.1 Influenza virion structure. The hemagglutinin (HA) and neuraminidase (NA) glycoproteins are imbedded in the lipid envelope. Lining the inside of the envelope is the matrix (M), and the genome is encoded by eight distinct RNA segments that are complexed with nucleoprotein (NP). The polymerase (P) is also a component of the virion.

Viral Life Cycle

Binding and internalization

The viral life cycle begins with binding to the cell surface. Hemagglutinin plays a central role in this step, by specifically binding sialic acid residues found on cell-surface sialylglycoproteins and gangliosides (see Section 11.3.1 of *Biochemstry* 5e). Hemagglutinin was crystallized in 1981 and was found to have an unusual structure. It is a homotrimer consisting of a stem-like region that projects from the cell surface, and a globular region on top. The crystal structure of HA bound to sialic acid has shown that sialic acid binds to a pocket of conserved amino acids of the globular region (Figure 11.2). In fact, these amino acids have remained unchanged since the influenza pandemic of 1968. The pocket is lined with amino acids that vary from strain to strain, and to which a number of neutralizing antibodies are directed. This suggests that antibody binding to these sites interferes with binding to sialic acid, thus abrogating infectivity.

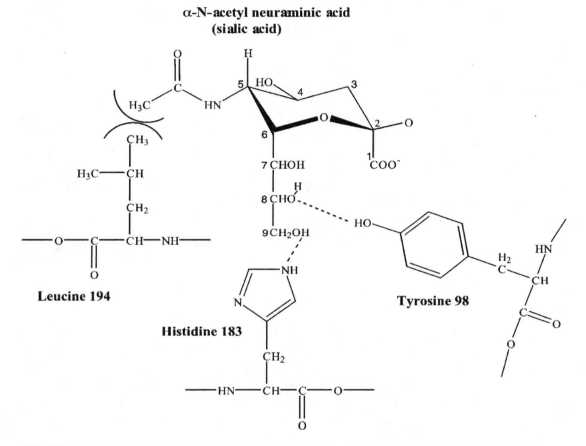

FIGURE 11.2 Schematic representation of hemagglutinin amino acid interactions with sialic acid. The three most important interactions are made with tyrosine 98, histidine 183, and leucine 194, as determined by crystallographic data and mutational analysis; however, other interactions contribute to binding. Tyrosine 98 and histidine 183 make hydrogen bonds with the hydroxyl groups at positions 8 and 9, respectively, and leucine 194 makes nonpolar interactions with the methyl group of the acetamido group [see references for more detail].

The virion is then internalized via the process of *endocytosis* and is incorporated into the endosome, where the acidic pH induces the ion channel activity of the M2 matrix protein. Acidification results in release of the viral genomic segments from the matrix, thus contributing to the disassembly of the virion. In addition, the low pH of the endosome induces a conformational change in HA that results in fusion of the viral envelope with the endosomal membrane, releasing the genome into the cytoplasm. Thus, HA carries out two vital functions: virion attachment to the cell surface and transfer of the genetic material into the cytoplasm.

Generation of new viral particles

Influenza virus is a negative-stranded RNA virus, meaning that the complement of the genome must be generated to serve as a template for protein synthesis. Thus, upon release into the cytoplasm the viral RNAs are transported into the nucleus where the viral RNA-dependent RNA polymerase generates mRNAs (positive-stranded) that are then used to generate the viral proteins. The viral polymerase also produces copies of the negative-stranded RNA genome by using a positive-stranded copy as template. The virus thus uses the same polymerase for both *transcription* and *replication,* exemplifying the remarkable efficiency of the genome. The genome segments and viral proteins then assemble at the inner surface of the cell membrane, and bud away from the cell. The enzymatic activity of neuraminidase plays a critical role at this point: without it the viral particles would remain bound to the cell via HA-sialic acid interactions.

Type A Influenza Viruses Are the Most Virulent

Influenza viruses are of the *Orthomyxoviridae* family, and they are classified into three types: A, B and C. Strains are grouped into one of these types according to antigenic determinants found on the viral nucleoprotein (NP) and matrix (M) proteins. They are further *subtyped* on the basis of antigenic determinants found on the HA and NA proteins. Influenza strains are designated by type, geographic site of origin, isolate number, year of isolation, and subtype. For example, A/Hong Kong/3/1957/H2N2 indicates the third isolate of a type A virus isolated in Hong Kong in 1957, which bears the H2 and N2 versions of hemagglutinin and neuraminidase, respectively. The type A viruses account for most epidemics in humans, in part due to the ability of HA and NA of this type to undergo antigenic variation.

Epidemiology

The transmission of influenza is influenced by several factors, including the particular virulence of the strain, and the degree of immunity among the population, as well as environmental conditions. Cases of influenza are most common in the fall through the winter months, with the spread often gaining momentum when children return to school, spreading the infection among themselves and bringing it home to their families. Outbreaks are recorded every year; however, any one community is usually struck every 1–3 years. Typically, between 10–20% of the populace is affected in an outbreak.

Pandemics have occurred periodically since 1889, with the most recent having occurred in 1968 (see Table 11.1). Pandemics coincide with an antigenic shift in the predominant circulating virus. This shift facilitates its spread, as the majority of the population is immunologically naïve to the new strain. Most strains circulating in the years following a pandemic are variants of the pandemic strain, and thus the population begins to acquire resistance to the subtype.

Although the factors affecting the virulence of a strain are not well understood, strain virulence is a factor in the outbreak of a pandemic. For example, a particularly virulent strain was responsible for the 1918–1919 outbreak, often killing by nightfall people who had felt well in the morning. In addition, the highest mortality rates were among young adults, an unusual feature of the disease. The factors behind the unusual characteristics of the 1918–1919 outbreak are still being investigated. Although pandemics are a striking manifestation of the devastation wrought by the virus, the cumulative mortality associated with the disease is greater in the years between pandemics.

Clinical Manifestations

Illness usually resolves after one week

The virus penetrates the airways in the form of droplets generated by coughs or sneezes from acutely infected individuals. The incubation time is fairly short, with symptoms occuring 18–72 hours after exposure. Symptoms include chills, fever (as high as 103°F), myalgia (particularly in the legs and lumbosacral area), headache, malaise, and fatigue. Most patients return to health after a week; however, fatigue and cough may continue for a longer period.

Complications

Complications are most often associated with those with chronic cardiovascular or pulmonary diseases, or the elderly. The most common complication is pneumonia, either as a consequence of the viral infection, or as a result of a secondary bacterial infection. The most common bacteria associated with secondary infections are *S. pneumoniae, S. aureus,* and *H. influenzae*. All three organisms infect the airways and may exploit the impaired bronchopulmonary defenses of the host in the wake of a viral infection. Death results from the acute inflammatory response to infection, resulting in edema and necrosis of pulmonary tissue, and eventually respiratory failure. Treatment for pneumonia includes aggressive oxygen therapy and antibiotic administration in cases of bacterial pneumonia. The antimicrobial regimen administered is typically broad-spectrum until the pathogen is identified. For example, in this case a combination of erythromycin and cefoxitin was first administered, and once the pathogen was identified as *S. pneumoniae,* the patient was switched to penicillin, to which this organism is generally susceptible.

Myocarditis, myositis, and encephalitis are very rare complications of influenza infection. Reye's syndrome is a hepatic complication that occurs in children, and is particularly associated with the type B viruses and with children who have taken certain drugs, such as the salicylates. Thus, children suspected of having the flu are not given aspirin. Reye's syndrome is a rare, albeit serious, complication, with a 50% fatality rate.

Diagnosis

The diagnosis of influenza may be made on the basis of symptoms and knowledge of an outbreak in the community. The illness may be distinguished from the common cold by the severity of symptoms and the high fever. A blood test or sputum sample may also be used to confirm influenza infection; however, this is rarely necessary.

Vaccination

Vaccination has become a widespread measure to prevent influenza infection in the United States, especially among high-risk groups. The most widely used vaccines are made up of inactivated influenza virus particles and consist of strains that circulated in the previous year's outbreak, and/or strains currently circulating in other geographic areas that are predicted to spread to the United States by the flu season. Three viral strains are chosen for each year's vaccine, and thus the vaccine is said to be *trivalent*. The vaccines are 50–80% effective when they closely match the circulating strains.

Treatment

Treatment is usually symptomatic, with patients advised to rest, drink plenty of fluids, and take nonsteroidal anti-inflammatories (NSAIDs) for the fever, headaches, and muscle aches.

If taken early in the infection the antiviral drugs *amantadine* and *rimantadine* can help reduce the duration and severity of the illness. These drugs act on the M2 matrix protein of type A viruses, inhibiting acidification of the virion, and thereby its disassembly. The neuraminidase inhibitors *zanamivir* and *oseltamivir* also reduce the severity of the illness and are active against both type A and type B viruses. These drugs are sialic acid derivatives and competitively inhibit neuraminidase. Although both the M2- and neuraminidase-inhibitor drugs are effective in uncomplicated cases of influenza, it is not yet known if they reduce the risk or severity of complications.

QUESTIONS

1. The affinity of HA for sialic acid is quite low, with a K_d of approximately 5 mM. How does efficient attachment occur with such a low affinity interaction?

2. Given that bacterial pneumonia is a common fatal complication of influenza infection, why do you think the influenza pandemic of 1918–1919 was so devastating in terms of loss of human life?

3. Mucus lines the respiratory tract and provides one of the first barriers of defense against inhaled pathogens. The mucins are a heavily glycosylated family of proteins that are the major constituents of mucus. How do these proteins act to avert infection?

4. Influenza viruses infect aquatic birds; however, these strains replicate poorly in humans and thus direct transmission from aquatic birds to humans has not been observed. Nevertheless, humans may acquire these strains from *intermediate hosts,* such as swine. Why is transmission to humans facilitated by passage through an intermediate host?

5. In 1977 there was a mild influenza pandemic caused by an H1N1 strain that predominantly affected young people, aged 19 or less. Why were young people particularly affected by this pandemic? [Hint: There was a mild epidemic in 1946–1947 caused by an H1N1 subtype.]

6. Two types of linkages between the terminal sialic acid residue and the penultimate galactose residue are found in nature: $\alpha(2,3)$ linkages and $\alpha(2,6)$ linkages (Figure 11.3). Influenza viruses isolated from humans preferentially bind $\alpha(2,6)$ linkages. International surveillance programs monitor influenza outbreaks and analyze the viruses for their antigenic properties in an effort to prevent the occurrence of a pandemic. To obtain sufficient material for analysis, virus isolated from patients is often grown in chicken eggs, cells that predominantly express the $\alpha(2,3)$ type of linkage. Do you think the virus purified from chicken eggs will accurately reflect the virus isolated from the infected patients?

7. From what do you think the name of the *hemagglutinin* molecule derived? [Hint: *hema* derives from the Greek for "blood"; "agglutinate" refers to the adhesion of parts.]

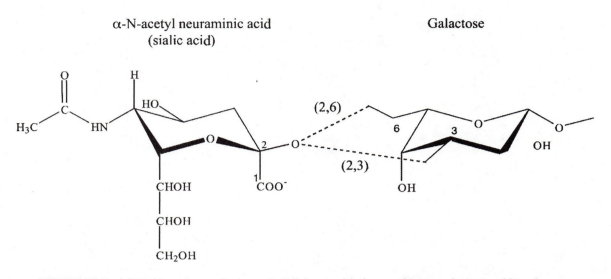

FIGURE 11.3 Sialic acid can form α(2,3) or α(2,6) linkages with the penultimate galactose residue of biological molecules.

FURTHER READING

1. Skehel, J. J., and Wiley, D. C. Receptor binding and membrane fusion in virus entry: The influenza hemagglutinin. *Annu. Rev. Biochem.* (2000) 69:531–569.

2. Weis, W., Brown, J. H., Cusack, S., Paulson, J. C., Skehel, J. J., and Wiley, D. Structure of the influenza virus haemagglutinin complexed with its receptor, sialic acid. *Nature* (1988) 333:426–431.

3. Zambon, M. C. The pathogenesis of influenza in humans. *Rev. Med. Virol.* (2001) 11(4):227–241.

4. Kandel, R., and Hartshorn, K. L. Prophylaxis and treatment of influenza virus infection. *BioDrugs* (2001) 15(5):303–323.

5. Stephenson, I., and Nicholson, K. G. Influenza: vaccination and treatment. *Eur. Respir. J.* (2001) 17(6):1282–1293.

For further information, see the following web sites:

Centers for Disease Control, Influenza: www.cdc.gov/ncidod/diseases/flu/fluvirus.htm

The World Health Organization, Influenza: http://who.int/emc/diseases/flu/

National Institute of Allergy and Infectious Diseases, Influenza: www.niaid.nih.gov/dmid/influenza/default.htm

CHAPTER 12

Lipids and Cell Membranes
An Error in Lipid Metabolism:
Cause of a Fatal Disease in Children

CASE HISTORY

Infant Appears Normal upon Delivery

A couple brings their 5-month-old son to the pediatrician because of irritability, manifest as persistent crying. This is the first child for this couple and the pregnancy was uneventful, as was the birth and first four months of life. The baby was born via a normal, spontaneous vaginal delivery, with apgar scores of 8 and 9. The birth weight was 7 pounds 8 ounces, the length was 20.5 inches, and the head circumference was 35 cm (60th percentile). Initial physical examination in the neonatal nursery was unremarkable, although there was a comment that it was difficult to see into the infant's eyes. He exhibited no organomegaly (the liver and spleen were of normal size). He had normal central nervous system examination, including normal suckling, rooting, grasp, and Moro reflexes. He had a normal cry and full motion of all extremities with normal muscle tone and mass. Neonatal blood tests revealed normal thyroid function and no evidence for phenylketonuria.

What are the manifestations of phenylketonuria and decreased thyroid hormone levels in infants?

What is the relevance of the difficulty in seeing the retina in this neonate?

Infant Thrived During the First Four Months of Life, but Parents Noticed Abnormal Behavior at Five Months

Mother and child were discharged from the hospital 48 hours after the birth. The infant was breast-fed and, after an initial weight loss of nearly 1 pound, he began to gain weight and exhibited no evidence of failing to thrive. He was waking twice per night to feed and would promptly go back to sleep afterward.

The parents reported that approximately five days ago he began to wake more frequently, and would respond with a vigorous startle to noise. At about the same time he began to cry more frequently and persistently without obvious cause, and was not calmed by a pacifier. In fact, he would only stop crying when being fed.

Examination Reveals Neurological Abnormalities and Cherry-Red Spot in Retina

He had no fever, rash, runny nose, cough, diarrhea, or change in urinary frequency. Physical examination revealed weight and height that were in the 60th percentile for his age, and normal head size without bulging fontanelles. The skin was warm and dry without rashes. The heart and lungs were normal, and the abdominal exam revealed a normal sized liver and spleen. Neurological examination was remarkable for an increase in muscle tone and exaggerated withdrawal response to touch. He exhibited jerking movements in response to auditory stimulation. He did not appear to visually track moving objects. His fundus (the back of the eye) was difficult to examine, but after pupillary dilation, macular pallor and a cherry-red spot were apparent.

When do infants typically begin tracking moving objects?

Is the absence of organomegaly significant?

Family History

Mother and father emigrated from the Ukraine two years ago and are Jewish. They are in good health, with no significant medical illnesses. They take no medicines and do not drink alcohol, smoke cigarettes, or use illicit drugs. The father works as a security guard in an office building and drives a cab part-time. The mother was a school teacher in the Ukraine but has not worked since she emigrated. Neither has a history of occupational exposures. They note no known family history of Tay-Sachs disease (TSD) or other significant neurological disease, but the father had a cousin who died in childhood of unknown causes.

Is the ethnicity of this family significant?

Diagnosis

A blood plasma assay for β-N-acetylhexosaminidase A activity revealed reduced activity, <0.1% of normal, and the infant was diagnosed with infantile TSD. DNA-testing confirmed the diagnosis, showing homozygosity of the +1273 TATC insertion mutation in the *HEX A* gene. The parents declined experimental or invasive treatment for their child. The child and parents were followed at home with supportive care and the parents were referred for genetic counseling.

What enzymatic reaction(s) is carried out by β-N-acetylhexosaminidase A?

Is TSD inherited in a recessive or dominant manner?

DISCUSSION

The Molecular Basis for Tay-Sachs Disease

Lysosomal storage diseases

TSD is one of a number of diseases classified as *lysosomal storage diseases* (see Section 26.16 of *Biochemistry* 5e). The lysosome is a membrane-enclosed intracellular organelle that breaks down all types of cellular macromolecules and recycles the building blocks into biosynthetic pathways. The organelle contains an array of degradative enzymes, and mutations in these account for over 30 distinct lysosomal storage diseases. Although they vary widely in their molecular defects and clinical manifestations, they are grouped together because they all result in the intralysosomal accumulation of an undegraded macromolecule.

Tay-Sachs disease is caused by defective or absent β-N-acetylhexosaminidase A

TSD is caused by an inherited mutation in the gene encoding the α subunit of a lysosomal enzyme called *β-N-acetylhexosaminidase A* (Hex A). Hex A removes a terminal sugar unit from a variety of macromolecules as a step in their degradation. The accumulation of one of the enzyme's substrates, a glycosphingolipid called ganglioside G_{M2} (Figure 12.1), results in distension and extensive vacuolization of the cell, which eventually leads to cell death. The highest concentration of gangliosides is found in neuronal cells, especially those of the central nervous system, and they are thus selectively destroyed in TSD.

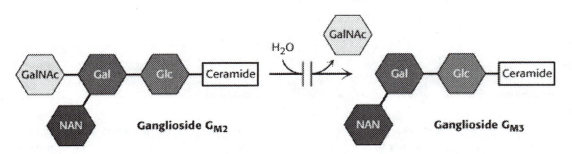

FIGURE 12.1 β-*N*-acetylhexosaminidase A (Hex A) catalyzes the removal of an *N*-acetylgalactosamine (GalNAC) residue from ganglioside G_{M2}, converting it into ganglioside G_{M3}. Gal: galactose, Glc: glucose, NAN: N-acetylneuraminate.

Mutation of either the α or β subunit can cause disease

Hex A is a *heterodimer*, formed by the association of an α subunit and a β subunit. A related disease, Sandhoff's Disease (SD), is caused by mutation of the β subunit. Those with SD are lacking Hex A activity, as well as that of a second enzyme, *β-N-acetylhexosaminidase B* (Hex B), which is a β subunit *homodimer*. The two isozymes are structurally very similar (the α and β subunits are 60% identical in sequence), but differ in some of their physicochemical properties, such as heat-stability and pI (isoelectric point: the pH at which the protein is neutral; see Section 4.1.4 of *Biochemistry* 5e). Hex A and Hex B have similar, but not identical, substrate specificities: while both can hydrolyze neutral substrates, Hex A is far more effective when hydrolyzing negatively charged substrates. It is important that removal of the terminal N-acetylgalactosamine of the negatively charged ganglioside G_{M2} is accomplished largely by Hex A, and thus it accumulates to roughly the same extent in TSD and SD. Consequently, the two diseases have almost identical clinical features.

Disease severity correlates with Hex A activity

There are three forms of TSD, the "classical" *infantile* form, the *juvenile* form and the *adult* form. A complete loss of Hex A activity causes the most severe, infantile form, with death often occurring by age 3. However, partial loss of Hex A activity leads to the milder forms with later onsets and longer survival times. Those with the juvenile form often survive into the teenage years, while survival well into the third or fourth decade is typical in those with the adult form. In vitro assays showed the infantile form to be associated with ≤0.1% of normal Hex A activity, while the juvenile and adult forms were associated with 0.5% and 2–4% of normal activities, respectively. Because clinically healthy individuals were found to have 11% and 20% of normal Hex A activity, it was concluded that below a critical threshold that lies between 5% and 10% of normal Hex A activity, disease characteristics become manifest. This concept is known as the *critical threshold hypothesis*. The infant described here has the infantile form and thus has a very poor prognosis.

Tay-Sachs disease-causing mutations

Although nearly 100 mutations causing TSD have been reported, three mutations account for 95% of the mutant alleles found in patients (Figure 12.2). The most common, and the one found in this infant, is a 4-base-pair (bp) insertion (TATC after codon 1273) in exon 11. It causes a frameshift and a premature stop codon, which causes degradation of the mRNA produced from this allele. The second most common mutation is a single-base substitution (+1G → C) at the 5′ end of the 12th intron. This causes a splicing defect and no normal mRNA is produced. These two mutations cause a complete elimination of normal α subunit mRNA, which results in the most severe, infantile form of the disease.

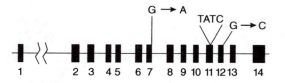

FIGURE 12.2 Schematic of *HEX A* gene structure and location of three most common mutations. The gene is located on chromosome 15 and has 14 exons (numbered). The most common mutations are a 4-bp insertion in the 11th exon, and single-base substitutions at the end of exon 7 and at the beginning of intron 12. The last 13 exons are clustered within approximately 15 Kb at the 3′ end of the gene, separated from the first exon by nearly 20 Kb.

The third commonly found mutation causes the adult form of the disease, and is the result of a missense mutation (805G → A), which causes a substitution of serine for glycine at amino acid 269 (269Gly → Ser). In this case, the mutation causes a reduction in Hex A activity via more than one mechanism: nucleotide 805 is the last in exon 7, and thus lies within the splice junction recognized by the splicing machinery. A splicing defect is thus one effect of the mutation, resulting in a reduced level of correctly spliced mRNA. Adding to this, even when correctly spliced, the mRNA gives rise to a protein carrying a detrimental amino acid substitution: the altered protein is unstable, and impaired in its ability to dimerize with the β subunit.

Epidemiology

TSD is most common amongst Ashkenazi Jews and French Canadians from southern Quebec, with 1 in 27 in these populations carrying a disease mutation. The carrier frequency is also elevated amongst Moroccan Jews and in the Cajun population of southwest Louisiana. The carrier frequency in the general population is approximately 1 in 250. The infantile form is by far the most common. Ashkenazi Jews derive from Eastern and Central Europe, and thus the infant described here, who is of Ukrainian descent, was in a high-risk group.

Inheritance Pattern

TSD is inherited in an *autosomal recessive* manner, and thus the disease will only occur if *both* copies of *HEX A* are faulty. The carrier status (one normal and one faulty copy) is asymptomatic; however, a carrier may pass the disease on to his offspring: if both parents are carriers, each of their children will have a 25% chance of having TSD, and a 50% chance of being a carrier (Figure 12.3).

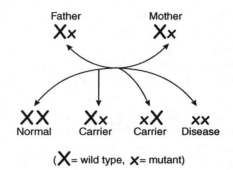

FIGURE 12.3 Inheritance pattern of TSD. The trait is inherited in an autosomal recessive manner. When both parents are carriers, each child will have a 25% chance of being healthy, a 25% chance of having the disease, and a 50% chance of being a carrier.

Disease Progression

Like the infant described here, a baby with infantile TSD appears healthy at birth and develops normally until between 3 and 6 months of age, at which time development of motor skills and mental abilities begins to slow. The first signs of the disease are typically decreased eye contact and, as was observed for this infant, an exaggerated startle response to noise. The infant subsequently loses any motor skills he had acquired, such as sitting up or crawling, and increasingly loses his vision. Upon examination at 5 months this infant showed no evidence of an ability to track objects, suggesting that loss of vision had already begun. Neurodegeneration is relentless, with progressive loss of physical and mental capacities. By 18 months seizures begin to occur, and a virtually complete vegetative state is reached in the late stages of disease. Death occurs by age 5, frequently due to bronchopneumonia caused by aspiration and an inability to cough.

Diagnosis

The clinical features in the affected infant as described in the case are usually sufficient to make the diagnosis. The cherry-red spot observed by retinal examination represents the fovea, a spot at the back of the retina that contains a high concentration of cones (light-sensitive cells). The cones are relatively unaffected while the remaining neuron-rich retina is largely destroyed and pale in comparison, hence the difficulty in observing the fundus and the prominance of the cherry-red spot. Indeed, even as a neonate his retina was noted as being difficult to observe, suggesting that some damage had already occurred. Several other gangliosidoses (including Sandhoff's disease) are also associated with macular pallor and a cherry-red spot.

Macrocephaly (head enlargement) due to lipid deposition in the brain may occur, but is not associated with hydrocephalus (fluid accumulation). Plasma levels of hexosaminidase activity may be used to confirm the diagnosis. Prenatal diagnosis is made by assaying hexosaminidase activity in amniotic fluid or from a chorion villus biopsy. Potential adult carriers are screened by measuring hexosaminidase level in plasma; however, a DNA-based diagnostic test is also recommended for individuals in high-risk groups.

Treatment

Unfortunately, physicians are virtually helpless in the face of TSD, as no effective treatment exists. Experimental therapeutic approaches have not met with much success to date. For example, enzyme replacement therapy and gene therapy have not been effective, perhaps because of the difficulty in delivering therapeutic agents to the brain. Bone marrow transplantation has also been attempted, as donor cells have been shown to serve as a source of the missing enzyme in other gangliosidoses, but has not yet proven effective in treating TSD. In addition, the risks of radiation to the brain and chronic immunosuppressive therapy, necessary steps in the transplantation procedure, are substantial. Given the limited effectiveness and invasiveness of current experimental treatments, many families, like this one, decline them.

Although the experimental treatments described above may prove useful in treating the forms of TSD with later onset, their effectiveness in treating the most common, infantile form is questionable. Reversing neurological damage that occurs during fetal development may not be possible, and hence significant effort has been focused on genetic screening and counseling. Screening programs have been very successful, reducing the incidence of the disease by close to 90% in some of the most high-risk groups.

QUESTIONS

1. What is the difference between TSD and Sandhoff's disease, at the molecular and clinical levels?

2. When investigators first sought to identify the defective enzyme in TSD, assays for Hex activity using an artificial substrate showed no difference between samples from Jewish patients with TSD and healthy individuals. How might this be explained?

3. Hex A, being a lumenal lysosomal enzyme, hydrolyzes ganglioside G_{M2} found on the *inner* leaflet of the lysosomal membrane. Do you think a related cytosolic enzyme carries out this reaction on ganglioside G_{M2} found in the *outer* leaflet?

4. In vivo, hydrolysis of ganglioside G_{M2} requires a cofactor called G_{M2} activator protein. In vitro, detergents can substitute for this cofactor. What function do you think G_{M2} activator protein carries out?

5. Despite the advent of DNA-based diagnostics, initial screening for TSD carrier status is still carried out by an enzymatic assay for Hex activity. Why? How is the level of Hex A activity distinguished from that of Hex B?

6. Describe the limitations of the enzymatic assay for Hex A activity.

7. Given that the infantile TSD is not evident at birth, with symptoms not appearing until between 3 and 6 months of age, is TSD considered to be a congenital defect?

FURTHER READING

1. Mahuran, D. J. Biochemical consequences of mutations causing the G_{M2} gangliosidoses. *Biochimica et Biophysica Acta* (1999) 1455:105–138.

2. Myerowitz, R. Tay-Sachs Disease-causing mutations and neutral polymorphisms in the Hex A gene. *Human Mutation* (1997) 9:195–208.

For further information, see the following web sites:

National Tay-Sachs and Allied Diseases Association: www.ntsad.org

National Center for Biotechnology Information, Online Inheritance in Man: www.ncbi.nlm.nih.gov/entrez/dispomim.cgi?id=272800

www.ntsad.org

National Center for Biotechnology Information, Genes and Disease: www.ncbi.nlm.nih.gov/disease/Tay_Sachs.htmFigure Legends

CHAPTER 13

Membrane Channels and Pumps
Ion Channels and the Cardiac Pump

M.C. is a 15-year-old male with a history of recurrent loss of consciousness (syncope). His first episode occurred at age 8 after jumping into a pool. He was rescued by a lifeguard who noticed he was having seizure-like movements. He was diagnosed with epilepsy and was monitored for recurrent seizures. Approximately one year later, after running up a flight of stairs, he had another episode of syncope, again with seizure activity. At this time he was treated with diphenylhydantoin (Dilantin) but continued to have rare episodes of loss of consciousness without seizure activity.

What causes syncope? What seems to trigger syncope in this patient?

What is the mode of action of diphenylhydantoin? Do you think it was benefiting this patient?

M.C.'s family history is remarkable considering he has a mother with a history of fainting, especially when "stressed," a maternal uncle who died suddenly at age 29 while playing basketball, and a maternal grandfather who died in his thirties of a "heart attack." He has two siblings who are well.

The only medicines he takes are diphenylhydantoin 300 mg per day and a multivitamin daily.

His physical examination is normal with the exception of mild gingival hyperplasia, considered to be a side-effect of long-term treatment with diphenylhydantoin. His resting electrocardiogram (Figure 13.1) demonstrated prolongation of the QT interval. His mother also had prolongation of the QT interval on her electrocardiogram.

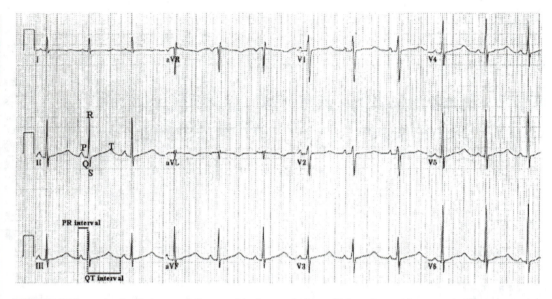

FIGURE 13.1 Twelve-lead electrocardiogram (ECG) recorded from this patient with congenital long QT syndrome. The ECG represents the summation of all the electrical activity occuring in the heart over time. The P wave is the ECG representation of atrial activation, the QRS complex, ventricular activation, and the T-wave ventricular recovery. The PR interval is the time from the onset of atrial activation to the onset of ventricular activation. The QT interval is the time from onset of activation of the ventricle to recovery or repolarization.

Diagnosis and Treatment

The diagnosis of the inherited long QT syndrome was made. Diphenylhydantoin was discontinued and he was started on a β-adrenergic blocking drug, long-acting propranolol 120 mg daily, and referred for implantation of an automatic defibrillator.

Is a prolonged QT interval a universal feature of long QT syndrome?

DISCUSSION

Action potentials are the basis for proper heart function

In Section 13.5.3 of *Biochemistry* 5e action potentials were discussed in terms of their role in transmitting nerve impulses; however, action potentials are also the basis for the organized contraction of heart myocytes that produces a heartbeat. As in the transmission of a nerve impulse, the well-orchestrated activity of a number of ion channels is necessary to generate an action potential in cardiac myocytes. The normal heartbeat begins with spontaneous *depolarization* of cells in the sinoatrial node, which activates or depolarizes cells in the atria and ventricles. This wave of depolarization brings about an orderly sequence of contraction of myocytes in the heart, and subsequent repolarization brings them back to the resting potential such that the process can be repeated.

Depolarizing Na^+ and Ca^{2+} currents flowing into the cell cytoplasm *activate* the myocytes (when the heart pumps), while repolarizing K^+ currents flowing out of the cell are active during the *recovery* phase (Figure 13.2). In addition, a reduction in depolarizing currents during the recovery phase contributes to the restoration of the resting intracellular negativity of the cardiac myocyte. The plateau (phase 2) of the action potential is a time of high membrane resistance (low current flow). Any perturbation of the balance between depolarizing and repolarizing currents during this phase can dramatically change the duration of the plateau and therefore the duration of the action potential.

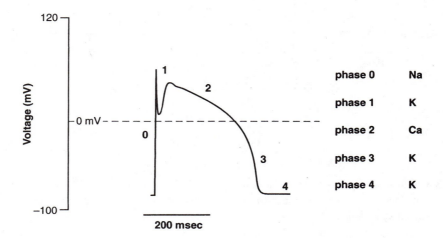

FIGURE 13.2 The phases of the ventricular action potential are inscribed as a result of the activity of a number of transmembrane glycoproteins known as ion channels. Although several channels are conducting current, often the activity of a single class of channel predominates during different phases of the action potential. Phase 0, or the rapid upstroke, is largely the result of activation of sodium channels; phase 1 occurs when sodium channels close and a class of potassium channels activate. The plateau, or phase 2, is the result of opening of calcium channels, and repolarization, or phase 3, is the result of activation of another class of potassium channels. The resting potential (called phase 4 in some cells) is near the reversal potential for potassium and is due to the activity of yet another class of potassium selective channels.

Long QT syndrome is a result of ion channel dysfunction

This child suffers from long QT syndrome (LQTS), a congenital cardiac disorder that is a result of ion channel dysfunction. The QT interval is the time elapsed during depolarization and recovery of the cardiac ventricles (Figure 13.3). Depolarization of the ventricles (reflected in the QRS complex of the ECG and phase 0 of the action potential) is very rapid, and hence any lengthening of the QT interval is usually a result of lengthening of the recovery phase (phases 1–3). During recovery Na^+ currents are inactive while K^+ currents are *active*, and hence a mutation that alters the relative activity of these currents would be expected to lengthen the QT interval. Indeed, of the six genetic loci linked to LQTS, five are components of Na^+ or K^+ channels (the sixth has yet to be identified). As expected, the mutations in the K^+ channel subunits cause reduced flow of ions through the channel, while those in the Na^+ channel result in a defect in inactivation of the channel, such that a depolarizing current continues to flow during the recovery phase.

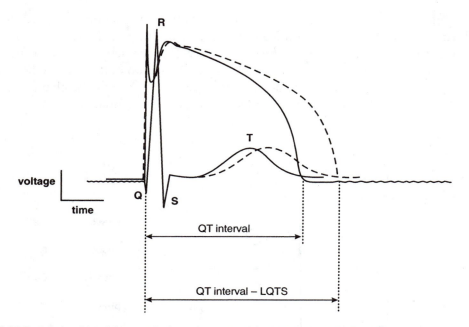

FIGURE 13.3 Relationship of the ventricular action potential ("AP") to the ECG. The QT interval is an index of the duration of the action potentials in ventricular myocytes. The QT interval and AP duration are prolonged in cases of the long QT syndrome.

Pathophysiology

QT prolongation can cause episodes of syncope, which may lead to life-threatening disturbances of the heart rhythm. In the setting of a long action potential with a sustained, relatively high plateau voltage, there is a possibility for reactivation of depolarizing currents before complete repolarization has occurred. This event is referred to as an early afterdepolarization (EAD) and can trigger a series of action potentials that can result in a specific type of repetitive and rapid activation of the ventricle (ventricular tachycardia) called *torsade de pointes* (because of its characteristic electrocardiographic appearance). When the rhythm is too rapid the heart cannot pump effectively, causing reduced blood flow to the brain, which causes syncope. Usually the heart rhythm will revert to normal within a minute, but it has the potential to degenerate into ventricular fibrillation and cardiac arrest.

In addition to the cardiac effects, there exists a rare form of LQTS associated with deafness. Patients with this form (called Jervell-Lange-Nielsen syndrome) inherit mutant copies of KvLQT1 or MinK (potassium channel subunits; see below) from *both* parents, and are thus homozygous for a mutant allele at one of these loci. This potassium channel is necessary for hearing to develop, and these patients are thus congenitally hearing impaired.

The molecular genetics of long QT syndrome

As mentioned above, LQTS has been associated with six genetic loci, five of which have been identified (Table 13.1). Four of these are potassium channel genes, while the fifth is a sodium channel gene. The KvLQT1 and HERG genes encode potassium channel α-subunits and are associated with the two most common types of inherited LQTS, LQT1, and LQT2. In addition, mutations in potassium channel accessory subunits have also been found to cause the syndrome. The Mink (KCNE1) gene product coassembles with KvLQT1, while the MiRP1 (KCNE2) gene product coassembles with HERG in the formation of functional potassium

TABLE 13.1 LQTS types. The mutated genes and currents affected are indicated for each type. The currents affected are either potassium (repolarizing) or sodium (depolarizing), and in the case of potassium currents, they are of either the rapidly activated type (Ik_r) or the slowly activated type (IK_s).

LQTS Type	Mutated Gene Common (Gene) name	Current Affected
LQT1	KvLQT1 (KCNQ1)	Potassium (IK_s)
LQT2	HERG (KCNH1)	Potassium (IK_r)
LQT3	HH1 (SCN5A)	Sodium (I_{Na})
LQT4	unknown	unknown
LQT5	MinK (KCNE1)	Potassium (IK_s)
LQT6	MiRP1 (KCNE2)	Potassium (Ik_r)

channels. KvLQT1/MinK and HERG/MiRP1 form distinct potassium channels that have distinct biophysical properties. KvLQT1/MinK forms a channel that activates slowly upon depolarization, and is thus responsible for the slowly activated component of the potassium current (IK_s), while the channel formed by HERG/MiRP1 is more rapidly activated and accounts for the rapidly activated component of the potassium current (IK_r). Finally, mutations in SCNA5, which encodes the cardiac sodium channel, have also been associated with LQTS. Genetic screening can be used to identify which of the five genes is affected, however, a negative result is obtained in approximately 50% of patients. This is because many mutations can cause the syndrome and only the most common are screened for; in addition, not all genes associated with the disease have been identified.

The six loci are each associated with a slightly different variant of the syndrome, called LQT1-6. In some cases they may be distinguished by differences in T-wave morphology on the ECG, as well as by the factors that trigger cardiac events. For example, cardiac events are more likely to occur while exercising in those with LQT1, while they typically occur during rest or sleep in those with LQT3.

Most cases of LQTS are acquired rather than congenital, the majority of which are caused by medications such as antiarrhythmics. However, evidence suggests that even some of these acquired forms have a genetic component: some mild mutations in ion channel genes may confer susceptibility to acquired LQTS.

Epidemiology

It has been estimated that up to one in 5000 people is affected with LQTS and that the disease causes 3000 deaths annually in the United States. Incidence is thought to be similar internationally. More females are diagnosed than males; however this may be because adult women, with longer QT intervals, are more easily diagnosed. In addition, mortality is greater in boys under 10 than girls, which may contribute to the apparently higher incidence in females. There is no evidence for race-related differences in incidence.

Diagnosis

LQTS is a life-threatening, yet treatable disorder, and thus early diagnosis is important. The two characteristic features of LQTS are episodes of syncope and prolongation of the

QT interval, however, the absence of one or even both features is sometimes observed in those with the syndrome. Sixty to seventy percent of patients exhibit a clearly elevated QT interval (>0.46 seconds, corrected for heart rate); however, many patients with the disease have borderline or even normal values. In these patients, repeated ECGs may reveal the prolongation, or the T-wave morphology may be abnormal and indicative of the syndrome.

Treatment

The mainstay of therapy for LQTS is treatment with β-adrenergic blocking agents or "β-blockers." The precise mechanism by which β-blockers act in LQTS in not certain; however these drugs blunt the effect of the sympathetic nervous system on the heart, and may inhibit the initiators of the abnormal heart rhythm. Treatment with β-blockers is effective for approximately 70% of patients. In addition, implantation of an automatic defibrillator is an option, and was recommended for this patient. This device is surgically implanted, and simply counts the heart rate. If the rate exceeds the value programmed into the defibrillator, a shock is delivered to the heart to restore normal heart rhythm.

In the future, therapy based on drugs tailored to the type of LQTS may become an option. For example, drugs that act to close the leaky sodium channel may be useful in treating LQT3, while drugs that keep potassium channels open may be effective in treating LQT1,2,5,6.

QUESTIONS

1. You are a physician treating two patients with LQTS. Both are heterozygous at the KvLQT1 locus, bearing one normal and one mutant copy. Patient 1 has a nonsense mutation in KvLQT1 that results in complete loss of expression of the gene, while patient 2 has a single missense mutation that causes expression of a defective version of KvLQT1. Interestingly, patient 2 experiences more frequent episodes of syncope, and his ECG revealed a more severe electrophysiological defect than patient 1. What might explain this paradox?

2. What causes deafness in some patients with LQTS?

3. You are a cardiologist and are asked to evaluate a patient with a history of syncope and a familial history of sudden death. You suspect LQTS; however, her QT interval is somewhat elevated but still within the normal range. What other tests might help you confirm the diagnosis of LQTS?

4. A heartbeat is produced when cardiac myocytes are activated and undergo an action potential, which results in orderly contraction of the myocardium. How might this electrical continuity between cells be achieved?

5. Can LQTS be diagnosed postmortem?

6. A mutational "hot spot" has been identified in the potassium channel KvLQT1 gene in many patients with LQTS. The mutation lies at codon 246, which normally encodes an alanine residue, but is frequently mutated to valine or glutamic acid in patients with the syndrome. Given that this alanine residue lies in the S6 transmembrane helix of the channel, how might these mutations affect channel function?

FURTHER READING

1. Splawski, I., Shen, J., Timothy, K. W., et. al. Spectrum of mutations in long QT syndrome genes. KVLQT1, HERG, SCNA5, KCNE1 and KCNE2. *Circulation* (2000) 102 (10):1178–1185.

2. Priori, S. G., Bloise R., and Crotti, L. The long QT syndrome. *Europace* (2001) 3:16–27.

3. Shalaby, F. Y., Levesque, P., Yang, W.-P., Little, W. A., Conder, M. L., Jenkins-West, T., and Blanar, M. A. Dominant-negative KvLQT1 mutations underlie the LQT1 form of long QT syndrome. *Circulation* (1997) 96:1733–1736.

For further information, see the following web sites:

National Center for Biotechnology Information:
www.ncbi.nlm.nih.gov/disease/LQT.html

Online Mendelian Inheritance in Man (OMIM), Long QT Syndrome 1:
www3.ncbi.nlm.nih.gov:80/htbin-post/Omim/dispmim?192500 [also see OMIM entries for LQT2,3,5 & 6]

Sudden Arrhythmia Death Syndromes Foundation: www.sads.org/

CHAPTER 14

Metabolism: Basic Concepts and Design
Simple Sugars Far and Wide, Yet Not a Grain to be Metabolized

A Diabetic Patient Is Referred to the ER with Signs of a Myocardial Infarction

A 57-year-old Caucasian female with a history of insulin-dependent diabetes mellitus presented to the emergency room (ER) complaining of five days of nausea and vomiting. She also noted that she was unable to lie flat due to breathlessness. Her primary care physician, who reported her oxygen saturation to be 85% on room air and her temperature to be 100.4°F, had referred her to the ER. A ventilation perfusion (V/Q) scan was done, which showed no evidence of pulmonary embolus. Admission laboratory studies revealed elevated creatine kinase and troponin I levels, suggestive of myocardial injury, so she was admitted to the coronary care unit (CCU) for further evaluation.

Her past medical history was remarkable for diabetes mellitus of 25 years' duration, requiring insulin. Her diabetes was complicated by coronary artery disease and a prior history of myocardial infarction, end-stage kidney disease requiring chronic ambulatory peritoneal dialysis (CAPD), autonomic neuropathy with orthostatic hypotension (low blood pressure when standing), and gastroparesis (delayed gastric emptying).

What is diabetes mellitus? What does the insulin-dependence suggest about the type of diabetes this patient suffers from?

Why did this patient not complain of chest pain, one of the most common symptoms of myocardial infarction?

Her Medical Regimen

Her medications on admission were aspirin, 325 mg daily, simvistatin (Zocor), 10 mg daily, metoprolol, 50 mg twice daily, metaclopramide (Reglan), 5 mg three times daily, Compazine, 10 mg every six hours as needed for nausea, iron (Niferex-150), 150 mg daily, sevelamer (Renagel), two tablets with each meal, calcitriol (Rocaltrol), 0.25 micrograms daily, Nephro-Vite, one tablet daily, erythropoietin, 5000 units injected subcutaneously two times a week, and citalopram (Celexa), 20 mg daily. Her insulin regimen is NPH, 8 units in the morning and 9 units in the evening injected subcutaneously, with regular insulin, 5 units and 3 units, injected subcutaneously in the morning and evening, respectively.

Physical Examination Suggests Cardiac Injury

Her physical examination on admission to the hospital was remarkable for a temperature of 99.8°F, blood pressure of 177/90, and heart rate of 103 beats per minute (bpm). Her respiratory rate was 22 per minute. She was drowsy, but arousable with deep stimulation. Her pupils were equal, round, and reactive to light, and the sclerae (whites of the eyes) were anicteric (no yellow coloration). She had a well-healed, carotid-end arterectomy scar on the left neck. Her carotids were of normal volume and there were no bruits. Her jugular venous pressure was elevated at approximately 12 cm of water. Her lungs were dull at bilateral bases and there were rales approximately one-quarter of the way up. Her heart exam revealed a diffuse and laterally displaced point of maximal impulse (PMI), suggestive of cardiac enlargement. She had a regular heart rate and rhythm, a normal S1, S2, and an S4 gallop (consistent with reduced compliance of the left ventricle) with no murmurs or rubs. On abdominal exam, she had normally active bowel sounds. In the left lower quadrant she had a Tenckhoff catheter (used for peritoneal dialysis) and the site was clean, dry, and intact. Her abdomen was soft, nontender, and nondistended. There was no hepatosplenomegaly. Her extremities had 1+ bilateral pedal edema. She was oriented to time, place, and person and had no obvious focal motor or sensory defects.

What are typical indicators of cardiac injury?

Is edema a common complication of diabetes? Does the 1+ grade indicate severe edema?

Laboratory and ECG Findings

Laboratory studies at the time of admission were remarkable for blood urea nitrogen (BUN) of 43 mg/dl, creatinine of 4.8 mg/dl, and glucose of 286 mg/dl. The total serum protein, 5.9 mg/dl, and albumin, 2.9 mg/dl, levels were low, but liver function tests revealed no abnormalities. She had mild anemia with a hematocrit of 34.4%, with normal white blood cell and platelet counts. Hemoglobin A1c level was 9.8% (upper limits of normal: 6.1%). She had a presenting creatine kinase of 89 IU/l with an elevated myocardial fraction (MB) of 32, representing an index of 36%. She had a presenting *troponin I level of 0.7 ng/l (upper limit of normal: 0.3 ng/l)*. Presenting electrocardiogram (ECG) showed sinus tachycardia at 100 bpm, with normal axis and normal intervals. She had left atrial enlargement. She had Q-waves in ECG leads II, III and AVF that had appeared since her last ECG, suggestive of inferior wall myocardial infarction of indeterminate age. Her chest radiogram showed small-to-moderate bilateral pleural effusions.

What do elevated levels of creatine kinase and troponin I indicate?

What is a normal blood-glucose level?

Treatment and Discharge

She underwent cardiac catheterization and coronary angiography, which revealed three-vessel coronary artery disease. She had a totally occluded right coronary artery with filling of the distal circulation by left-to-right collateral vessels. The left coronary system revealed a totally occluded circumflex artery distal to the takeoff of a large obtuse margin vessel that had diffuse narrowings. The left anterior descending vessel was also a small caliber vessel that had diffuse atherosclerosis. A left ventriculogram revealed a moderately severe reduction in ejection fraction of 35% (>55% is normal) with an akinetic inferior wall. In light of the diffuse nature of the atherosclerosis and the small caliber coronary arteries she was not a candidate for bypass surgery. She was prescribed oral aspirin and coumarin to prevent further thrombi, and was discharged from the hospital on day 3.

What is coronary angiography, how is it carried out, and what does it reveal?

What is the collateral circulation and how does it help those with heart failure?

What is atherosclerosis and is it common among diabetics?

DISCUSSION

Diabetes Mellitus: Definition and Types

Diabetes mellitus (DM) is a disturbance of carbohydrate metabolism that results from insufficient production of the pancreatic hormone, *insulin,* and/or peripheral tissue resistance to insulin's action. High blood-glucose levels normally stimulate the secretion of insulin, which in turn stimulates glucose uptake by cells of peripheral tissues (especially skeletal muscle and liver). The reduced cellular-glucose uptake that occurs in diabetes results in *hyperglycemia* (high blood-glucose level), the major cause of the most serious complications. Thus, in diabetes, carbohydrate metabolism is restricted by reduced *accessibility of substrate,* one of the principal control mechanisms of metabolic pathways (see Section 14.3.3 of *Biochemistry* 5e).

DM is divided into two major categories, based on the cause of hyperglycemia. *Type 1* is characterized by a loss of insulin-producing cells of the pancreas (β cells), usually as a result of autoimmune destruction. Type 1 usually has a juvenile onset (before age 25), and patients are dependent on the administration of exogenous insulin; thus type 1 diabetics are *insulin dependent.* This patient, with severe complications and long-standing insulin dependence, would be classified as a type 1 diabetic.

Type 2 diabetes is characterized predominantly by *insulin resistance,* meaning that although insulin is produced by the pancreas, target cells respond poorly, most likely due to defective signaling through the *insulin receptor* found on the surface of target cells. The pancreas typically responds by overproducing insulin, and the β cells often eventually fail, leading to insufficient insulin production. Thus, although type 2 DM may initially be controlled with dietary and lifestyle changes, and oral hypoglycemics, patients often ultimately become insulin dependent. Type 2 DM usually has a later onset (typically over age 40) and is clinically less severe than type 1.

Epidemiology

Type 2 diabetes is by far the more common form, accounting for approximately 90% of cases worldwide. One hundred and thirty five million people are estimated to have the disease globally, including 16 million in the United States alone (~6% of the population). There is considerable variation in incidence with geographic area, with the United States having an intermediate level. The incidence of type 1 diabetes also varies with geographic area, with Finland having the highest rate (35/100,00 per year) and countries of the Pacific Rim having the lowest (1–3/100,00 per year). The incidence of DM is roughly equal in men and women in most age groups, but is slightly higher in men over the age of 60.

Diagnosis

The recommended diagnostic criterion for DM is a fasting plasma glucose level of ≥126 mg/dl (<110 mg/dl is normal). Type 1 diabetes typically has a sudden onset (over a period of a few days or weeks) with symptoms of thirst, excessive urination, fatigue, weight loss and a propensity for infections. Similar symptoms occur in type 2 diabetics, but the onset is typically more gradual. Indeed, symptoms often go unnoticed for several years, as studies have shown that most type 2 diabetics have had the disease for at least four years prior to diagnosis.

Etiology

Although the mechanism is largely unknown, autoimmune destruction of the β cells of the pancreas is the principal cause of type 1 DM (see Section 36.6.2 of *Biochemistry* 5e). There are certain genetic predisposing factors, such as particular alleles of genes that encode the class II MHC molecules (involved in the immune response), but an environmental "trigger" is thought to initiate β-cell destruction. Several environmental stimuli have been proposed as triggers for type 1 DM, but none have definitively been confirmed. Possible candidates include viral infections and exposure to bovine milk. Although genetic factors are involved in determining susceptibility to type 1 DM, most individuals with genetic susceptibility factors do not get the disease, and most type 1 diabetics do not have first-degree relatives with the disease.

Conversely, type 2 diabetes has a strong genetic component: if both parents have type 2 DM, the risk of the disease in each of their children reaches nearly 40%. In addition, obesity is a strong determinant: between 80% and 90% of those with the disease are obese. Adipocytes release substances that may cause insulin resistance, thus contributing to the pathophysiology of the disease.

Treatment

Hyperglycemia lies at the root of the most serious complications of diabetes, and hence controlling the blood sugar is the primary focus of therapy. In addition, as diabetes typically affects numerous organs, diabetics routinely follow complex drug regimens. A case in point, this patient takes 13 different medications related to her condition, and requires chronic ambulatory dialysis.

Insulin regimens typically make use of both long-acting insulin, such as the NPH insulin used by this patient, as well as the shorter-acting, regular insulin. The long-acting insulin provides a fairly constant "baseline" level of insulin in the systemic circulation, while the shorter-acting regular insulin is taken to coincide with the rise in blood-sugar level that occurs with meals. This regimen is thus designed to mimic the normal pattern of insulin secretion that occurs in a healthy individual, attempting to avoid large swings in the level of serum glucose. Alternatively, some patients receive constant infusions of insulin via implanted pumps.

Complications

The vascular complications of diabetes are the most troublesome

There are *acute* and *chronic* complications of diabetes; however, the chronic complications are the most serious, accounting for most of the morbidity and mortality associated with the disease. Of the chronic complications, the *vascular* complications are the most important and are caused by chronic hyperglycemia. Over time, hyperglycemia can cause the narrowing of blood vessels, reducing blood flow through them and thus causing necrosis of the tissues they supply. The small blood vessels of the retina and kidney are often affected, as are those feeding the nerves of the extremities and the autonomic nervous system; thus retinopathy, nephropathy, and neuropathy are common *microvascular* complications. Larger blood vessels are also often affected, which can cause *macrovascular* complications such as myocardial infarctions or strokes. The macrovascular complications in diabetics are generally the result of the accelerated rate of *atherosclerosis* in diabetics. Atherosclerosis is the deposition of fatty plaques within blood vessels, which narrows them and renders them susceptible to complete occlusion by a thrombus (blood clot).

The laboratory studies performed on this patient upon admission revealed hyperglycemia and an elevated hemoglobin A1c (HbA1c) level, indicating poor control of blood glucose, the most important factor in the development of complications. Indeed, with end-stage kidney disease, autonomic neuropathy, and a history of coronary artery disease and myocardial infarction, she presented with some of the classic features of the disease.

Myocardial infarction and impaired cardiac performance

This patient presented at the ER with several indicators of a *myocardial infarction*. A myocardial infarction occurs when there is necrosis of part of the myocardium due to an interruption in blood flow that supplies it. It usually occurs when an atherosclerotic plaque ruptures, with ensuing thrombosis at the site, which occludes the blood vessel. The portion of the myocardium supplied by the artery soon dies, producing a number of symptoms and signs that may be used to diagnose the condition.

Impaired contractile performance of the heart may produce low cardiac output and pulmonary venous congestion, resulting in symptoms such as nausea and breathlessness that this patient experienced. A number of compensatory mechanisms are recruited in order to maintain cardiac output, including an increase in sympathetic nervous system tone, which results in elevated heart rate (HR) and blood pressure. The reduced contractility of the heart reduces the stroke volume (SV), the amount of blood pumped with each beat (SV = end diastolic volume [EDV] − end systolic volume [ESV]). Blood accumulates in the ventricles, with the increased pressure backing up into the veins, which accounts for the findings of elevated venous pressure and cardiac enlargement on physical examination. Cardiac dilation, with concomitant increase in EDV, serves to maintain the stroke volume, and together with the increased heart rate, maintains cardiac output (CO = HR × SV).

Diagnosis and treatment of myocardial infarction

Myocardial infarction is typically diagnosed by a combination of clinical history; characteristic electrocardiographic (ECG) changes; and elevation in the serum levels of cardiac enzymes, creatine kinase (MB isoform), troponin I, and lactate dehydrogenase (LDH). The spatial distribution of ECG changes can localize the region of the heart that is infarcted, in this case the inferior, or diaphragmatic, surface of the heart. This was confirmed by the lack of movement (akinetic segment) of the inferior wall on the ventriculogram. Chest pain, a common symptom of MI, is often absent in diabetic patients, possibly because of neuropathy.

Typically, a patient who is experiencing myocardial infarction is aggressively treated with pharmacological or mechanical measures that are designed to open the occluded coronary artery. However, such maneuvers are performed only if symptoms are relatively short lived, and in this case the patient had been experiencing symptoms for five days. Thus, diagnostic cardiac catheterization and angiography were performed instead. The presence of diffuse coronary artery disease and an akinetic inferior wall with collaterals suggested that the right coronary occlusion was old, and therefore no attempt was made to open the vessel. Upon discharge, patients are typically prescribed antithrombotic medications, such as aspirin or coumarin, to prevent the recurrence of thrombosis.

QUESTIONS

1. This patient developed diabetes at age 32. Is this typical for type 1 diabetes mellitus?

2. In diabetes, the reduced accessibility of glucose signals the starved state, which stimulates the mobilization of energy from stored lipid reserves. Fatty acid breakdown can lead to an excess of acetyl-CoA (see Section 22.2 of *Biochemistry* 5e), which is unable to enter the citric acid cycle due to a shortage of oxaloacetate (see Figure 17.15 of *Biochemistry* 5e). As a result, there is a shift to an alternative pathway in which acetyl-CoA is used in the generation of *ketone bodies*. What are the steps in the formation of the ketone bodies (acetoacetate, D-3-hydroxybutyrate, and acetone) from acetyl-CoA, and how do they cause ketoacidosis, a life-threatening complication of diabetes?

3. In vivo, insulin is secreted into the hepatic portal vein, and thus the liver is exposed to the highest concentration of the hormone. In contrast, exogenous insulin is administered subcutaneously in diabetic patients, at a site distant from the liver. What might be the effect of this?

4. Do the antithrombotic drugs, coumarin and aspirin, act via the same mechanism?

5. Elevated creatine kinase level in the serum is an indicator of MI. Why does the myocardium express high levels of this enzyme?

6. What is hemoglobin A1c, and what does its level reflect?

FURTHER READING

1. Cooper. M. E., Bonnet, F., Oldfield, M., and Jandeleit-Dahm, K. Mechanisms of diabetic vasculopathy: An overview. *American Journal of Hypertension* (2001) 14:475-486.

2. Kirpichnikov, D., and Sowers, J. R. Diabetes mellitus and diabetes-associated vascular disease. *Trends in Endocrinology and Metabolism* (2001) 12(5):225-230.

3. Kahn, C. R. New concepts in the pathogenesis of diabetes mellitus. *Adv. Intern. Med.* (1996) 41:285-321.

For further information, see the following web sites:

American Diabetes Association: www.diabetes.org

National Institute of Diabetes and Digestive and Kidney Diseases: www.niddk.nih.gov

Centers for Disease Control, Diabetes Public Health Resource: www.cdc.gov/diabetes/faqs.htm

National Center for Biotechnology Information, Online Mendelian Inheritance in Man, Diabetes Mellitus, type 1: www3.ncbi.nlm.nih.gov:80/htbin-post/Omim/dispmim?222100

CHAPTER 15

Signal Transduction Pathways
A Historical Case of Bipolar Illness

CASE HISTORY

A Chief Executive with the Blues

A 51-year-old man is seen by his private physician with complaints of despondency and the blues. He has had a history of what he describes as episodic "melancholia" since his youth. The episodes of the blues alternate with periods of extreme energy and success in his professional life.

What is the clinical spectrum of bipolar illness?

What are the criteria for making this diagnosis?

A Life Marked by Successes, Tragedy, and Depression

There was a well-known family history of depression, and his personal life was marked by a number of tragedies. His father had a history of depression, often withdrawing from his family and friends. Several other paternal relatives had similar disturbances of mood. He often found himself at odds with his father, who was a farmer and carpenter, because he preferred to read rather than work in the fields. He had an older sister, and a younger brother who died in infancy. When he was 9 years old his mother died, an event that he believes has affected him psychologically for his entire life. About a year later his father remarried, and he described his relationship with his stepmother as caring and respectful. When he was 19 his older sister died during childbirth. He describes this as a devastating loss followed by a long period of depression of his mood.

What are the symptoms of depression?

What are the symptoms of mania?

At 26 years of age he was engaged to be married, but unfortunately his fiancée died of complications of typhoid fever. This again prompted a period of profound depression of mood and withdrawal from the rest of his family, and he buried himself in his work. He had a number of jobs in his youth, including running a store, surveying, and working in a post office. He ultimately received his degree in law and settled on life as a public servant, running energetic and successful campaigns for the state legislature.

Several years later, after a tumultuous relationship, he was married, and over the course of the marriage had four children, only one of whom survived. The death of one of his sons was particularly painful and again prompted a period of profoundly depressed mood with uncontrolled weeping, withdrawal, sleep disturbance, and loss of appetite.

The only significant illness or injury that he had suffered was a head injury, having been kicked and rendered unconscious by a mule. Upon regaining consciousness, he began to experience petit-mal (unresponsiveness without seizure-like movements) epileptic seizures. He took no medicines or herbs regularly, did not smoke, and would drink only socially. He did suffer from constipation throughout his life.

At present his mood was good. He felt vital with a good, but not excessive, energy level and normal libido. He was sleeping well, had a good appetite, and had not noticed any significant change in his weight. He denied any slowed or restless movements, feelings of worthlessness or excessive guilt, trouble in thinking, concentrating, or making decisions. He was not excessively anxious, irritable, or impulsive. He did not have recurrent thoughts of death or suicide.

What His Physical Examination Might Have Revealed

On physical examination he was 6 feet 4 inches tall and weighed 180 pounds. He was thin but normally developed. His blood pressure was 140/60 mmHg, with a heart rate of 74 beats per minute. His head, ears, eyes, nose, and throat examination revealed no abnormalities. His neck was supple and there was no thyromegaly and no cervical lymphadenopathy. The chest was resonant to percussion and clear to auscultation. The heart examination revealed a fourth heart sound and a murmur of aortic regurgitation. His abdomen was soft, nontender, with active bowel sounds, and there was no organomegaly. His extremities were characterized by long fingers (arachnodactyly) with normal pulses and no edema. The neurological examination revealed orientation to person, place, and circumstance. He could name the current U.S. president and his predecessor. His attention span was normal, and long- and short-term memory was intact. He had no focal sensory or motor deficits and his gait was normal. Deep tendon reflexes were symmetrical and normally active.

Does the physical examination suggest any other diagnoses?

The Diagnosis and Treatment Then and Now

The diagnosis that was rendered by his doctors to the patient, Abraham Lincoln, was that he suffered from nothing more than "hypochondria." The patient accepted this diagnosis and believed that he now understood his disease; indeed, he had been quoted as saying, "I have within the last few days been making a most discreditable exhibition of myself in the way of hypochrondrism . . . ."

There was no specific therapy prescribed for this patient. Indeed, over the years, Lincoln was able to develop a core of self-confidence that supported him through the most agonizing moments of his life and the life of our nation. Lincoln had sage advice for those suffering with melancholia: "spend leisure time with your spouse, or try wrestling on the living room rug with your children. Laugh at your own shortcomings, and bond closely with your friends. Commit yourself to meaningful goals, and enjoy verbalizing your ideas and aspirations. Light entertainment, i.e., the theater, is always a good diversion. If none of these untangle your web, you may decide to involve yourself in a national dilemma."

Were he alive today Lincoln may have been diagnosed with bipolar disorder II (major depression plus hypomania). He might have been treated with lithium carbonate for mood stabilization and intensive psychotherapy and family therapy. Recently, other pharmacotherapeutics have been used with success in the treatment of bipolar disorder, including anticonvulsants such as valproic acid. Many patients are treated with combination regimens of antipsychotic drugs and mood stabilizers. Antidepressants are generally avoided because of the manicogenic properties of these agents.

DISCUSSION

Bipolar Disorder Is a Mood Disorder

Cycles of elation and depression characterize the disorder

Bipolar disorder is a *mood disorder,* a class of mental disorders that also includes depressive disorders. Mood disorders are characterized by pervasive mood and behavioral disturbances that are typically recurrent. Bipolar disorders are distinguished from the depressive disorders by the presence of *manic (or hypomanic)* affective episodes, which are periods when the individual feels extremely energetic, elated, and uninhibited. These alternate with periods of depression, during which the individual feels intense sadness and hopelessness. The syndrome of cyclic episodes of depression and mania has been recognized for over 2000 years; however, not until 1978 did *The Diagnostic and Statistical Manual of Mental Disorders (DSM)* include bipolar disorder as a distinct diagnosis.

Joy and sadness in response to life events are normal and must be distinguished from the abnormal responses that occur in those with mood disorders. The responses are disproportionately intense in those with mood disorders and last for longer than the expected amount of time. In addition, an episode may occur spontaneously, in the absence of any trigger. In those with mood disorders, the mood disturbances are usually severe enough to interfere with the ability to function socially or in a work or school setting. Bipolar disorder is a chronic disease and must be managed throughout one's lifetime.

Epidemiology

Mood disorders are the most common psychiatric disorders, with the depressive disorders accounting for the majority of cases. Approximately 20% of women and 12% of men are expected to suffer from some type of mood disorder in their lifetime. Bipolar disorder is estimated to affect 2% of the general population. Both sexes are equally affected, and the disorder is found across socioeconomic strata; however, it is somewhat more common in upper socioeconomic classes. Gender and cultural factors appear to influence clinical manifestations; for example, depressive episodes typically predominate in women, whereas mania often predominates in men. In addition, self-reproach typifies depressive episodes among Anglo-Saxon cultures, whereas manic episodes are particularly extreme in Mediterranean cultures.

Bipolar disorder may strike in the teens, 20s, or 30s, but it most often strikes in the 20s. Over two million Americans are estimated to suffer from bipolar disorder.

Risk Factors

Childhood loss of a parent has not been shown to increase the chance of developing a mood disorder; however, if a mood disorder strikes, it may strike at a younger age and lead to more serious clinical manifestations. Genetics also play a role, although the genes involved have yet to be identified. There is an 80% concordance rate between monozygotic twins, and two-thirds of patients have a close relative with depression or bipolar disorder. The data accumulated so far indicate that a number of genes are likely to be involved, and that both genetic and environmental factors contribute to susceptibility. Although mood disorders occur in all personality types, extroverted individuals seem to be at increased risk for bipolar disorder.

Clinical Manifestations

Alternating shifts of mania and depression characterize the disorder

The disorder is characterized by alternating shifts of depression and mania. Both physical and psychological stressors can induce an affective episode, in particular, a separation, such as death of a loved one or a divorce. The onset of an affective episode can occur over a period of days or weeks (sometimes even hours) and typically lasts for 3–6 months. A day or two of little sleep often coincides with the shift from a depressive to a manic episode, and the shift may be experimentally induced by sleep deprivation. Many patients feel well between episodes, especially early in the course of the disorder; however, as many as one-third experience residual symptoms between episodes. Rarely, some patients experience persistent, unremitting symptoms despite treatment.

There are subtypes of bipolar disorder

There are three types of bipolar disorder, which differ in the extent of the manic phase. Type I is associated with frank depressive and manic phases, whereas type II is associated with somewhat more moderate manic episodes termed *hypomanic* phases. Hypomanic phases are characterized by increased confidence, optimism, and psychomotor activity, but no psychosis. In type III the manic phases are still more subtle, and may be difficult to recognize. Various *mixed states,* in which features of both depression and mania are manifested within a single episode, may be superimposed upon these types. For example, a manic episode may be interrupted by brief periods of tearfulness.

The depressive and manic phases

The *depressive* phase is characterized by depressed psychomotor activity, fatigue, sadness, self-reproach, apathy, and an inability to enjoy life's pleasures. The patient often feels that life is no longer worth living, and many contemplate suicide during these episodes. In extreme cases patients may fall into a depressive stupor.

The *manic* phase is characterized by extreme confidence, which is often delusional, bringing convictions of inordinate wealth, power, and intelligence. The patient feels as though he can do anything. He is often unable to concentrate and leaps from one thought to the next at such a rapid pace that it is difficult to make sense of what he is saying. Auditory and visual hallucinations may occur. The patient often loses a sense of judgment and may exhibit reckless behavior. The need for sleep decreases and psychomotor activity increases, often to a frenetic pace, and in extreme cases may reach a point of senseless agitation known as *delirious mania.* This is a rare, albeit life-threatening complication, which may lead to death from sheer physical exhaustion.

In the absence of treatment the disease tends to worsen, with the affective episodes becoming more severe and more frequent. *Rapid cycling,* defined as four or more episodes within a year, becomes more common as the disease progresses in untreated cases, and is more prevalent among women.

Suicide

Bipolar disorder brings an increased risk of suicide. It has been estimated that 15–25% of untreated or inadequately treated patients attempt suicide. Suicide attempts are more common early in the course of the disorder, especially during the first episode; however, the attempts become more serious later on. More women attempt suicide than men; however, equal numbers of men and women succeed. There is also a high rate of accidental death in those with bipolar disorder.

Etiology

The etiology of bipolar disorder is unknown. The clear genetic component to the disorder has led many researchers to search for a molecular defect; however, the underlying mechanism(s) of bipolar disorder has remained elusive. Abnormalities in neural pathways involving serotonin, norepinephrine, dopamine, and acetylcholine have been implicated, and imaging studies have indicated that irregularities in various parts of the brain are associated with the disorder. Investigations into the mechanisms of action of the drugs used to treat the disorder have identified some candidate molecules involved in mediating the therapeutic effects (see below), and these studies may also shed some light on the disease mechanism.

Diagnosis

Diagnosis often brings a sense of relief to patients. Knowing that their mood swings are due to an identifiable illness that may be treated brings the realization that they are not "crazy" and that concrete steps may be taken to control the disorder. This, in itself, represents a major advance since the time of Abraham Lincoln, who was diagnosed with *hypochondria,* which is defined as the depression of spirits centered on imaginary ailments. Fortunately, psychiatry has moved forward in the intervening years, and mental illnesses are now recognized as having a biochemical basis, and are not simply "in the mind," as was once believed. Although unrelated to his mental condition, Lincoln had arachnodactyly, was exceptionally tall, and had cardiac abnormalities consistent with Marfan syndrome (see Chapter 28).

As with other mental illnesses, diagnosis of bipolar disorder may not be made on the basis of a physiological change, but rather depends upon a description of the symptoms and the family history, if it is available. It often begins with depression, which switches to bipolar disorder after a few years. The occurrence of major manic (or hypomanic) and depressive episodes, as defined in the *DSM,* is the major diagnostic criterion for bipolar disorder. In addition, various brain imaging techniques may be helpful, such as magnetic resonance imaging (MRI) and positron emission tomography (PET); however, these are more useful in supporting a diagnosis than in making it or ruling it out. Distinguishing between depression and bipolar disease is often a challenge as patients often present during a depressive episode and do not recognize the mania as part of the disorder.

Treatment

Combination of drug and psychosocial therapy is most effective

Most patients with bipolar disorder can expect to benefit from the available drug treatments. The medications are able to control the mood swings such that many patients can return to productive lives. Medications are taken for extensive periods to stabilize moods, with higher dosages or additional drugs given during affective episodes. In addition to medications, psychosocial therapy, in particular "talk therapy," has proven useful in treating those with bipolar disorder, stabilizing moods, and reducing the frequency of hospitalization.

Lithium is the mainstay of therapy

Lithium (Li$^+$, in the form of lithium carbonate) is the mainstay of therapy for treating the manic episodes of bipolar disorder. It is a naturally occurring alkali metal and rapidly enters the bloodstream from the gastrointestinal tract, reaching its peak concentration in the serum after about 90 minutes. Ninety-five percent is excreted in the urine within 24 hours. Steady state is not reached until after 4–6 days of treatment, and there is a delay of approximately two weeks in its antimanic activity. Thus, other antipsychotics must often be taken in the initial part of treatment. At lower dosages lithium is also useful as a prophylactic, reducing the chance of recurrence of manic and, to a lesser extent, depressive episodes. Lithium is effective in roughly two-thirds of patients. Noncompliance with the drug regimen is common because of the side effects of nausea, thirst, polyuria, and weight gain, among others. *Valproic acid* is a good alternative to lithium, and appears to be particularly effective in treating patients with rapid cycling. Another anticonvulsant, *carbamazepine,* may be clinically effective as well, although it is not yet approved by the FDA for treatment of bipolar disorder.

Mechanism of Drug Action

Elucidation of mechanism may lead to improved pharmacological agents

The mechanism of action of lithium is unknown; however, a number of molecular targets for the drug have been identified. Distinguishing which of these, if any, is involved in the neuropsychiatric effects of lithium has yet to be determined. Identifying the molecular target(s) of lithium that is responsible for its neuropsychiatric effect is clinically significant because it will allow for the development of more effective and better-tolerated therapies. Lithium has adverse side effects, and its therapeutic dose is close to the toxic level. Furthermore, it is not effective in all patients, and thus improved drug therapies would have great therapeutic value. In addition, identification of the mechanism underlying lithium's action may shed some light on the pathogenesis of the disorder, of which little is known.

Inositol depletion hypothesis

One of the most persuasive hypotheses is that lithium acts through *depletion of inositol.* This would ultimately limit the level of phosphatidyl inositol 4,5-bisphosphate (PIP$_2$), a precursor of inositol 1,4,5-trisphosphate (IP$_3$), an important cellular signaling molecule (see Section 15.2 of *Biochemistry* 5e). A number of neurotransmitters are known to signal through G-protein-coupled receptors to activate phospholipase C and the phosphoinositide pathway. Lithium has been shown to uncompetitively inhibit inositol monophosphatase (IMPase) with a K$_i$ of 0.8 mM, which is in the therapeutic range for treatment of bipolar disorder. This enzyme dephosphorylates inositol monophosphate (IMP), regenerating inositol (Figure 15.1), and is thus important in inositol recycling. Lithium likely inhibits the enzyme by displacing one of the magnesium ions necessary for activity.

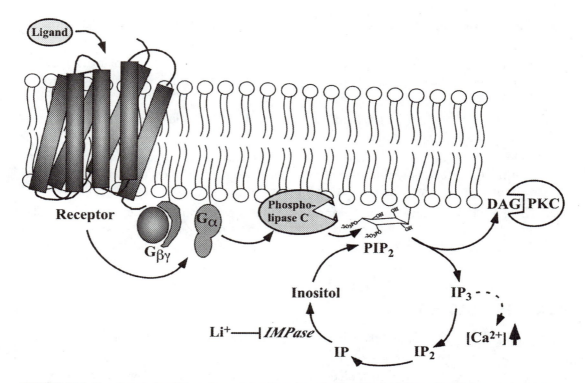

FIGURE 15.1 The phosphoinositide pathway. It has been hypothesized that inhibition of inositol monophosphatase (IMPase) by lithium accounts for its therapeutic effect in bipolar disorder. IMPase dephosphorylates inositol monophosphate (IP) to generate inositol, and thus inhibition of this enzyme would inhibit recycling of inositol for use in the generation of PIP_2, a critical component of the pathway. PIP_2: phosphatidyl inositol 4,5-bisphosphate; DAG: diacylglycerol; PKC: protein kinase C; IP_3: inositol 1,4,5-trisphosphate.

Other drugs may act via the same pathway

Thus, perhaps lithium normalizes overactive signaling through the phosphoinositide pathway, thereby reducing the symptoms of the disorder. Interestingly, a recent study (see Williams, et al., 2002) has indicated that carbamazepine and valproic acid also act to deplete inositol, and thus, these structurally diverse drugs may all act via a common mechanism to stabilize moods in these patients. Other targets of lithium include glycogen synthase kinase-3β and a large family of structurally related phosphomonoesterases. The contribution of these, and perhaps other as yet unidentified targets, to the pathogenesis of bipolar disorder has yet to be determined.

QUESTIONS

1. Why is aberrant signal transduction an appealing hypothesis for the underlying mechanism of bipolar disorder?

2. Lithium has been shown to have little effect on the psychiatric state of patients without mental disorders. Assuming that lithium's therapeutic effect comes from reducing excessive signaling through the phosphoinositide pathway via inhibition of IMPase, why might you expect a minimal effect of the drug on healthy individuals?

3. Unlike most drugs, lithium appears to have multiple, unrelated targets. What property of lithium do you think might account for this?

4. Elevation of the G-protein subunit, $G\alpha_s$, has been associated with bipolar disorder, especially the manic episodes. Which enzymatic activities might you expect to be affected by $G\alpha_s$ elevation?

5. Why is the central nervous system (CNS) particularly dependent upon *recycling* of inositol to meet its inositol needs?

6. Preliminary trials suggest that the drug *tamoxifen,* an antiestrogen commonly used to treat breast cancer, is useful in treating bipolar disorder. What is the proposed mechanism of action of the drug in treating bipolar disorder?

7. There is a delayed onset of action for the two most common drugs used to treat bipolar disorder, lithium and valproic acid. Given that these drugs are likely to act by modulating the activity of signal transduction pathways, what might account for this delay?

FURTHER READING

1. Shutes, M. H. *Lincoln and the doctors: a medical narrative of the life of Abraham Lincoln.* Pioneer Press, New York, 1933.

2. Phiel, C. J., and Klein, P. S. Molecular targets of lithium action. *Annu. Rev. Pharmacol. Toxicol.* (2001) 41:789–813.

3. Williams, R. S., Cheng, L., Mudge, A. W., and Harwood, A. J. A common mechanism of action for three mood-stabilizing drugs. *Nature* (2002) 417(6886):292–295.

4. Bezchlibnyk, Y., and Young, L. T. The neurobiology of bipolar disorder: focus on signal transduction pathways and the regulation of gene expression. *Can. J. Psychiatry* (2002) 47(2):135–148.

5. Ghaemi, S. N., Ko, J. Y., and Goodwin, F. K. "Cade's disease" and beyond: misdiagnosis, antidepressant use, and a proposed definition for bipolar spectrum disorder. *Can. J. Psychiatry* (2002) 47(2):125–134.

6. Kato, T. Molecular genetics of bipolar disorder. *Neurosci. Res.* (2001) 40(2):105–113.

7. Sachs, G. S., Printz, D. J., Kahn, D. A., Carpenter, D., Docherty, J. P. The Expert Consensus Guideline Series: Medication Treatment of Bipolar Disorder 2000. *Postgrad. Med. Special Report* (2000):1–104.

For further information, see the following web sites:

Depression Awareness, Recognition, and Treatment (D/ART) Program, National Institute of Mental Health: www.nimh.nih.gov

National Depressive and Manic Depressive Association: www.ndmda.org

National Alliance for the Mentally Ill: www.nami.org

CHAPTER 16

Glycolysis and Gluconeogenesis
African Sleeping Sickness:
Time for a Wake-Up Call

CASE HISTORY

Patient Presents to the ER with Fever and Diarrhea

A 22-year-old college student came into the ER complaining of a three-day history of malaise, fever, and diarrhea. Three days ago she returned from a two-month trip as an exchange student in East Africa, in Uganda, Kenya, and the Sudan. An anthropology major, she was carrying out research on hunting practices among tribal peoples in remote, rural areas. She had faithfully taken chloroquine, 300 mg weekly, for malaria prophylaxis, beginning one week before departure from the United States, and had continued the prescribed regimen upon her return. She had no significant medical history and had never been hospitalized. The previous day she had developed a throbbing headache that was unresponsive to acetaminophen. Several of the students who participated in the exchange program experienced episodes of diarrhea over the course of the trip, although she had not.

She does not recall having a rash and denies muscle and joint pains or stiff neck. She denied abdominal pain or fullness and does not recall having a fever during the trip. She did not complain of itching and had no history of edema.

What is the potential significance of her recent travel history?

Physical Examination and Laboratory Results

Physical examination revealed a healthy appearing young woman with blood pressure 140/75 mmHg, pulse 95 per minute, temperature 39.5°C orally, and respirations 15 per minute. Detailed examination of the skin revealed a 3-cm raised, red lesion on the back of her left arm, but no rashes. On questioning she remembered having been bitten by a "horse fly" several days before her return to the United States. The abdomen was soft and nontender, with mild enlargement of the spleen. Her cervical and axillary lymph nodes were enlarged, mobile, rubbery, and minimally tender.

Initial laboratory results were: hemoglobin, 12.5 mg/dl; hematocrit, 33%; white blood cell count, 5000 /mm^3 with a left shift; platelets, 200,000/mm^3; glucose, 79 mg/dl; sodium, 120 mEq/l; potassium, 4.2 mEq/l; and chloride, 90 mEq/l. Spinal fluid was obtained from a lumbar puncture, and was found to contain normal glucose and protein levels and no evidence of trypanosomes. The opening pressure was normal.

Does the raised lesion suggest any possible diagnoses?

What is the significance of these laboratory values?

Diagnosis and Treatment

Examination of a thick smear of peripheral blood stained with Wright's stain revealed spindle-shaped flagellates. Detailed examination of these organisms revealed the presence of a central nucleus and a kinetoplast at one end, features characteristic of African trypanosomes. The patient was diagnosed with stage I human African trypanosomiasis (HAT), and specimens were sent for analysis.

The patient was admitted to the hospital and 1 g of suramin was administered by slow, intravenous infusion. Additional doses were administered on days 3 and 7, with gradual improvement of her symptoms. She did not develop any neurological or psychiatric symptoms. She was discharged on day 10, apparently fully recovered, but returned for the final suramin infusions on days 14 and 21. The specimens sent for analysis revealed the presence of *Trypanosoma brucei rhodesiense*, a subspecies of African trypanosomes. At one-year follow-up the patient was asymptomatic and refused lumbar puncture.

What is suramin? Does it have any potentially harmful side effects?

What other drugs may be used to treat human African trypanosomiasis?

Few pharmaceutical companies are currently developing new drugs for the treatment of HAT. Why?

DISCUSSION

Etiology of African Sleeping Sickness

Caused by two Trypanosoma brucei *subspecies*

African sleeping sickness, or HAT, is caused by a protozoan hemoflagellate of the Trypanosoma brucei *complex. T. brucei* is a kinetoplastid, meaning that it possesses a subcellular organelle known as a *kinetoplast*. The kinetoplast is a region of the mitochondria that contains DNA and is visible using staining techniques such as Giemsa or Wright staining. Only three types of kinetoplastids cause disease in humans—*T. brucei, T. cruzi,* and *Leishmania.*

HAT is transmitted via an insect vector, the tse tse fly of the genus *Glossina.* The disease may be caused by either of two morphologically indistinguishable trypanosome subspecies: *Trypanosoma brucei rhodesiense* or *Trypanosoma brucei gambiense.* They may be distinguished by the geographic area in which they were acquired and by the clinical features of the diseases they cause. *T. brucei rhodesiense* is prevalent in East African countries and the disease typically follows an acute course that can lead to death in a matter of months or even weeks.

Conversely, infection with *T. brucei gambiense,* which is found in West Africa, causes a protracted illness that may last for years; however, it too is fatal if left untreated.

Antelopes are a reservoir for T. b. rhodesiense

Humans are the only reservoir for *T. b. gambiense;* however, deer and antelopes found in the savannah and woodland areas of central and Eastern Africa serve as reservoirs for *T. b. rhodesiense* and thus proximity to infected animals can result in human infection. The parasite is adapted to these game animals and does not harm them; however, infection in humans is lethal. Infection is also fatal in cattle, taking an economic toll by hindering the development of livestock and dairy industries. Hunters, game wardens, as well as tourists visiting game reserves, are at increased risk for becoming infected; however, only 21 cases of East African trypanosomiasis in travelers to Africa have been reported to the CDC since 1967. There is no prophylactic treatment for HAT and thus wearing protective clothing to guard against the insect bite is the best defense.

History of the disease

Although the symptoms of the disease have been recognized for centuries, the parasite was not identified until 1903 by two British physicians working in Gambia. In the same year the tse tse fly was identified as the disease vector. The disease was gradually brought under control in the first half of the 20th century thanks to systematic surveillance and drug treatment, as well as vector control. The disease was virtually eliminated in the 1960s; however, since then the disease has returned and has again become a threat to public health. This has largely been attributed to the limited financial resources, weak health systems, political instability, and war in the countries where it is prevalent. Of the 36 countries affected by HAT, 22 are among the least developed in the world.

Epidemiology

HAT is found exclusively in sub-Saharan Africa. Between 1940 and 1960 the disease incidence dropped dramatically, from approximately 60,000 cases annually to virtually none. These statistics are believed to be quite accurate because of systematic screening of the population that was carried out during that time. Since the mid-1970s increasing numbers of cases have been reported; however, screening for infection is no longer a widespread practice and thus the number of reported cases are thought to be a vast underestimate of the actual number. African sleeping sickness tends to occur in *foci,* such that one village may be ravaged by the disease while a neighboring community is unaffected, and thus comprehensive national screening programs are necessary to accurately assess disease prevalence.

Forty-five thousand cases were reported in 1999; however, the true number of cases is thought to be between 300,000 and 500,000. The Democratic Republic of the Congo, Angola, Uganda, and Sudan are the most seriously affected. *T. b. rhodesiense* is found in Kenya as well as eastern parts of The Congo and Uganda, all areas visited by this patient.

Microbiology of *T. brucei*

Life cycle

T. brucei multiplies in extracellular bodily fluids, such as the blood, lymph, and cerebrospinal fluid (CSF). Humans become infected through the (painful) bite of several species of the genus

Glossina tsetse fly. Transmission may also occur through contaminated blood or from mother to fetus through the placenta. The parasite (in the *trypomastigote* stage) replicates in the glucose-rich medium of the blood. It may then be transmitted to the fly as it takes a blood meal from an infected person. In the fly it passes through several developmental stages, from the *procyclic* to the *epimastigote* stages in the midgut, and finally to the *metacyclic* stage (the infective stage) in the salivary glands. The parasite may then be returned to a mammalian host through insect saliva.

Antigenic variation allows evasion of the immune system

Infection induces a vigorous humoral immune response; however, the parasite has evolved the ability to change its surface antigen (the variable surface glycoprotein, *VSG*), thus allowing it to continue to replicate. The cycles of antigenic variation are believed to account for the cyclic nature of the fevers that often occur with the disease. The lysis of cells as a result of immune destruction releases toxic substances that induce inflammation and fever. During the afebrile phase that follows, the new clone replicates until the immune system can again mobilize against the "novel" intruder.

Parasites depend upon glycolysis in human phase

The cell differentiation patterns that occur as the parasite progresses through its life cycle are accompanied by variations in parasite metabolism. In the bloodstream form, trypanosomes depend entirely on glycolysis for ATP, relying on the ample levels of glucose available in the blood, while in the insect it switches to oxidative metabolism and utilizes proline as its primary source of energy. The dependence of the parasite on glycolysis in the human phase has made this process an attractive target for the development of new drugs. In trypanosomes, most of the reactions in the glycolytic pathway take place in specialized membrane-bound organelles related to peroxisomes called *glycosomes*.

Glycosomes

Each trypanosome has been estimated to contain 200–300 glycosomes, and they occupy ~4% of the cell volume. The flux through their glycolytic pathway is approximately 20 times higher than that through most cells. Although it was formerly believed that the high glycolytic flux was a result of the concentration of glycolytic enzymes in a fairly small space, this view is now being challenged. Evidence now indicates that glycolytic flux is not limited by the diffusion of intermediates from one enzyme to the next, but rather by the rate of catalysis of the enzymes in the pathway. However, computer modeling has indicated that confinement of the glycolytic enzymes to the glycosome prevents the accumulation of metabolites to potentially toxic levels, and allows for the rapid onset of ATP production following a period of starvation.

Glycolytic pathway is divided into glycosomal and cytosolic parts

The glycolytic pathway is divided into two parts in trypanosomes, with the enzymes involved in converting glucose into 3-phosphoglycerate localized to the glycosome, while the enzymes that convert 3-phosphoglycerate into pyruvate are found in the cytosol (Figure 16.1). Thus, the consumption and production of ATP in the glycosome is balanced. Similarly, the redox balance of NADH/NAD$^+$ is unaffected: NADH produced in the reaction catalyzed by glyceraldehyde-3-phosphate dehydrogenase is consumed in the generation of glycerol-3-phosphate from dihydroxyacetone phosphate (via glycerol-3-phosphate dehydrogenase). The mitochondrion participates in this pathway by providing the oxidase (glycerol-3-phosphate oxidase) that, with the help of a molecular shuttle, transfers the two electrons from NADH to molecular oxygen in the mitochondrion. Under anaerobic conditions glycerol-3-phosphate is converted into glycerol via the activity of glycerol kinase.

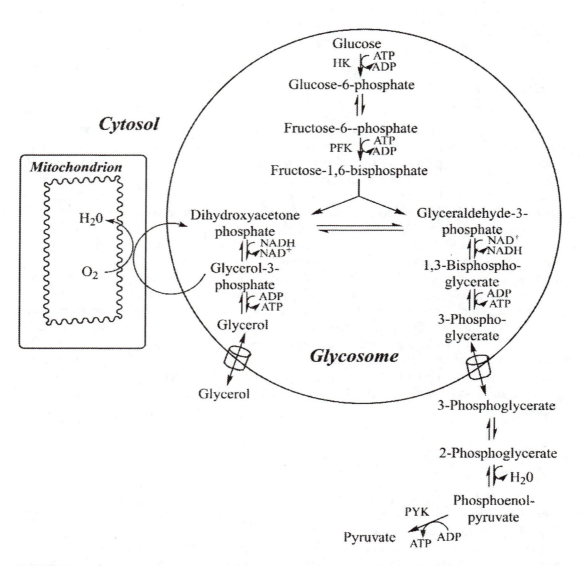

FIGURE 16.1 The *T. brucei* glycolytic pathway. The reactions leading to the generation of 3-phosphoglycerate occur in the glycosome, whereas the remaining steps take place in the cytosol. Pyruvate is the final product and is excreted into the bloodstream. Under aerobic conditions the electrons from NADH are shuttled to the mitochondrion from glycerol-3-phosphate to regenerate NAD$^+$; however, under anaerobic conditions glycerol-3-phosphate is converted into glycerol via the activity of glycerol kinase. Under anaerobic conditions glycerol and pyruvate are produced in equimolar amounts. HK: hexokinase; PFK: phosphofructokinase; PYK: pyruvate kinase.

Glycolytic enzymes are attractive targets for chemotherapy

If sufficient parasite vs. host specificity could be achieved, the glycolytic enzymes would be ideal targets for antitrypanosomatid chemotherapy. Studies have shown that glyceraldehyde-3-phosphate dehydrogenase, phosphoglycerate kinase, and glycerol-3-phosphate dehydrogenase are all partially rate controlling, and thus would be particularly good targets. Indeed, adenine analogs have been found to effectively inhibit parasite glyceraldehyde-3-phosphate dehydrogenase without affecting the human enzyme, and in addition were shown to kill parasites in vitro (see Aronov, et al., 1999). These results validate the approach of using glycolysis as a target for antitrypanosomatid chemotherapy, and provide the basis for further studies to test the in vivo effectiveness of these agents.

Clinical Features

A localized infection at the site of the fly bite may occur, with the development of a sore, or *trypanosomal chancre,* within a week or so of the bite. At the same time fevers develop, often cyclically, with several days of fever followed by an afebrile period. Swollen and tender lymph nodes (lymphadenopathy) are typical, particularly the cervical lymph nodes in the neck, which may be visible. Maculopapular rashes may also occur. Muscle aches, joint pain and a generalized malaise characterize this stage of the illness, which is known as *stage I.* Mild leukocytosis, thrombocytopenia, and anemia often accompany this stage, as do high levels of immunoglobulins, particularly those of the IgM class.

In *stage II,* which can occur from months to years after the bite, the parasite crosses the blood–brain barrier and invades the central nervous system (CNS). Once disease reaches this stage it is more serious and more difficult to treat. This stage is characterized by protean neurological manifestations. The patient becomes confused, and speech may become indistinct. This stage is also characterized by disturbance of the circadian rhythm, with the patient typically being listless and dazed during the day (hence the name *sleeping sickness*) and restless and agitated at night. Ataxia is common, as are tremors and fasciculations. When untreated, neurological damage progresses, eventually leading to coma and death. Death typically occurs in a matter of 9–12 months for *T. b. rhodesiense;* however, infection with *T. b. gambiense* is typically chronic, and can last for years before claiming a life.

Diagnosis

Although the clinical features and travel history may suggest the etiology, diagnosis rests upon visualization of the parasite. A sample from the chancre may reveal parasites, and blood samples may also contain evidence of trypanosomes. Lymph node or bone marrow aspirates may also reveal the presence of parasites. In the late stages of disease (stage II) trypanosomes may be found in the cerebrospinal fluid. Other indicators of stage II HAT include increased pressure of the CSF, or elevated numbers of blood cells or protein levels within it. Differentiation between *T. b. rhodesiense* and *T. b. gambiense* may not be made morphologically. Instead, a number of tests based on antibody recognition of subspecies-specific antigens are available, and in addition, inoculation into mice or rats will cause parasitemia in the case of *T. b. rhodesiense,* but not *T. b. gambiense* because of host specificities.

Treatment

Currently used drugs are inadequate

Treatment options for HAT are currently, and always have been, inadequate. Suramin and pentamidine, developed in the first part of the 20th century, are recommended for stage I disease, but are ineffective in cases of CNS involvement because they inefficiently cross the blood–brain barrier. In addition, these drugs are toxic, causing complications such as liver and kidney dysfunction, and hypotension. The introduction in 1949 of melarsoprol, a drug active against stage II disease, was groundbreaking; however, it also is highly toxic, causing reactive encephalopathy in up to 18% of patients. The drug is an arsenic-based compound, and is fatal in 5–10% of those treated. Fortunately, this patient had only stage I disease, and suramin proved effective and did not produce side effects. The mechanism of action of suramin is unknown; however, it affects the function of a number of proteins. The drug DL-α-difluoromethylornithine (DFMO), an inhibitor of polyamine biosynthesis, was introduced in 1990, but it is active only against *T. b. gambiense* and is very expensive. Combinations of these antitrypanosomatid agents are now being tested in an attempt to lower toxicity and maximize effectiveness.

Eflornithine

The history of another drug, eflornithine, illustrates an important problem in providing drugs to developing countries. Eflornithine was developed in the 1970s and its potential in treating *T. brucei* was soon recognized, bringing dramatic recoveries. It was found to be active against both stages of disease of *T. b. gambiense* infection, and was remarkably effective, curing more than 90% of patients with stage II disease. However, following several corporate mergers, drug production was halted in 1999. The discovery that eflornithine was effective in removing facial hair in women resurrected the drug and production was reinitiated, and has led to a collaboration between the pharmaceutical industry and public health organizations to provide and distribute the drug to patients in West Africa. Thus, although eflornithine had the potential to save thousands of African lives, production ceased until it was discovered that it could be marketed in the Western world to treat a cosmetic problem.

Economic barriers to new drug development for HAT

Many medications are too costly for the Third World market, providing little incentive for their production and distribution. Further, the cost of new drug development is enormous, with few pharmaceutical companies willing to risk the cost for a drug that is unlikely to be profitable. Thus, although basic science research has provided some promising avenues for new chemotherapeutic agents for HAT, none is currently in Phases I–III of clinical development, and consequently no new medications are likely to enter the market in the next decade.

QUESTIONS

1. This patient was treated with suramin (Figure 16.2) for stage I HAT. This drug is not used for stage II disease because it is unable to cross the blood–brain barrier in sufficient quantities to be effective in the CNS. What property of the drug do you think renders it unable to cross the blood–brain barrier?

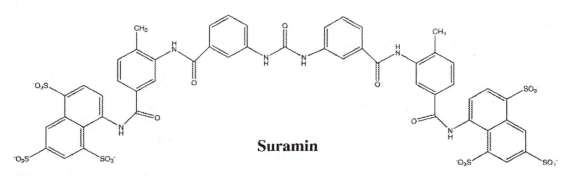

FIGURE 16.2 Structure of suramin.

2. Do you think vaccine development is a viable approach to controlling African sleeping sickness?

3. The glycosome is found in all kinetoplastids, and is thus thought to be evolutionarily ancient, predating multicellular organisms with glucose-rich bloodstreams. Why has this been used as an argument *against* the hypothesis that glycosomes evolved to increase glycolytic flux?

4. Treatment of *T. brucei* with glycerol and salicylhydroxamic acid (an inhibitor of glycerol-3-phosphate oxidase) kills the parasite. Why?

5. In many organisms phosphofructokinase is a key regulatory enzyme in the glycolytic pathway, and its activity is affected by the ATP/AMP ratio (as well as other effectors). *T. brucei* phosphofructokinase does not appear to be sensitive to the levels of these nucleotides. What might explain this?

6. Inhibition of one of the glycolytic enzymes of *T. brucei* can be lethal to the parasite as a consequence of reduced flux through the pathway and inhibition of ATP synthesis such that ATP levels are too low to sustain life. However, this may be difficult to achieve as most enzymes are not rate-limiting and inhibition would have to be close to complete for there to be any effect. How else could inhibition of a glycolytic enzyme result in death of the parasite?

7. Would inhibitors of components of the parasite oxidative transport chain in the mitochondrion be useful in treating HAT?

FURTHER READING

1. Aronov, A. M., Suresh, S., et al. Structure-based design of submicromolar, biologically active inhibitors of trypanosomatid glyceraldehyde-3-phosphate dehydrogenase. *Proc. Natl. Acad. Sci. U.S.A.* (1999) 96:4273–4278.

2. Michels, P. A. M., Hannaert, V., and Bringaud, F. Metabolic aspects of glycosomes in trypanosomatidae—new data and views. *Parasitol. Today* (2000) 16(11): 482–489.

3. Keiser, J., Stich, A., and Burri, C. New drugs for the treatment of human African trypanosomiasis: research and development. *Trends in Parasitol.* (2001) 17(1):42–49.

For further information, see the following web sites:

Centers for Disease Control, African Trypanosomiasis:
www.cdc.gov/ncidod/dpd/parasites/trypanosomiasis/default.htm

World Health Organization, African trypanosomiasis:
www.who.int/emc/diseases/tryp/index.html

CHAPTER 17

The Citric Acid Cycle
The Danger of Two Carbon Fragments

CASE HISTORY

Patient in Alcoholic Withdrawal Is Brought to the ER

A 56-year-old man was admitted to the hospital with acute alcohol withdrawal. He was brought into the ER 12 hours earlier when he was found on the street in a stuporous state. His admission blood alcohol level was 270 mg/dl, and toxicology screening revealed no illicit drugs. He was somnolent but easily arousable, and belligerent when stirred.

Does the alcohol level indicate anything about the chronicity of his alcohol abuse?

Patient Has a History of Alcoholism

He began drinking beer and wine in his early teens. While he was working as an automobile mechanic he would drink approximately a case of beer and/or several bottles of wine each day. More recently he drank whiskey or vodka and admitted to drinking at least a fifth (a typical sized bottle, 4/5 of a liter) each day depending on his finances. He was homeless, having lost his job five years earlier as a result of repeatedly coming to work intoxicated. He had been married earlier in life but was divorced shortly thereafter—another casualty of his drinking. He did not drive, having lost his license for driving while intoxicated. He smoked cigarettes when he had the opportunity.

He had previously been treated in an in-patient detoxification program, had taken disulfiram (Antabuse), and had attended Alcoholics Anonymous (AA), but always returned to drinking. He has had multiple blackouts and has been treated for withdrawal, but he denied a history of delirium tremens (DTs), hallucinations, or convulsions. He denied a history of gastrointestinal bleeding or dark stools. He did not have known liver or heart disease; a history of jaundice; or numbness, tingling, or pain in the legs and feet. He had been hospitalized only for alcohol detoxification and denied any other significant medical illnesses.

What are the criteria for the diagnosis of alcoholism?

What are the main treatment strategies for alcoholism?

What are DTs?

Why was he asked about darkening of his stools?

What would numbness, pain, or parasthesias represent in such a patient?

Physical Examination

The physical examination was remarkable for a well-developed but unkempt man who intermittently slept during the examination. He had a strong smell of alcohol on his breath. Examination of his skin revealed multiple spider angiomata (branching nests of dilated blood vessels) on the chest and neck. The skin and sclera were not icteric (yellow). The blood pressure was 122/58 mmHg and the heart rate was 94 beats per minute (bpm) and irregular, the respiratory rate was 16 per minute and regular, and his temperature was 36°C orally.

His neck was supple with no distention of the jugular veins. The thyroid was not enlarged, and the carotid arterial pulsations were of normal volume without bruits. The chest examination revealed rales at the bases of both lungs that cleared with deep inspiration. His heart examination revealed a laterally displaced point of maximal impulse (PMI) with frequent extrasystoles and a grade II/VI systolic ejection murmur. The abdomen was soft and nontender with active bowel sounds, and no fluid wave or other evidence for ascites. The liver was enlarged and nodular, and the spleen was not palpable. Rectal examination revealed external hemorrhoids but no masses, and scant amounts of dark guaiac positive stool.

Neurological examination revealed that he was oriented to self, month, year, and place. He had a short-term memory deficit; he was unable to remember three objects at one minute, although the testing was complicated by residual alcohol intoxication. He had normal muscle mass and strength, and responded to pinprick in all the extremities.

What are spider angiomata and what do they signify?

What is ascites and why is it observed in alcoholics?

What are the common neurological problems exhibited by alcoholics?

What is the significance of the consistency of the liver on examination?

Alcoholics are predisposed to hemorrhoids. Why?

Laboratory Evaluation and Radiogram

The laboratory evaluation was remarkable for anemia, with a hematocrit of 33% with macrocytosis (increased red blood cell size). The white blood cell count was normal. Liver function tests (LFTs) were abnormal, with an aspartate amino transferase (AST) of 90 IU/l (upper limit of normal [ULN]: 31 IU/l), an alanine amino transferase (ALT) of 84 IU/l (ULN: 31 IU/l), total bilirubin (TB) of 1.2 mg/dl (ULN: 1.2 mg/dl), and an alkaline phosphatase of 128 IU/L (ULN: 120 IU/l). Renal function tests were abnormal, with a blood urea nitrogen (BUN) of 38 mg/dl (ULN: 20 mg/dl) and a creatinine of 0.9 mg/dl (ULN: 1.4 mg/dl). His blood sugar was low at 62 mg/dl, serum lactate was 27.3 mg/dl (normal range: 4.5–19.8 mg/dl) and uric acid was 9.4 mg/dl (normal range: 4.1–8.8 mg/dl). He had a mild coagulopathy, with a prothrombin time (PT) of 16 seconds (ULN: 13 seconds), with an international normalized ratio (INR) of 1.2. The chest radiogram revealed no evidence for infiltrates in the lungs and a large heart with a cardiothoracic ratio of 18/32.

What is the significance of the macrocytic anemia?

Why are the liver function tests only mildly abnormal?

What is the significance of the abnormal blood coagulation tests?

Why was the blood sugar low?

Patient Develops DTs Despite Treatment

He was given intravenous normal saline with dextrose supplemented with thiamine. Over the course of his ER stay he became less somnolent and more agitated. He was treated with intravenous diazepam (Valium) but continued to be agitated. He then experienced a convulsion for which he was given a dose of diphenylhydantoin (Dilantin). He was transiently more confused after the seizure, his skin became diffusely erythematous, his heart rate increased to 120 bpm, and he became tremulous. He was treated with more intravenous diazepam, and admitted with a diagnosis of impending DTs.

DISCUSSION

Alcoholism

Effects of alcoholism are devastating

Alcoholism is a chronic disease characterized by the consumption of excessive amounts of alcohol, repeated attempts to stop drinking, and continuation despite devastating social and occupational consequences. Alcoholism is both a psychological and physical addiction. Many alcoholics become divorced and lose their jobs, and many eventually become homeless, like this patient. Although homeless populations are notoriously difficult to sample, it has been estimated that over 60% of homeless adults in Western industrialized urban areas are alcoholics. Alcoholism is a prevalent problem. It has been estimated that 14 million people in the United States—1 in 13 adults—are alcoholics or abuse alcohol (the diagnostic criteria for alcoholism are hotly debated, and physical addiction may not be a necessary requisite).

Alcoholism affects life expectancy, cutting approximately 15 years from the life spans of both men and women. Death occurs through motor vehicle and other accidents; acts of violence (suicide and homicide); and medical conditions, such as cirrhosis, heart failure, and certain types of cancer. Alcoholism clearly takes a heavy economic toll: alcohol-related problems, including accidents, loss of productivity, medical problems, and crime, are estimated to cost $300 billion annually in the United States.

Pathogenesis

Excess NADH impacts several metabolic pathways

A number of metabolic processes are affected by excessive consumption of alcohol (ethanol). Ethanol is metabolized in the liver by *alcohol dehydrogenase (ADH)*, which oxidizes ethanol to acetaldehyde. This toxic compound is then oxidized to acetate via the enzyme *aldehyde dehydrogenase (ALDH)* (Figure 17.1). An important feature of these reactions is that NADH is produced, which decreases the NAD⁺/NADH ratio in the cell, accounting for a number of the pathological effects of alcoholism.

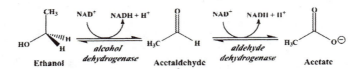

FIGURE 17.1 Metabolism of ethanol in the liver produces NADH.

Impaired gluconeogenesis may cause hypoglycemia

The increased level of NADH impairs the process of *gluconeogenesis* because the elevated NADH level inhibits the formation of pyruvate from lactate (see Section 16.3 of *Biochemistry* 5e). This can lead to hypoglycemia, especially in the setting of chronic malnutrition, which is common among alcoholics. Most alcoholics get much of their calories from alcohol alone, and damage to the gastrointestinal tract causes reduced absorption of nutrients. Blood-glucose levels can become dangerously low, especially when an alcoholic binge occurs within or after a period of low food intake, and may lead to death from hypoglycemic coma. In addition, lactate levels may build up, which may lead to lactic acidosis.

Citric acid cycle is inhibited

The elevated levels of NADH also act to inhibit the *citric acid cycle*. Two enzymes in the cycle, *isocitrate dehydrogenase* and *α-ketoglutarate dehydrogenase,* are inhibited by NADH (see Section 17.2.2 of *Biochemistry* 5e), leading to the accumulation of acetyl CoA (Figure 17.2). This is exacerbated by the conversion of acetate to acetyl CoA via the activity of *thiokinase*. The build-up of acetyl CoA has several downstream effects. First, it may be converted into ketone bodies (see Section 22.3.5 of *Biochemistry* 5e), thus exacerbating acidosis. It may also be converted to acetaldehyde, a reactive compound that can impair protein function, via the activity of aldehyde dehydrogenase. Acetyl CoA is also the precursor of fatty acid biosynthesis (see Section 22.4.1 of *Biochemistry* 5e), and much of it is diverted into this pathway, leading to excess formation of triacylglycerols and eventually to hepatic steatosis (fatty liver), which can lead to liver dysfunction, cirrhosis, and liver failure. Inhibition of fatty acid oxidation (an NAD⁺-dependent process) also contributes to the development of hepatic steatosis. Liver failure is one of the most serious complications of alcoholism.

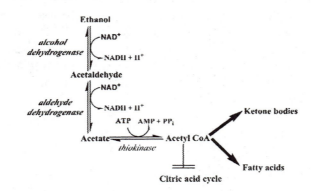

FIGURE 17.2 Inhibition of isocitrate dyhydrogenase and α-ketoglutarate dehydrogenase inhibits entry of acetyl CoA into the citric acid cycle. Acetyl CoA is instead diverted into alternative pathways, ketone body formation and fatty acid synthesis.

Oxidative stress may contribute to pathogenesis

Finally, an alternative pathway for the oxidation of ethanol occurs in the endoplasmic reticulum, via a process known as the *microsomal ethanol-oxidizing system*. Acetaldehyde and acetate are produced via this pathway as in the pathway catalyzed by ADH and ALDH; however, NADPH, an important cellular reducing agent, is consumed. This may lead to oxidative damage and precipitate the development of cirrhosis.

Clinical Features

CNS effects

Alcoholism is associated with a number of chronic and acute medical problems. Acute problems include acidosis, hypoglycemia, and hepatic failure, as well as withdrawal and DTs (see below), and these must be treated as medical emergencies. The chronic problems include central nervous system (CNS), cardiovascular, and hepatic disturbances. Peripheral neuropathy, perhaps as a result of thiamine deficiency, is seen in 5–15% of alcoholics, and causes symptoms of tingling and numbness of the extremities. Blackouts are periods when all or part of a day spent drinking is completely forgotten, and are common among alcoholics, including this patient. Alcoholics also often have cognitive problems, with impairment of both short- and long-term memory, and may exhibit a range of psychiatric symptoms during an alcoholic binge or upon withdrawal from one. Extreme anxiety, sadness, delusions, and hallucinations are among the psychiatric problems that may occur.

Gastrointestinal effects

Alcoholics also suffer from gastrointestinal problems, largely as a result of damage to the gastric mucosa and increased acid production. Gastritis is common, and poor absorption of nutrients as a result of damage to the gastric mucosa exacerbates the malnutrition associated with alcoholics. This may lead to damage to the tissue and gastrointestinal bleeding, which may be manifested as darkened stools. However, among the gastrointestinal disturbances, liver dysfunction predominates, with cirrhosis, a life-threatening complication, occurring in 15–20% of alcoholics. Cirrhosis in alcoholics occurs in the setting of hepatic steatosis, or "fatty liver." As will be seen in later chapters (see Chapters 22 and 23), cirrhosis can occur in response to a number of disorders; however, alcoholism is the most frequent cause in Western industrialized countries.

The nodular liver that was palpated on physical examination of this patient is typical of a cirrhotic liver, indicating that he had fairly advanced liver damage. The fairly mild abnormalities in liver function tests belie the underlying liver dysfunction in this case. AST and ALT are present at high concentrations in hepatocytes, and their elevation in the serum generally correlates with necrosis of hepatic tissue. However, when liver damage is advanced, few hepatocytes remain, and thus LFTs will not necessarily reflect the degree of liver damage. Spider angiomata and hemorrhoids often occur together with cirrhosis.

Cardiovascular and hematological effects

Cardiovascular effects of alcoholism include hypertension, cardiomyopathy, and arrhythmias. Enlargement of the heart and cardiac arrhythmias such as atrial fibrillation may predispose to intracardiac thrombi and increased risk of thromboembolism. Blood coagulation defects due to reduced synthesis of coagulation factors in the liver may reduce the risk of embolic complication but enhance the chance of severe bleeding. The hematopoietic system is also affected, with macrocytic anemia occurring in a number of patients as a result of folate or pyridoxine deficiencies. Interestingly, chronic, moderate alcohol consumption (1–2 drinks daily) has been found to have beneficial cardiovascular effects, reducing the risks of cardiovascular mortality and ischemic stroke.

Delirium tremens

This patient developed DTs, a response to withdrawal from ethanol. Ethanol acts through the γ-aminobutyric acid A ($GABA_A$) and N-methyl-D-aspartate (NMDA) receptors as a CNS depressant, and withdrawal results in effects opposite to those of the drug. Thus, autonomous nervous system activity increases, with increased heart rate, respiratory rate, and body temperature, and the patient often becomes agitated and anxious. Seizures may occur, as well as other neurological disturbances such as delusions and hallucinations. Although symptoms may occur within a few hours of withdrawal, the peak may not occur until between 48 and 72 hours later. Symptoms generally improve by day 4 or 5.

DTs are potentially fatal and thus early recognition and treatment are critical. Mortality is close to 35% if left untreated, but only ~5% if recognized and treated early. Intravenous administration of benzodiazepines, such as the diazepam used on this patient, is standard treatment, as they are very effective and act quickly, typically within minutes of intravenous administration. These drugs act on the $GABA_A$ receptor, enhancing GABA binding and depressing CNS activity.

Epidemiology

The life-time risk of developing alcoholism in most Western countries is between 10% and 15% for men, and ~5% for women. Alcoholism is approximately 2.5 times more prevalent in men than women, and it occurs in all socioeconomic strata and in individuals of all races. There appears to be a genetic component to risk of alcoholism, as the children of alcoholics have been shown to be at higher risk for developing the disease. In addition, studies in twins have shown a greater concordance of drinking habits between identical twins compared to fraternal twins. The genetic factors appear to be related to the response to alcohol: those with a predilection to alcoholism appear to be relatively resistant to the effects of alcohol, as measured by factors such as lower impairment of alcohol-induced psychomotor skills and lower alcohol-induced secretion of certain hormones. In addition, certain personality types, such as risk-seekers and antisocial personalities, have been linked to the risk of developing alcoholism.

Diagnosis

The clinical diagnosis of alcoholism is not based on the amount or frequency of alcohol consumption, but rather on the *development of problems* associated with its use. Thus, the diagnosis is best made through a detailed history that includes information on marital and occupational status, accidents, and medical problems. Patients may not recognize that they are alcoholic, or may deny it if they do, and thus various screening tests, such as the CAGE or MAST questionnaires, have been devised to assist in diagnosing alcoholism.

A number of physical signs and laboratory tests may support a diagnosis of alcoholism. Abnormal LFTs are an indicator, as is an elevated mean corpuscular volume (MCV). Serum uric acid, lactate, and triacylglycerols may be elevated. Hypertension and tachycardia may also occur. Levels of γ-glutamyl transferase (GGT) may be used as a marker for alcoholism because it is induced by ethanol; however, this is not a specific marker as it is induced by a variety of drugs, and its level is elevated in a number of hepato-biliary diseases. Carbohydrate-deficient transferrin (CDT) is another marker that appears to be more sensitive; however, this test is more expensive to perform.

Unlike most cases, diagnosis of this patient was clear-cut. He had a clear history of alcoholism and presented with high blood alcohol and symptoms of withdrawal, and he subsequently developed DTs.

Treatment

Education and counseling are the mainstays of treatment for alcoholism. The AA 12-step approach to changing behavior has helped many alcoholics recover. Complete abstinence is recommended because few alcoholics (<10%) are able to return to moderate drinking habits. Drug treatments include opiate antagonists, such as naltrexone, which bind opiate receptors, thus inhibiting ethanol binding and signaling through the receptors. Ethanol binding to the opiate receptors contributes to the mood-altering effects of alcohol consumption. Although the data are still preliminary, the opiate antagonists are believed to reduce the craving for alcohol. Disulfiram (Antabuse), an irreversible inhibitor of ALDH, may also be helpful. Use of this drug results in the accumulation of acetaldehyde (see Figure 17.1) when alcohol is consumed. This toxic compound results in unpleasant (and potentially dangerous) symptoms such as facial flushing, rapid heart and respiratory rates, and throbbing headache, hence the effectiveness of the drug.

QUESTIONS

1. The two genetic rat lines, ANA (alcohol-avoiding) and AA (alcohol-preferring), have been used in experimental studies. Hepatic enzyme activities of ADH and ALDH were measured in these lines, and one was found to have *higher* ADH and *lower* ALDH activity than the other. Do you think it was the ANA or the AA line?

2. Why are alcoholics particularly susceptible to gout?

3. Why is ethanol often given as an antidote to poisoning by ethylene glycol (a common constituent of antifreeze) or methanol (found in antifreeze, paint thinner, windshield-wiper fluid, and other consumer products)?

4. What is fetal alcohol syndrome (FAS)?

5. What is the "French paradox," and how does alcohol feature in it?

6. The ratio of lactate to pyruvate is affected by ingestion of a large dose of alcohol. How is the ratio affected and what is the biochemical basis for the change in equilibrium?

7. How is ethanol absorbed in the gastrointestinal (GI) tract?

FURTHER READING

1. Lieber, C. Hepatic, metabolic and nutritional disorders of alcoholism: from pathogenesis to therapy. *Crit. Rev. Clin. Lab. Sci.* (2000) 37(6):551–584.

2. Chang, P. H., and Steinberg, M. B. Alcohol withdrawal. *Med. Clin. North Am.* (2001) 85(5):1191–1212.

3. Wiese, J. G., Shlipak, M. G., and Browner, W. S. The alcohol hangover. *Ann. Intern. Med.* (2000) 132(11):897–902.

4. Eaton, S., Record, C. O., and Bartlett, K. Multiple biochemical effects in the pathogenesis of alcoholic fatty liver. *Eur. J. Clin. Invest.* (1997) 27(9):719–722.

For further information, see the following web sites:

National Institute on Alcohol Abuse and Alcoholism: http://www.niaaa.nih.gov/

National Council on Alcoholism and Drug Dependence: http://www.ncadd.org/

American Council on Alcoholism: http://www.aca-usa.org/

Alcoholics Anonymous: http://www.alcoholics-anonymous.org/

CHAPTER 18

Oxidative Phosphorylation
A Lazy Eye Plus

Woman Presents to ER with Dizziness

A 24-year-old African-American woman presented to the emergency room with dizziness and a heart rate of 25–30 beats per minute (bpm). She had been well until one week prior to this hospital visit when she began to feel "drained," without much energy, but she did not have any dizziness until today. She denied syncope but felt as if she would pass out if she had to stand for more than a few seconds. She did not complain of chest pain or shortness of breath. There was no known prior history of heart disease.

What is the cause of this patient's lightheadedness? Why do the symptoms increase with standing?

She Has a History of Opthalmoplegia

Her past medical history was remarkable for droopy eyelids and a "lazy eye" since childhood, for which she was evaluated by an ophthalmologist. She was found to have external ophthalmoplegia on eye examination without evidence for visual loss. She has one daughter, age 4, who was the product of a normal gestation and delivery, and who is well without medical problems. She has two sisters who are in their thirties, also without significant medical illness. Her mother is 55 years old and has hypertension; her father, who does not live in the same city, is 57 years old and without significant medical illnesses.

What is the significance of the eye findings? What is ophthalmoplegia?

Does the absence of a family history of similar symptoms help in the diagnosis?

She lives with her mother and daughter, and is not currently working outside the home. She has a high school education, and before the birth of her daughter she worked as a teacher's aid. She never held a job where she was exposed to noxious chemicals or dust. Her blood-lead levels had been tested when she was younger and they were never elevated. She smokes one half of a pack of cigarettes per day, but does not drink alcohol nor use illicit drugs. The review of systems revealed no significant history of muscle weakness, skin rashes, arthritis or arthralgia, gastrointestinal symptoms, or tick bites.

What are the entities that we are trying to rule out in the review of systems?

Physical Examination Reveals Opthalmoplegia and Cardiac Abnormalities

On physical examination her blood pressure was 159/89 mmHg and pulse was 32 per minute and slightly irregular. The respiratory rate was 20 per minute and the temperature was 36.5°C. She was an anxious, fully alert, oriented, cooperative, mildly obese woman in no acute distress. The skin was dry without rashes. The head, eyes, ears, nose, and throat examination revealed limitations in extreme lateral eye movements bilaterally, with a mild exotropia on the right. She had lid lag bilaterally. The neck revealed no jugular venous distention, normal carotid artery pulsations without bruits, and no thyromegaly. She had full range of motion of her neck without stiffness. Her chest was barrel-shaped, resonant to percussion, and clear to auscultation. The heart exam revealed a slow but regular rhythm, a persistently split-second heart sound, and a systolic ejection murmur. Her abdomen was soft and nontender with normally active bowel sounds. There was no clubbing, cyanosis, or edema of her extremities. Examination of the joints revealed no abnormalities. Neurological evaluation revealed external ophthalmoplegia as described previously, but no other evidence for cranial nerve defects. There were no focal motor or sensory defects and gait, balance, and coordination were normal.

Laboratory Results Reveal Elevated Creatine Kinase and Lactate

Laboratory examination revealed normal renal function tests (blood urea nitrogen and creatinine) and serum electrolytes. The serum glucose was normal, as was the total cholesterol. The creatine kinase (CK) was mildly elevated at 209 IU/l (upper limit of normal [ULN]: 150 IU/l) without an elevation in the cardiac specific (MB) fraction, and the serum lactate level was elevated at 4.1 mM (ULN: 1.8 mM). Thyroid function tests were normal, serum ferritin was 24 ng/ml (normal range: 10–130 ng/ml), and the rapid plasma reagin (RPR) test for syphilis, hepatitis B surface antigen, anti-hepatitis C antibody, and lyme titers were all negative. Antinuclear antibody (ANA) tests were low titer with a nonspecific staining pattern. The white blood cell, differential, and platelet counts were normal. She was mildly anemic, with a hematocrit of 33.6%, and was negative in a screen for sickle-cell anemia. She underwent a lumbar puncture and examination of the cerebrospinal fluid (CSF), and the only abnormality was an elevation in the protein level to 61 mg/l (ULN: 45 mg/l).

Electrocardiogram Reveals Heart Block

The chest radiogram showed cardiomegaly (enlargement of the heart silhouette) without other abnormalities. The electrocardiogram (ECG) demonstrated complete atrioventricular heart block. The ventricular response rate was approximately 25 per minute with a sinus rate of approximately 80 per minute (Figure 18.1).

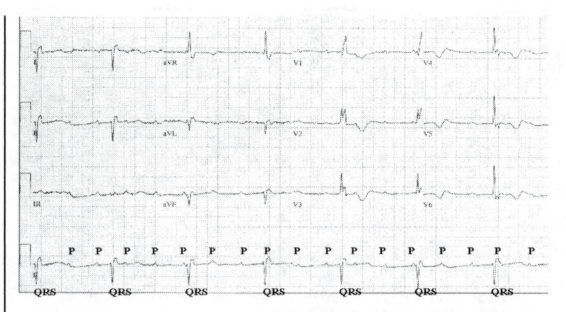

FIGURE 18.1 Electrocardiogram demonstrating complete heart block. The ECG signature of atrial depolarization (P wave) has no temporal relationship to the signature of ventricular depolarization (QRS complex). The effective heart rate is that produced by ventricular depolarization and therefore contraction.

What is atrioventricular heart block, and how is it treated?

What is the significance of the elevation in creatine kinase?

Why were tests for Lyme disease performed?

Diagnosis and Treatment

The diagnosis of Kearns-Sayre syndrome (KSS) was made on the basis of the findings of ophthalmoplegia and complete heart block. A mild skeletal muscle myopathy may also have been present, but the patient did not suffer from symptoms of encephalopathy or retinitis, other manifestations of this mitochondrial DNA disease.

In view of her symptoms and profoundly slow heart rate she received a temporary cardiac pacemaker with prompt resolution of her dizziness. She was admitted to the coronary care unit (CCU) for close observation, and on the following day a permanent pacemaker was implanted.

What is a cardiac pacemaker and what does it do?

She underwent a skeletal muscle biopsy, which revealed ragged red fibers with subsarcolemmal accumulations of morphologically abnormal mitochondria. Mitochondrial DNA (mtDNA) analysis revealed a nearly 5-Kb deletion. She did not have significant visual symptoms or muscle weakness, and no further therapy (e.g., creatine or coenzyme Q10) was recommended. She was scheduled for routine follow-up in the pacemaker and neuromuscular diseases clinics.

> *Why is KSS not typically an inherited problem?*
>
> *How is mitochondrial DNA inherited?*
>
> *Are the skeletal muscle biopsy findings specific for KSS?*
>
> *What are creatine and coenzyme Q10, and why might they be useful in the treatment of KSS?*

DISCUSSION

The Mitochondrion

The mitochondrion is distinctive in that it is the only subcellular organelle to have its own DNA, carrying a 16,569 base pair circular genome that encodes 37 genes. The 13 protein-encoding genes are components of the electron-transport chain, with the remainder encoding tRNAs and rRNAs needed for mitochondrial protein synthesis. There are between 2 and 10 copies of the mitochondrial genome per mitochondrion. The remaining components of the mitochondrion are encoded by nuclear genes; in fact, the large majority—close to 3000—of mitochondrial proteins are nuclear encoded.

The most important role of the mitochondria is the generation of ATP; the citric acid cycle and oxidative phosphorylation, from which most cellular energy is derived, take place there. Most cells have hundreds of mitochondria, and cell types with the highest energy needs typically carry the largest numbers. For example, the mitochondria make up close to 60% of the volume of extraocular muscle cells and 40% of the volume of cardiac muscle cells. Although ATP production is the predominant role of the mitochondrion, it also carries out other functions, some of which are cell-type specific. For example , dihydroorotate dehydrogenase, an important enzyme in pyrimidine biosynthesis, is found in the mitochondria, and the urea cycle takes place in liver cell mitochondria. Thus, rather than being a uniform ATP-generating organelle, the mitochondrion in fact performs myriad functions, many of which are tailored to the cell type in which it is found.

Mitochondrial Diseases

Caused by mutation in mitochondrial or nuclear DNA

Mitochondrial dysfunction may be caused by mutations in either mitochondrial DNA (mtDNA), or in mitochondrial genes encoded within nuclear DNA (nDNA), and the mutations may be either sporadic or inherited. Mitochondrial diseases may be *heteroplasmic,* meaning that wild-type (wt) and mutant mitochondria coexist within the same cell, or *homoplasmic,* when the mitochondria within a cell are identical. Homoplasmic diseases are usually the result of relatively benign mutations, as any mutation that severely compromises mitochondrial function is embryonically lethal. As opposed to nDNA inheritance, mtDNA is inherited solely from the mother, and thus many mitochondrial diseases are transmitted maternally.

Mitochondrial diseases have widespread and variable clinical features

There are several hallmarks of mitochondrial diseases, the first being that their effects are widespread, reflecting the dependence of many organs and tissues on oxidative phosphorylation. The central and peripheral nervous systems are usually affected, as are the eyes, muscles, kidneys, and endocrine glands. A second feature is that the clinical features associated with a particular mutation can be highly variable. This has been attributed to the heteroplasmic nature of many of the mitochondrial diseases, which results in the various bodily tissues and organs carrying differing proportions of defective mitochondria. The particular energy requirement of a tissue, as well as any specialized functions the mitochondria carry out there, will determine the clinical outcome of a shortage of functional mitochondria. Thus, patients carrying the *identical* mutation may have dramatically different clinical manifestations, depending upon the distribution of defective mitochondria throughout the body. This property of mitochondrial diseases has made it very difficult to associate a particular mutation with a clinical phenotype.

Delayed onset is typical

A third feature of mitochondrial diseases is that they typically have a delayed onset, and once symptoms appear they usually progressively worsen. This is also thought to stem from the heteroplasmic nature of most mitochondrial diseases. Because wt and defective mitochondria coexist within a cell, the wt organelles may often supply the energy needs of the cell for a time, despite overall reduced cellular mitochondrial function. However, the burden of reduced mitochondrial function tends to increase with time: reduced ATP production compromises the efficiency of many cellular processes, some of which influence the maintenance of functioning mitochondria. Once a certain *threshold* of defective mitochondria is reached, the minimum energy requirement of the cell will not be met and it will die. This will increase the burden on neighboring cells within a tissue, which in turn may begin to decline and eventually die.

Pathophysiology

Defective mitochondria are thought to cause disease via three main mechanisms. First, a shortage of ATP, needed by the cell for numerous vital processes, will often result in cellular dysfunction, and eventually lead to cell death. Second, defects in components of the electron transport chain will often result in increased production of reactive oxygen species (ROS). As seen in Section 18.3 of *Biochemistry* 5e, during oxidative phosphorylation electrons are passed from NADH through a series of complexes and are ultimately transferred to molecular oxygen, which reacts with protons to generate water. However, occasionally the electrons diverge from their path and react with molecular oxygen to form ROS, which can damage cellular components such as proteins, lipids, and DNA, eventually leading to cell death. Finally, inappropriate activation of the *apoptosis* (programmed cell death) pathway, in which the mitochondria play an important role (see Section 18.6.6 of *Biochemistry* 5e), may occur in defective mitochondria. Although all three mechanisms contribute to the pathophysiology of mitochondrial diseases, the primary defect is believed to be impaired ATP production.

Clinical features

The mitochondrial diseases are a heterogeneous group of disorders; however, tissues with high-energy needs are most commonly affected, leading to a set of symptoms commonly observed in patients with these disorders. However, usually only a subset of these will be present in any individual patient, making specific diagnoses challenging. Patients often present with the onset of muscle weakness or cardiac abnormalities, reflecting the dependence of skeletal and cardiac muscle cells on a plentiful supply of ATP. In addition, encephalopathy is often observed, with symptoms such as seizures, stroke-like events, migraines, and dementia. Neuronal cells also have high and fluctuating energy needs. Endocrine disorders are also common; indeed, the prevalence of diabetes mellitus is particularly high in patients with mitochondrial diseases. Other features associated with mitochondrial diseases are visual and hearing impairment, short stature, opthalmoplegia, and renal tubular disease.

Kearns-Sayre Syndrome

Molecular defect

The most common molecular defect in KSS is a large deletion of the mitochondrial genome that ranges in size from 1.3 to 8 Kb, but is typically a 4.9 Kb deletion such as the one exhibited by this patient (Figure 18.2). It comprises genes for components of complexes I, III, IV, and V of the electron-transport chain, as well as some tRNA genes. The mechanism of deletion is unknown; however, recombination, or slippage during DNA replication, are possibilities, as the deletion is usually flanked by directly repeated sequences. The mutation is usually sporadic, and is usually thought to occur during embryonic development. This explains why this patient had no family history of the disease. KSS is extremely rare, with only a few hundred cases having been described in the literature. It does not appear to have any predilections for gender or race.

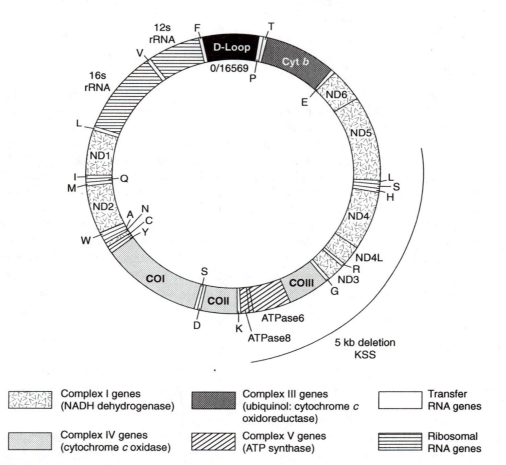

FIGURE 18.2 Schematic representation of the mitochondrial DNA with the various genes and complexes indicated. The D-loop is a regulatory region and does not encode genes. The location of the 5-Kb deletion common in those with KSS is indicated.

Clinical manifestations

The typical clinical manifestations of KSS are (1) onset prior to age 20, (2) progressive opthalmoplegia, and (3) atypical pigmentation of the retina. In addition, patients may exhibit cardiac conduction defects and elevated protein levels in the cerebrospinal fluid (CSF). In addition, any of the clinical manifestations associated with mitochondrial diseases in general may occur in KSS patients (see above). KSS is a progressive disease with a fairly poor prognosis; death usually occurs in the third or fourth decade of life.

This patient presented with two of the three cardinal clinical features of KSS, as well as some other commonly observed manifestations. She had opthalmoplegia since childhood, and the physical examination revealed extreme deficits in eye movement. The cerebrospinal fluid exhibited elevated protein levels, indicating a central nervous system abnormality, and the skeletal muscle biopsy revealed "ragged red fibers," which are a hallmark of a severe defect in oxidative phosphorylation. The elevated CK level revealed by the laboratory evaluation had already hinted at a skeletal muscle myopathy. This enzyme is found in high concentrations in skeletal and cardiac muscle cells, and its elevation in the serum is an indicator of cell necrosis. The elevation was found not to be due to the cardiac-specific isozyme (MB isozyme), implicating skeletal muscle as the source. Elevated lactate levels are a consequence of a defect in oxidative phosphorylation—with this pathway blocked, the pyruvate produced from glycolysis is instead converted to lactate (see Section 16.1.9 of *Biochemistry* 5e). Elevation of pyruvate levels is also often observed.

The heart examination and electrocardiogram revealed abnormalities consistent with a cardiac conduction defect, or *heart block,* a significant cause of mortality in KSS patients. Heart block is a disorder in which contractions of the atria are not coordinated with contractions of the ventricles. A heartbeat is normally triggered by depolarization of cells of the sino-atrial node, found in the right atrium. The depolarization spreads through the atria (causing them to contract) and onto the atrial-ventricular node, which then triggers contraction of the ventricles. Complete heart block refers to complete electrical, and therefore contractile, dissociation between atria and ventricles, such that the ventricles contract at their own natural rhythm of 20–40 beats per minute. This does not provide sufficient oxygenated blood to the brain and muscles, and leads to episodes of syncope (fainting) and overall weakness. This cardiac disorder thus explains this patient's presenting symptoms: she felt dizzy and drained, and as though she could pass out at any moment. The implantation of the cardiac pacemaker, which provides the electrical impulses needed to cause regular ventricular contraction, resulted in prompt resolution of her symptoms.

Diagnosis

Diagnosis may be made on the basis of clinical manifestations and laboratory results; however, given the variability characteristic of these diseases, identification of the DNA lesion is usually necessary for definitive diagnosis. Identification of the specific DNA lesion is not usually helpful in predicting prognosis as this varies dramatically, even among individuals carrying the identical mutation; however, it is important for genetic counseling.

Treatment

No treatment has yet been found to alter the course of KSS, however certain vitamin and cofactor supplements may nevertheless be recommended. These are intended to favor the generation of ATP, or to scavenge free radicals (such as ROS), thus slowing the progression of the disease. For example, coenzyme Q10 may be recommended, as it participates both in the citric acid cycle and oxidative phosphorylation (see Section 18.3 of *Biochemistry* 5e), as well as acting as a free radical scavenger. Creatine supplementation may build up bodily stores of creatine phosphate, a readily mobilized source of cellular ATP (see Section 14.1.5 of *Biochemistry* 5e). Eating several small meals a day is thought to be beneficial, and avoidance of fasting and any type of physiological stress, especially cold, is recommended.

QUESTIONS

1. Why are diseases that are caused by mutations in mtDNA maternally inherited?

2. How do mitochondria segregate during cell division? How might this pertain to the high degree of variability observed in the clinical manifestations of KSS (and other diseases caused by mutations in mtDNA) between individuals?

3. Why is the serum lactate/pyruvate (L/P) ratio useful in identifying the molecular defect involved in a mitochondrial disease?

4. Why is mitochondrial DNA particularly susceptible to mutation?

5. Why is mitochondrial DNA frequently used in forensics analyses?

6. Why are neurons particularly affected by defective mitochondria?

7. Why have researchers speculated that aging is the result of the accumulation of mtDNA mutations?

8. How does the inheritance pattern of a disease caused by an mtDNA mutation differ from that of an X-linked disorder?

FURTHER READING

1. Moraes, C. T., Di Mauro, S., Zeviani, M., et al. Mitochondrial DNA deletions in progressive external ophthalmoplegia and Kearns-Sayre syndrome. *N. Engl. J. Med.* (1989) 320:1293–1299.

2. Porteous, W. K., et al. Bioenergetic consequences of accumulating the common 4977-bp mitochondrial DNA deletion. *Eur. J. Biochem* (1998) 257:192–201.

3. Chinnery, P. F., and Turnbull, D. M. Mitochondrial DNA mutations in the pathogenesis of human disease. *Mol. Med. Today.* (2000) 6(11):425–432.

For further information, see the following web sites:

OMIM, Kearns-Sayre Syndrome:
http://www.ncbi.nlm.nih.gov/entrez/dispomim.cgi?id=530000

NINDS information page, Kearns-Sayre Syndrome:
http://www.ninds.nih.gov/health_and_medical/disorders/kearns_sayre.htm

e-medicine, Kearns-Sayre Syndrome: http://www.emedicine.com/ped/topic2763.htm

United Mitochondrial Disease Foundation: www.umdf.org

CHAPTER 19

The Light Reactions of Photosynthesis
A Little Sun Will Do You Good

CASE HISTORY

15-Month-Old Presents with Loss of Motor Milestones

In mid-April, a 15-month-old African-American girl was brought by her mother to the pediatrician for her check-up. She had been well and developing normally until approximately 13 months of age. After initially standing and walking at 10 months, she had been increasingly hesitant to bear weight on her legs, and by 13 months she had stopped walking altogether. She returned to crawling, although her mother reported that in the past few weeks she crawled only when strongly motivated to do so, such as when a favorite toy was out of reach. The development of her fine motor skills was unaffected. She was able to pick up small objects with the pincer grip, she could feed herself with a spoon, and could twist lids off of jars.

She was born at term (birth weight 3.90 kg) and had been breast-fed until she reached 11 months. Since then she had been consuming approximately 700 ml/day of a soy-based "health food" beverage, which was not supplemented with vitamin D or calcium. She also consumed home-prepared vegetable and meat-based foods, but no dairy products, and her mother described her as a "good eater." Due to relatively cold weather, she had spent almost all of her time indoors over the past six months. The patient's mother was well-educated and informed and concerned about her daughter's development.

What do the growth curves suggest?

Is the ethnicity of this child significant?

Is the time of year significant?

Physical Examination

Upon examination the patient was alert and responsive. Her axillary temperature was 36.6°C. Her weight was 8.4 kg and her length was 71.5 cm, compared with 9.2 kg and 70.0 cm at her 12-month check-up (see Figure 19.1). She appeared quite thin and had reduced muscle bulk. She was weak and could not sit up unsupported, but she attempted to reach for objects placed within her vicinity. Her reflexes were normal. All immunizations were up to date.

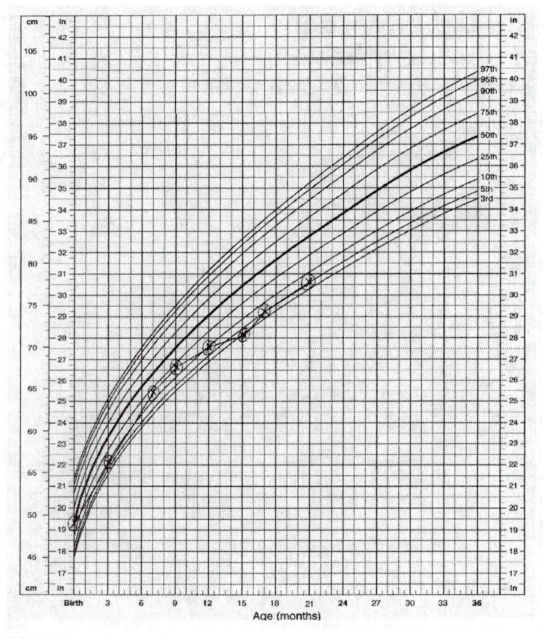

FIGURE 19.1 Length growth chart showing the marked effect of vitamin D deficiency on growth, as well as the dramatic recovery upon diagnosis and vitamin D supplementation.

Laboratory Evaluation

Radiographs revealed three notable skeletal findings: a lumbar kyphoscoliosis (a spinal deformity), enlargement of the costal cartilages (manifest as a row of bony protrusions at the junction of the ribs and their cartilages), and flaring of the wrists. Laboratory tests revealed a low serum phosphorus of 2.0 mg/dl (normal range: 4.5–6.9 mg/dl), a normal calcium level of 9.2 mg/dl (normal range: 8.0–10.3 mg/dl), and very high alkaline phosphatase levels 2100 IU/l (upper limit of normal: 120 IU/l). Serum vitamin D levels (as 25-hydroxyvitamin D) were low at 6.5 pg/ml (normal range: 8.9–46.7 pg/ml), and the parathyroid hormone level was elevated at 97 pg/ml (normal range: 10–65 pg/ml).

Diagnosis and Treatment

Based on these observations, the patient was diagnosed with rickets as a result of vitamin D nutrient deficiency and inadequate exposure to sunlight. The patient was prescribed 1000 IU/day of vitamin D_2 for 10 weeks, followed by 200 IU/day thereafter and a diet rich in calcium. Within two weeks of the start of treatment, she began to show signs of increased strength and was walking within four months of treatment. Follow-up radiographs at four months showed marked improvement in bone mineralization and skeletal abnormalities.

Why was she treated with calcium?

What foods are rich in vitamin D?

DISCUSSION

Light Participates in a Number of Physiological Processes

Sunlight consists of a range of electromagnetic radiation, mostly in the infrared and visible spectrums; however, it also contains ultraviolet (UV) light, which is important biochemically because it is high enough in energy to elicit photochemical reactions in skin. As discussed in Chapter 19 of *Biochemistry* 5e, light is critical to the process of photosynthesis, whereby plants derive energy in the form of ATP; although we do not use light to generate ATP, it is important to many other physiological processes in mammals. Our eyes and brains process light such that we can see, and mental acuity, body temperature, and secretion of a number of hormones vary in circadian patterns in response to light/dark cycles. Light also affects mood and behavior via mechanisms that are currently being investigated.

Light is also used in a number of medical therapies, such as in treating jaundice in newborns. Upon delivery, the immature livers of some infants are unable to effectively clear *bilirubin*, a breakdown product of hemoglobin, from their bodies—a process that was previously carried out by the mother in utero. Ultraviolet light converts bilirubin to a more soluble form, which may then be excreted. Another therapy, *photodynamic therapy*, makes use of drugs called photosensitizers to absorb light and utilize the energy to destroy malignant cells. Light therapy is also used in the treatment of a number of skin conditions, such as psoriasis. Finally, as will be discussed here, the generation of the active form of vitamin D in skin depends upon the absorption of light. Rickets, a disease characterized by skeletal abnormalities, weakness, and poor growth in children, results from vitamin D deficiency and occurs when dietary vitamin D intake or exposure to sunlight is low.

Although light is beneficial to human health in many respects, its hazards have become evident in the recent past. Light can damage DNA and stimulate the formation of reactive oxygen species (ROS), thereby increasing the risk of skin cancers. Thus, although a little sun will indeed do you good, a lot of sun will not do any good, causing painful sunburns and skin cancer.

Vitamin D

The symptoms of rickets have been recognized for centuries; however, it was not until 1645 and 1650 that the first scientific reports of the disease were made. Subsequently, in 1919, the lack of a trace dietary component was found to be the cause of rickets, and shortly thereafter it was discovered that a substance equivalent to this dietary component was generated by exposure of skin to UV light. The chemical structures of vitamin D_2 and vitamin D_3 were determined in Germany in 1932 and 1936, respectively. In the United States milk has been fortified with vitamin D since the 1920s; however, vitamin D fortification did not become widespread in Canada until the 1960s and, consequently, rickets remained a significant problem there until then.

Despite its name, vitamin D is not strictly a vitamin because it may be synthesized by the body via the action of UV light on *7-dehydrocholesterol* in the skin (see Section 26.4.7 of *Biochemistry* 5e). There are two forms of vitamin D, *vitamin D_2* (or ergocalciferol), which is found in plants and irradiated yeast, and *vitamin D_3* (or cholecalciferol), found in animals (Figure 19.2). Most foods have a low content of vitamin D and thus vitamin D deficiency was common in previous centuries. Today a number of foods, such as milk, are fortified with vitamin D, and thus the deficiency is less common. However, as was highlighted by the patient described here, a number of health-food beverages are not supplemented with this important nutrient, which can lead to a dangerous deficiency in young children when they are substituted for milk.

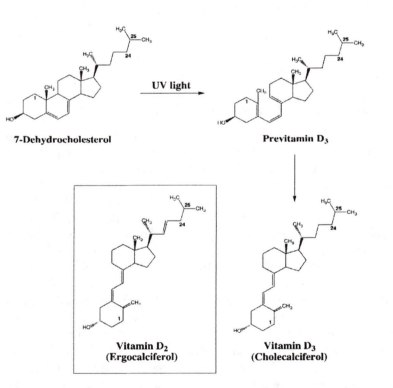

FIGURE 19.2 7-dehydrocholesterol is converted into previtamin D_3 by UV light. Previtamin D_3 then isomerizes to vitamin D_3. Vitamin D_3 is activated by hydroxylation at C_1 and C_{25}, and hydroxylation of C_{24} appears to be the first step in its inactivation. The structure of vitamin D_2 is shown for comparison.

Metabolism of Vitamin D$_3$

The first step in the formation of vitamin D$_3$ is the photolysis of one of the rings of 7-dehydrocholesterol (see Figure 26.29 of *Biochemistry* 5e). Like chlorophyll, the ring (a cyclohexadiene unit) possesses alternating single and double bonds, which makes it an effective photoreceptor. Lysis of the ring produces *previtamin D$_3$*, which is rapidly isomerized to *vitamin D$_3$*. When the entire body is exposed to enough light to result in a mild sunburn, the increase in serum vitamin D is equivalent to consumption of 10,000 to 25,000 international units (IU), which is 50–125 times the recommended daily allowance (200 IU).

Vitamin D$_3$ (as well as vitamin D$_2$ taken in from the diet) is inactive and must be converted to the active form by hydroxylases in the liver and kidneys. The vitamin passes from the epidermis into the circulation and is hydroxylated at C$_{25}$ by the hepatic enzyme, vitamin D-25 hydroxylase, generating *25-hydroxyvitamin D$_3$*. With a half-life of approximately 21 days, this metabolite is very stable and serves as a storage form of 1,25-hydroxyvitamin D$_3$, the most active form of the vitamin (see below).

25-hydroxyvitamin D$_3$ is bound by a transport protein, vitamin D–binding protein, and transported from the liver through the circulation to the kidney, where the hydroxylation of C$_1$ by 25-hydroxyvitamin D-1α-hydroxylase yields *1,25-dihydroxyvitamin D$_3$*, the most active metabolite. Although the formation of 25-hydroxyvitamin D$_3$ is regulated by feedback inhibition, the subsequent hydroxylation of C$_1$ is the most important regulated step. Low serum calcium stimulates the secretion of parathyroid hormone (PTH), which activates 25-hydroxyvitamin D-1α-hydroxylase. Thus, low serum calcium levels stimulate the generation of 1,25-dihydroxyvitamin D$_3$, which ultimately acts to increase calcium and phosphorus absorption from the intestine and their release from bone. The kidney also contains vitamin D-24-hydroxylase, which hydroxylates C$_{24}$ of both 25-hydroxyvitamin D$_3$ and 1,25-dihydroxyvitamin D$_3$, which are believed to be steps in the inactivation of the vitamin.

Physiological Effects of Vitamin D

Most of the biological activity of vitamin D is mediated through its receptor, the vitamin D receptor (VDR), which is a nuclear hormone receptor. 1,25-dihydroxyvitamin D diffuses into the nucleus and binds the receptor, thereby increasing receptor affinity for a DNA regulatory sequence, the vitamin D responsive element (VDRE), and stimulating gene transcription. Although a number of cell types possess the VDR, the principal target tissues of vitamin D appear to be the intestine, the bone, and the kidneys. It stimulates the absorption of calcium and phosphorus from the intestine, as well as their resorption from bone, thus increasing their serum concentrations. In addition, vitamin D is thought to affect the handling of calcium and phosphorus by the kidneys; however, these processes are not yet well understood.

Rickets

Clinical features

Vitamin D deficiency in children results in *rickets,* a disease characterized by bone deformities that result from poor mineralization. Both calcium and phosphate are important in the calcification of bone, and thus vitamin D deficiency interferes with their normal development and growth in children. The defect is most apparent at the *metaphysis* or the growing portion of the long bones, and impaired growth of the bones is reflected in reduced or lack of growth of the child. Bones become weak, and deformities such as bowed legs and knock-knees are not uncommon; the bones also become brittle and susceptible to fractures.

Vertebral softening and *kyphoscoliosis* are typical findings, as are other bone distortions, such as the flaring wrists observed in this child. Enlargement of the *costal cartilages,* which connect the ribs to the sternum (breast bone), is another common finding. The prominence of these costochondro (rib-cartilage) junctions gives a beaded appearance that is often referred to as the *rachitic* (afflicted with rickets) *rosary.* In addition to bone, the mineralization of epiphesial cartilage is also defective in rickets. Symptoms of *hypocalcemia,* such as seizures and cardiac abnormalities, may also occur in some cases.

Common laboratory features are low serum phosphorus levels, low-to-normal serum calcium levels, elevated parathyroid hormone levels, and low 25-hydroxyvitamin D levels. The low calcium absorption that results from the deficiency results in elevated PTH levels, which act to increase bone resorption, thus normalizing serum calcium levels. However, PTH stimulates elimination of phosphorus by the kidneys into the urine. The high serum alkaline phosphatase level is a result of active osteoblasts (immature bone cells), which are stimulated by the loss of bone mass. The diagnosis is made on the basis of laboratory, and especially radiographic, findings showing poor bone mineralization.

Etiology

Rickets is caused by poor calcification of the osteoid (uncalcified bone matrix), and this most frequently results from deficiency of vitamin D or of its metabolites. Lack of sunlight and/or dietary deficiency is the most common etiology. Although less common, calcium or phosphorus deficiency may also cause rickets.

Treatment

This patient's dramatic response to vitamin D supplementation is typical, with healing of bones and improvement in symptoms occuring in a matter of weeks. Most children make a full recovery and there are few long-term effects, although children who have had rickets appear to be at increased risk of degenerative joint disease in adulthood.

QUESTIONS

1. Melanin, a pigment found in skin, hair, and eyes absorbs broadly in the visible and UV spectra, but most strongly in the UV range. How would you expect the rate of vitamin D_3 synthesis in a dark-skinned person to compare with that in a light-skinned person?

2. Rickets reached epidemic proportions in many urban areas in Northern Europe and the United States during the industrial revolution. Can you think of a reason why?

3. Breast milk contains little vitamin D and thus breast-fed infants are at risk for developing rickets. Can you think of an evolutionary explanation for the low level of this important nutrient in human milk?

4. Is vitamin D a vitamin? Is it a hormone?

5. How does parathyroid hormone collaborate with vitamin D to increase serum calcium levels?

6. What would be the effect of renal failure on vitamin D metabolism?

FURTHER READING

1. Chesney, R. W. Vitamin D deficiency and rickets. *Rev. Endocr. Metab. Disord.* (2001) 2(2):145–151.

2. Joiner, T. A., Foster, C., and Shope, T. The many faces of vitamin D deficiency rickets. *Pediatr. Rev.* (2000) 21(9):296–302.

3. Miller, W. L., and Portale, A. A. Genetics of vitamin D biosynthesis and its disorders. *Best Pract. Res. Clin. Endocrinol. Metab.* (2001) 15(1):95–109.

4. Christakos, S., Raval-Pandya, M., Wernyj, R. P., and Yang, W. Genomic mechanisms involved in the pleiotropic actions of 1,25-dihydroxyvitamin D_3. *Biochem. J.* (1996) 316(Pt. 2):361–371.

5. Tomashek, K. M., Nesby, S., Scanlon, K. S., Cogswell, M. E., et al. Nutritional rickets in Georgia. *Pediatrics* (2001) 107(4):e45.

For further information, see the following web sites:

American Society of Photobiology: www.pol-us.net/ASP_Home/index.shtml

National Institutes of Health, Clinical Center, Vitamin D: www.cc.nih.gov/ccc/supplements/vitd.html

Agency for Health Care Research and Quality, Vitamin supplementation to prevent rickets in breast-fed babies: www.ahcpr.gov/child/rickets.htm

CDC, Morbidity and Mortality Weekly Report, March 30, 2001 50(12):224–227. Severe malnutrition among young children—Georgia, January 1997–June 1999. www.cdc.gov/mmwr/preview/mmwrhtml/mm5012a3.htm

CHAPTER 20

Pentose Phosphate Pathway and the Calvin Cycle
Bad Oysters, Bad Chicken, or Bad Enzyme?

CASE HISTORY

Young Man with Gastrointestinal Symptoms Is Admitted to the Hospital

A 22-year-old Caucasian man was admitted to the hospital with fevers, chills, anorexia, and abdominal and low-back pain. He was well until three days ago when he began to experience nausea, diarrhea, and fevers. Several days before he had been to a cookout and had eaten raw oysters and barbecued chicken. He was seen one day ago in an urgent care clinic with diarrhea. He was able to keep fluid and foods down, so he was managed conservatively with oral hydration and imodium. A complete blood count reportedly showed a normal white-blood-cell count (WBC) and normal hematocrit. A stool sample was cultured.

After returning home he developed worsening abdominal pain and more frequent liquid bowel movements with traces of blood on the toilet paper. The abdominal pain became more severe, and he began to develop severe bilateral flank pain. He took acetaminophen, but continued to have fevers, sweats, and shaking chills. He noticed that his urine was dark and foul smelling. His friends took him to the emergency room after he complained of feeling anxious and lightheaded.

What is the likely cause of his anxiety and lightheadedness?

What is the significance of blood in the stool?

Patient History Is Unremarkable

His past medical history was unremarkable. He was generally healthy, although as an infant he required an additional three-day stay in the hospital for neonatal jaundice. There was no prior history of significant gastrointestinal illnesses. He was unaware of anyone else at the barbecue having developed similar symptoms. He has seasonal allergies, for which he takes fexofenadine, 10 mg daily, on an as-needed basis. He recently graduated from college and works as a loan officer in a bank. He drinks socially, and does not smoke or use recreational drugs. He is of

Italian descent, and the family history is unremarkable. His parents are both in their late forties and are healthy, and he has three younger sisters who are all well. His mother's parents are both dead; his maternal grandfather was killed in an automobile accident and maternal grandmother died of complications of uterine cancer. His father's parents are alive and both have hypertension but no other medical illnesses.

Physical Examination

On physical examination he was acutely ill. His skin was pale and his mucous membranes were dry. The blood pressure was 92/58 mmHg, with a heart rate of 118 beats per minute. The respiratory rate was 26 per minute and unlabored. He had a temperature of 38.6°C. His head, eyes, ears, nose, and throat examination revealed conjuctival pallor without scleral icterus, and the throat was erythematous without an exudate. His neck was supple. The chest was resonant and clear, and the heart examination revealed a rapid heart rate with a flow murmur but was otherwise normal. His abdomen was diffusely tender without peritoneal signs. He had active bowel sounds and no enlargement of the liver or the spleen, although both were tender when palpated. Rectal examination revealed no masses and scant amounts of guaiac-positive stool (positive for blood). The neurological examination revealed no focal abnormalities.

Is the absence of organomegaly significant in this patient?

What does the pallor suggest?

Laboratory Evaluation, ECG, and Chest Radiogram

The laboratory evaluation was remarkable for anemia, with a hemoglobin concentration of 8.9 g/dl and a hematocrit of 29%. The WBC was elevated at 16,200 per mm³, with 72% polymorphonuclear leukocytes, 6% band forms, 18% lymphocytes, 2% eosinophils, and 2% basophils (representing a "left shift" or shift to early WBC forms). The peripheral blood smear demonstrated cell fragments, microspherocytes, and eccentrocytes or "bite cells," with no reticulocytes (nucleated red blood cells [RBCs]). Crystal violet staining revealed the presence of Heinz bodies in the red blood cells.

Serum electrolytes and glucose were normal, the blood urea nitrogen (BUN) was elevated at 30 mg/dl with a normal creatinine of 1.0 mg/dl. Liver function tests were mildly elevated, with an alanine amino transferase of 50 IU/l (upper limit of normal [ULN]: 31 IU/l) and an aspartate amino transferase of 44 IU/l (ULN: 31 U/l), lactate dehydrogenase (LDH) was 358 IU/ml, alkaline phosphatase of 146 IU/l (ULN: 120 IU/l), and the total bilirubin was 2.2 mg/dl (ULN: 1.2 mg/dl), all of which was unconjugated. Serum haptoglobin levels were normal. The urine was clear and dark and had a positive dipstick test for bilirubin and hemoglobin pigments, with trace protein and no glucose or ketones.

An electrocardiogram revealed sinus tachycardia with a rate of 120 beats per minute without other abnormalities. The chest radiogram was normal and the flat plate radiogram of the abdomen revealed no intraperitoneal air.

What might be the cause of the sudden development of anemia?

What are Heinz bodies?

What does the abnormality of red-blood-cell shape signify?

Why are the liver function tests abnormal?

What is intraperitoneal air, and what does it mean?

Hemolytic anemias are typically associated with a decrease in serum haptoglobin. What is haptoglobin and why is it normal in this case?

What are reticulocytes and why are they not elevated in this case?

Diagnosis and Treatment

The patient was treated with vigorous intravenous fluid replacement, with an increase in the blood pressure and a decrease in the heart rate. The stool cultures taken the day before were found to be growing a *Salmonella* species that ultimately turned out to be *S. enteritis,* and he was treated with ciprofloxacin, 500 mg, every 12 hours for five days. The diarrhea and abdominal symptoms resolved but the anemia persisted and, in fact, worsened, with the lowest hematocrit recorded three days after admission, at 21%. At the same time reticulocytes began to appear in the peripheral blood. He refused a transfusion, but his hematocrit slowly increased. A Coombs (antiglobulin) test was negative. A positive fluorescent spot test was consistent with the diagnosis of glucose-6 phosphate dehydrogenase (G6PD) deficiency. A DNA test revealed a cytosine-to-thymine substitution at nucleotide 563, consistent with the Mediterranean G6PD variant.

What are the clinical presentations of G6PD deficiency?

What is the cause of acute hemolytic episodes in these patients?

Patients with anemia are often treated with iron, but this patient was not. Why?

What is a Coombs test and what does it signify?

DISCUSSION

Biochemical Activity of G6PD

G6PD is a universally expressed housekeeping enzyme that catalyzes the first step in the pentose phosphate pathway (Figure 20.1), the dehydrogenation of glucose-6-phosphate to generate 6-phosphoglucuno-δ-lactone (see Section 20.3.1 of *Biochemistry* 5e). The enzyme exists in a dimer-tetramer equilibrium and consists of identical subunits of 514 amino acids, corresponding to a molecular weight of 59.3 kDa each. The step catalyzed by G6PD is virtually irreversible and is rate limiting, and thus serves as an important control point in the pathway. The pathway produces NADPH, which is utilized in a number of biosynthetic pathways.

The enzyme is regulated by the availability of $NADP^+$, and thus as NADPH is consumed in biosynthetic pathways, the rise in $NADP^+$ level stimulates the enzyme such that NADPH is replenished. High $NADP^+$ concentrations favor the formation of dimers and tetramers, the active forms of the enzyme. Interestingly, each monomer appears to have *two* $NADP^+$-binding sites—the *coenzyme* site and the *structural* site. The latter is found at the dimer interface and is thought to stabilize the formation of the dimer. Comparison of *G6PD* gene sequences from different organisms has revealed a highly conserved 8-amino-acid (RIDHYLGK) substrate binding site from amino acids 198–205, and a dinucleotide-binding fingerprint GxxGDLx (residues 38–44), the $NADP^+$ coenzyme binding site.

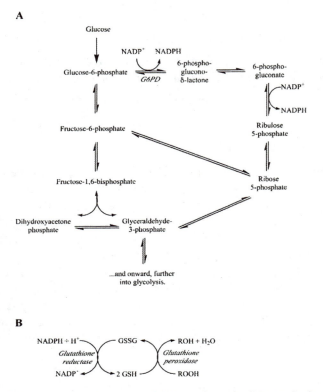

FIGURE 20.1 (A) *Glucose-6-phosphate dehydrogenase (G6PD)* catalyzes the first step in the pentose phosphate pathway. The pathway is coordinated with glycolysis through interconversion of intermediates between the two pathways. (B) NADPH is an important product of the pathway and is used in the reduction of harmful peroxides via the activities of glutathione reductase and glutathione peroxidase.

Physiological Role of G6PD

The pentose phosphate pathway is crucial, as it *generates NADPH,* a reducing agent important in many biosynthetic reactions (see Table 20.4 of *Biochemistry* 5e). In addition, it *maintains glutathione in the reduced form,* which combats damage to cellular macromolecules from reactive oxygen species (ROS). The pathway also serves as a *source of pentose sugars* needed for nucleotide biosynthesis, and is thus particularly important in rapidly dividing cells.

Interestingly, the pentose phosphate pathway and glycolytic pathways are linked: glucose-6-phosphate lies at the intersection of both pathways (see Figure 20.1), and intermediates from each pathway can be funneled into the other depending upon the needs of the cell. Thus, for example, when the need for NADPH predominates, the oxidative phase of the pentose phosphate pathway will be favored (see Figure 20.19 of *Biochemistry* 5e). The ribose-5-phosphate generated in this phase, rather than proceeding through the nonoxidative phase, will be shunted through the glycolytic pathway, and back to glucose-6-phosphate such that the process may be repeated, thus enhancing NADPH production.

Molecular Genetics of G6PD Deficiency

More than one hundred mutations in human *G6PD* have been described, most of which are single amino acid substitutions. The mutations are spread throughout the coding region; however, there is a cluster of mutations in a region of the gene that corresponds to the dimer-binding interface, and some of these interfere with binding of the structural NADP+ (see above). The mutations in this region are among the most severe, causing dramatically reduced enzyme activity.

The normal enzyme is designated G6PD B, and is the most common variant observed in all population groups that have been studied. Approximately 20% of people of African heritage carry the A+ mutant, which is caused by a single amino acid substitution (asparagine to aspartate at amino acid 126). G6PD A+ is functionally normal but is distinguished from G6PD B by a greater electrophoretic mobility. A number of the double missense mutations are found in association with the A+ mutation, indicating that the second mutation arose subsequent to the 126Asn → Asp substitution. Although the A+ mutation alone does not affect enzyme function, it may contribute to destabilization of the enzyme when present together with other mutations.

The A− variant has the same electrophoretic mobility as the A+ variant; however, it is the product of two mutations (126Asn → Asp and usually 68Val → Met) and is functionally impaired. The A− variant is found in 10–15% of African Americans and individuals from western and central Africa. *G6PD Mediterranean* is the most common abnormal variant found in Caucasians, and is the result of substitution of phenylalanine for serine at residue 188 (188Ser → Phe). This is the variant expressed in this patient. Its electrophoretic mobility is identical to that of G6PD B. Several variants are also found among certain Asian populations. Erythrocytes carrying the A− variant have between 5% and 15% of normal G6PD activity, whereas those with the Mediterranean variant have barely detectable enzyme activity. The World Health Organization has classified G6PD variants according to percentage of normal enzyme activity and extent of hemolytic anemia in patients. The A− variant is a class 3 variant (> 10% activity), whereas the Mediterranean variant is in class 2 (< 10% activity). Both variants are unstable and susceptible to proteolytic degradation.

Epidemiology

G6PD deficiency is an X-linked (X q28) disorder and predominantly affects men; however, in certain areas with high gene frequencies homozygous females are found and are affected by the disorder as well. Heterozygous females may also be symptomatic due to inactivation of the normal X chromosome. It has been estimated that G6PD deficiency affects over 400 million people worldwide. The highest gene frequency of mutant *G6PD* has been identified in Kurdish Jews (0.7). Mutations causing G6PD deficiencies are also prevalent (~0.1) in individuals from Africa, the Mediterranean, and Southeast Asia. This patient, who is of Italian descent, was thus in a high-risk group. In contrast, the frequency of *G6PD* mutations is about three orders of magnitude lower in Middle and Northern Europe (0.0005). Selective pressure due to the protection afforded against malaria is the likely cause of the particularly high prevalence of G6PD deficiency in tropical and subtropical regions of the world. Indeed, this hypothesis has been supported by laboratory studies showing poor growth of the malaria parasite in G6PD-deficient cells.

Pathogenesis

The most clinically significant feature of the deficiency is reduced cellular capacity for combating oxidative stress as a consequence of lower NADPH and reduced glutathione levels. The deficiency predominantly affects erythrocytes because the pentose phosphate pathway is the major source of NADPH in this cell type. There is also some evidence for mildly impaired liver function as a result of the deficiency as well.

In normal individuals oxidative stress stimulates the pentose phosphate pathway, generating sufficient NADPH and reduced glutathione to cope with the insult, but in patients with G6PD deficiency oxygen radicals are inefficiently reduced, leading to damage to all types of cellular macromolecules. In particular, the primary protein constituent of erythrocytes—hemoglobin—is not maintained in the reduced state, and it consequently forms large disulfide-linked aggregates referred to as *Heinz bodies*. The presence of Heinz bodies is associated with increased cell fragility and susceptibility to lysis. Older erythrocytes are most susceptible because G6PD activity decreases with time, with the activity decreasing by roughly 50% over the ~120-day life span of a normal red blood cell. In the most frequently observed G6PD mutants—*G6PD A*– and *G6PD Mediterranean*—reduced protein stability hastens the time-dependent decrease in enzyme activity, and older erythrocytes are disproportionally destroyed in a hemolytic crisis.

Clinical Features

Episodic hemolytic anemia and neonatal jaundice are most common features

G6PD deficiency is associated with a range of clinical features: whereas most individuals are asymptomatic, severe cases may be associated with chronic nonspherocytic hemolytic anemia (CNSHA). Patients with the deficiency are at increased risk for neonatal jaundice and episodic hemolytic anemia precipitated by infection, drugs, or fava beans. A common trigger is infection, either viral or bacterial, and although the underlying mechanism is unknown, it has been suggested that reactive oxygen species produced during inflammation may be involved. Thus, this patient's *S. enteritis* infection was the likely precipitant of the hemolytic crisis in this case. In addition, various drugs (such as some antimalarials and sulfonamides) and other substances that generate oxygen radicals may trigger a crisis. Indeed, investigations into the hemolytic effects of primaquine (an antimalarial) in certain patients led to the identification of G6PD deficiency in the 1950s. Fava beans produce pro-oxidant glycosides, and thus consumption of fava beans has been associated with the onset of a crisis as well. Not all variants of G6PD are associated with sensitivity to fava beans; however, those that are cause what is referred to as *favism*.

Neonatal jaundice is also common, and it has been estimated that as many as one third of those with neonatal jaundice have G6PD deficiency. This has been attributed to an increased susceptibility of neonatal erythrocytes to hemolysis, as well as to suboptimal liver function as a result of the deficiency and the immaturity of the organ in the neonatal period. Kernicterus, a serious complication of neonatal jaundice that can cause brain damage, is increased in incidence in all populations with high frequencies of mutant *G6PD* alleles.

Laboratory and clinical features of a hemolytic crisis

The massive destruction of erythrocytes that occurs during a hemolytic crisis has a number of characteristic signs and symptoms. The decreased plasma hemoglobin and increased level of its metabolite—bilirubin—are typical, and jaundice may be apparent in extreme cases. Haptoglobulin is an abundant serum protein that binds free hemoglobin to form a complex that is rapidly removed from the circulation by phagocytes. Its level is generally decreased during a hemolytic crisis; however, in this case the level was found to be in the normal range. This likely reflects the competing effects of hemolysis and systemic infection in this patient, which would tend to offset each other. Haptoglobin is an *acute phase reactant,* one of a group of plasma proteins whose level rises upon tissue injury, inflammation, or malignancy; its serum concentration typically increases in the face of infection. Splenomegaly is often observed during an acute hemolytic crisis because the spleen is important in the clearance of damaged erythrocytes.

When the hemoglobin concentration in the bloodstream becomes very high it is filtered through the renal glomeruli and passes into the urine, causing *hemoglobinurea*. In this patient the elevated BUN level in the setting of a normal serum creatinine most likely reflects dehydration rather than kidney dysfunction. Liver function tests may be elevated, reflecting damage to the organ as a result of the deficiency or hemolysis.

Damage to erythrocytic cells can be visualized in a blood smear as cellular fragments, "bite cells" (cratered erythrocytes formed by removal of Heinz bodies), and spherocytes, and typically the *reticulocyte* count is elevated, a reflection of increased bone marrow red-blood-cell production in response to the crisis. However there is a delay between hemolysis and the subsequent production of reticulocytes in the bone marrow and their appearance in the circulation, accounting for the initial normal reticulocyte count in this patient. A subsequent test three days after admission showed the characteristic elevation in reticulocyte levels.

This patient thus presented with a number of features of an acute hemolytic crisis, which was supported by the laboratory tests. Lightheadedness is a result of anemia and dehydration, with associated reduction in blood pressure and compromised oxygen and nutrient delivery to the brain, and tachycardia is a compensatory cardiac response. Once the gastrointestinal infection resolved, the hematocrit continued to fall, a common feature in those with the Mediterranean G6PD variant. Hemolytic crises in those with the A− variant are generally self-limiting because the newly formed reticulocytes have sufficient G6PD activity to maintain cellular integrity; however, the enzyme activity is so low in patients with the Mediterranean variant that a blood transfusion is usually recommended to restore normal erythrocyte levels and forestall complications. This patient did not wish to be transfused, but fortunately his hematocrit nevertheless began to rise and he recovered.

Diagnosis

The most commonly used test for diagnosing G6PD deficiency is the *fluorescence spot test,* which is based on the conversion of NADP⁺ to NADPH and visualization of the fluorescence emitted as a result of this conversion under UV light. The test is not always reliable and may yield a false-positive result when a patient is severely anemic, and a false-negative result if the reticulocyte count is high. In these cases quantitative assessment of G6PD activity must be made. In addition, the test is not reliable in identifying heterozygous females. The G6PD activity may fall in the normal range in this subgroup; thus molecular analysis is necessary to identify the defect.

Treatment

There is no treatment for G6PD deficiency; thus avoidance of precipitating factors is the primary recommendation. Hemolytic crises may be treated with intravenous fluids and oxygen, and occasionally with blood transfusion; however, in most cases they are self-limiting. Bone marrow transplantation is a feasible strategy in the most severe forms of the enzyme deficiency. Gene therapy may also prove to be a feasible option in the future. Indeed, retroviral infection with a retrovirus that harbors human *G6PD* produced lifelong expression of the enzyme in recipient mice.

QUESTIONS

1. Females heterozygous for the *G6PD* gene were found to have two distinct populations of erythrocytes, one that contains normal G6PD (and no mutant enzyme), and the other that contains mutant G6PD (and no wild type). How can this be explained?

2. Most mutations in the *G6PD* gene are single amino acid substitutions, and very few large deletions, frameshift mutations and nonsense mutations have been described. In addition, mutations in the 8-amino-acid substrate binding site have not been described. What might be the explanation for this?

3. The pentose phosphate pathway is a major source of cellular NADPH; however, another cellular enzyme activity also produces NADPH. What is it and what reaction does it catalyze?

4. Is iron-deficiency anemia a *hemolytic anemia?* Briefly describe the underlying mechanism of iron-deficiency anemia.

5. An African American man in the midst of a drug-induced acute hemolytic crisis is admitted to the hospital. He has a history of neonatal jaundice and you suspect G6PD deficiency. You order the fluorescent spot test, which comes back negative. Is it possible that the result is a false negative? Why?

6. The enzyme *catalase* (see Section 18.3.6 of *Biochemistry* 5e), like glutathione peroxidase (see Section 20.5.1 of *Biochemistry* 5e) can dispose of harmful cellular peroxides. Both enzymes are found in erythrocytes. Do you think catalase plays a major role in the disposal of peroxides in the erythrocytes of G6PD-deficient patients undergoing a hemolytic crisis? Under conditions when the patient is not in a crisis?

FURTHER READING

1. Au, S. W. N, Gover, S., Lam, V. M. S., and Adams, M. J. Human glucose-6-phosphate dehydrogenase: The crystal structure reveals a structural $NADP^+$ molecule and provides insights into enzyme deficiency. *Structure* (2000) 8:293–303.

2. Mehta, A., Mason, P. J., and Vulliamy, T. J. Glucose-6-phosphate dehydrogenase deficiency. *Best Pract. Res. Clin. Haematol.* (2000) 13:21–38.

3. Ruwende, C., and Hill, A. Glucose-6-phosphate dehydrogenase deficiency and malaria. *J. Mol. Med.* (1998) 76(8):581–588.

For further information, see the following web sites:

The G6PD deficiency homepage: http://www.rialto.com/g6pd/

OMIM: http://www.ncbi.nlm.nih.gov/htbin-post/Omim/dispmim?305900

CHAPTER 21

Glycogen Metabolism
Muscle Aches and Tea-Colored Urine

CASE HISTORY

Patient with Nausea and Lethargy Is Admitted Following Overexertion

A 15-year-old boy was admitted to the hospital with nausea, vomiting, lethargy, and dark-colored urine. He had a history of easy fatigability, muscle cramps, and difficulty keeping up with children his age. He avoided physical activity because of these symptoms.

Two days prior to admission, while playing with friends after school, he developed muscle stiffness, cramping, and fatigue with bursts of exercise. However, he continued playing because he experienced a "second wind" and easing of the muscular symptoms. He remained continuously active for several hours but developed extreme fatigue upon returning home. He did not go to school the following day because of diffuse myalgias (aching of the muscles) and a low-grade fever, for which he was given a single dose of acetaminophen, 500 mg orally. He complained of nausea and refused to eat but did drink water and juices that he initially was able to keep down. On the morning of admission he was difficult to rouse from sleep and complained of worsening nausea, and vomited. His mother noticed that after he had gone to the bathroom his urine was tea-colored, so she brought him to the emergency department.

What are the causes of exercise-induced muscle cramps in a child of this age?

Is the "second wind" significant?

What are the causes of darkening of the urine?

Patient History

He was the oldest of three children born by normal spontaneous vaginal delivery. He was of average height and weight throughout his childhood, and attained all developmental milestones on time. He had no known medical illnesses and had received all routine immunizations without difficulty. He was an above average student and did not exhibit any behavioral problems. He took no medicines regularly and did not smoke, use alcohol, or use illicit drugs. His younger brother and sister, ages 12 and 10 years, had similar but milder difficulties with physical activity. The parents were healthy, without limitations in physical activity, but not particularly athletic; both had sedentary jobs. There was no known family history of neuromuscular diseases.

Physical Examination

On physical examination he was lethargic but easily arousable, and appeared acutely ill. He was normally developed for his age. His blood pressure was 104/58 mmHg, heart rate was 110 beats per minute (bpm), respirations were 22 per minute, and his temperature was 38°C. The head, eyes, ears, nose, and throat examination was normal. Neck was supple without nuchal rigidity (stiff neck), the thyroid was normal, and there were no palpable cervical lymph nodes. His chest was resonant to percussion and clear to auscultation. The heart examination was normal. He had normally active bowel sounds and no abdominal tenderness or organomegaly. He had no costovertebral angle tenderness. Examination of the musculoskeletal system revealed normal muscle tone and bulk with diffuse mild tenderness. There was no atrophy or fasciculations in any of the muscle groups. Neurological examination was normal.

What is nuchal rigidity, and what does its presence signify?

Is the absence of enlargement of the liver significant?

What are muscle fasciculations and what is their significance?

Laboratory Evaluation Reveals Rhabdomyolysis

Initial laboratory evaluation was remarkable for a normal complete blood count. Serum chemistries revealed a potassium of 5.2 mEq/L, blood urea nitrogen (BUN) of 45 mg/dl, creatinine of 2.0 mg/dl, creatine kinase (CK) of 2114 IU/L (upper limits of normal 195 IU/L), aldolase of 44.1 IU/L (upper limits of normal 7.6 IU/L), and a mild elevation in uric acid (8.0 mg/dl). Serum glucose, cholesterol, triglycerides, and liver function tests were normal. The serum and urine myoglobin levels were elevated. Examination of the urine revealed pigmented casts. Arterial blood gas revealed a pH of 7.36, pCO_2 (partial pressure of CO_2) of 44 mmHg, and pO_2 of 92 mmHg on room air. An electrocardiogram revealed sinus tachycardia but was otherwise normal. A diagnosis of rhabdomyolysis was made and he was given intravenous fluids, bicarbonate, and mannitol to induce an alkaline osmotic diuresis.

What is rhabdomyolysis, and why is raising the blood pH and inducing a diuresis helpful in the management of this problem?

What are creatine kinase and aldolase, and what is the significance of their elevation?

What do casts in the urine signify?

Intravenous fluids produced significant improvement in the lethargy and nausea. The muscle tenderness and weakness subsided. The levels of CK and aldolase decreased to 310 IU/L and 10 IU/L, respectively, over the next four days, and myoglobin was no longer detectable in the urine. The serum potassium, uric acid, urea nitrogen, and creatinine all normalized.

Diagnosis and Treatment

He underwent further evaluation for a presumed skeletal myopathy. An ischemic exercise test showed flat lactate and pyruvate curves and a normal rise in ammonia (basal 79 and maximum 280 micrograms/dl). He underwent a skeletal muscle biopsy, and the morphology of the muscle was normal, although rare areas of necrosis were observed. Ultrastructural changes included an increase in intracellular glycogen, and staining for myophosphorylase was negative. He was diagnosed with glycogen storage disease type V (McArdle disease). He was treated with a high-protein diet and it was recommended that he avoid severe bursts of activity, but that he perform moderate exercise with ample time for warm-up.

What is the purpose of the high-protein diet?

A diet rich in fats has also been suggested for patients with McArdle disease. What is the rationale?

Why is an adequate warm-up prior to exercise important in these patients?

DISCUSSION

Glycogen Storage Diseases

McArdle disease is but one of a number of *glycogen storage diseases* (see Table 21.1 of *Biochemistry* 5e), so called because they are caused by defects in glycogen metabolism, leading to the accumulation, or *storage*, of glycogen in bodily tissues. The liver and muscle are the most seriously affected by these disorders because they possess the most glycogen, and their functions depend upon its use. The liver uses glycogen to maintain blood glucose at a constant level, while muscle cells depend upon glycogen as a readily mobilized source of energy. Defects in liver glycogen metabolic enzymes lead to hypoglycemia and cirrhosis, whereas defects in muscle enzymes typically result in exercise intolerance and muscle atrophy.

Myophosphorylase Deficiency Is the Cause of McArdle Disease

Glycogen breakdown is initiated by *glycogen phosphorylase,* which removes 1,4-glucosyl residues from the outer branches of glycogen, producing glucose-1-phosphate (see Section 21.1.1 of *Biochemistry* 5e). The glycogen phosphorylase enzyme is a homodimer, consisting of two identical subunits of 97 kDa each. Mature skeletal muscle has a single glycogen phosphorylase isoenzyme, *myophosphorylase (MPL)*, whose gene resides on chromosome 11q13. McArdle disease (glycogen storage disease type V) is caused by genetic defects in the myophosphorylase gene (*PYGM*) that result in little or no gene expression. The molecular defects in *PYGM* are heterogeneous, with the most common mutation in the United States being a nonsense mutation in exon 1, at position 49 (R49X), but missense and frameshift mutations have also been described. Over 30 distinct mutations have been described in patients with McArdle disease. The disease is inherited in an autosomal-recessive manner; however, symptomatic heterozygotes have been reported.

Clinical Manifestations of McArdle Disease

Disease is fairly benign

McArdle disease is relatively benign and is characterized by exercise-induced muscle cramping, stiffness, pain, and fatigue that promptly resolves with rest. Fatigue and weakness during exercise are due to the unavailability of glycogen as an energy source for muscle contraction (Figure 21.1). In a matter of a few seconds the available supply of ATP and creatine phosphate are exhausted (see Figure 14.7 of *Biochemistry* 5e), and the cell must rely upon metabolic processes for the production of ATP for its energy needs. Glycogen, an *intracellular* source of glucose, serves as an important early source of energy during intense activity, until extracellular sources may be mobilized. The second wind experienced by this patient while playing with his friends is typical of those with McArdle disease, and reflects the mobilization of *extramuscular* energy sources (especially glucose and fatty acids) triggered by hormonal and neural stimuli.

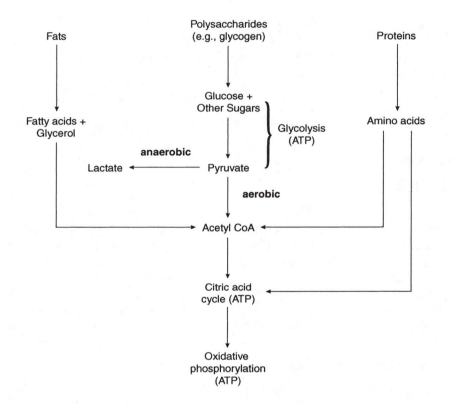

FIGURE 21.1 Energy-yielding metabolic pathways. Glycogen serves as a source of glucose for the energy-yielding processes of glycolysis (2 ATP/glucose molecule), the citric acid cycle (2 ATP/glucose), and oxidative phosphorylation (26 ATP/glucose). Under anaerobic conditions lactate is formed from pyruvate.

Disease progresses through phases

Typically, the onset of symptoms occurs at puberty, and there are even well-documented cases with symptomatic onset in adulthood. The reason for the relative absence of symptoms in young children is unclear. The disease may occur in phases, with signs and symptoms in late childhood and adolescence consisting of intermittent dark urine due to *myoglobinuria,* an indicator of muscle destruction (see below). In early adulthood muscular symptoms become more prominent, with cramping and pain or stiffness on exertion, occasionally followed by transient myoglobinuria. Later in life (fourth or fifth decade), exertional muscular symptoms give way to more persistent and progressive weakness and wasting of muscle, and at this stage myoglobinuria is rare.

Rhabdomyolysis

Rhabdomyolysis is a life-threatening complication of McArdle disease that results from the breakdown of skeletal muscle and the spillage of muscle-cell constituents into the bloodstream. It is typically triggered by overexertion, as in this case. Upon muscle injury, myoglobin, aldolase, potassium, creatine kinase, and uric acid leak into the bloodstream. The elevated potassium level (hyperkaelemia) may lead to life-threatening cardiac rhythm disturbances, and the high serum myoglobin may lead to acute renal failure (ARF) if left untreated. This patient did not exhibit serious cardiac symptoms; however, the tea-colored urine noticed by his mother is the result of the pigmented myoglobin in his urine. Myoglobin in the glomerular filtrate may precipitate, causing tubular obstruction and damage to the kidney. The administration of intravenous fluids and the induction of alkaline diuresis helps clear myoglobin from the kidney, thus preventing the development of ARF. Fortunately, he responded well to therapy and the signs and symptoms of kidney malfunction receded: his BUN and creatinine levels normalized, and the nausea and lethargy subsided.

Incidence

Glycogen storage diseases are rare, with 2–3 cases per 100,000, internationally. McArdle disease is one of the more common forms; however, it is still extremely rare, with only a few hundred cases having been reported in the United States.

Diagnosis

McArdle disease is rare and, when uncomplicated, is primarily characterized by fatigue and cramps during exercise, which may not immediately strike one as out of the ordinary. The disease is thus thought to be underdiagnosed, especially in the early stages. This patient, however, presented with rhabdomyolysis, a common complication of McArdle disease. Although there are other causes of this disorder, such as injury, alcohol abuse, or the ingestion of certain medicines or toxic substances, this patient (as well as some family members) had a history of exercise intolerance, suggesting an inherited metabolic disorder. The results of the ischemic exercise test suggested a muscle glycogen storage disease: the failure of pyruvate (the product of *glycolysis;* see Section 16.1 of *Biochemistry* 5e) and lactate to rise upon ischemic exercise would result from a shortage of glucose, secondary to an inability to breakdown glycogen (see Figure 21.1). The definitive diagnosis of McArdle disease may be made only biochemically, or by muscle biopsy with a demonstration of a reduction or absence of MPL immunoreactivity, and increased muscle glycogen stores.

Treatment

There is currently no treatment for McArdle disease; however, the symptoms may be managed with dietary and exercise regimens. Although intense exercise that would necessitate glycogen as an energy source is not recommended, moderate, regular exercise is thought to be beneficial. Maintaining cardiovascular fitness will allow for effective transport of extramuscular sources of energy such as fatty acids and glucose, as well as maintaining the activities of mitochondrial enzymes involved in the citric acid cycle and oxidative phosphorylation. A high-protein diet is sometimes recommended and may improve endurance during exercise in some cases. The metabolism of amino acids produces metabolic intermediates that may enter the citric acid cycle and thus generate energy (see Section 23.5 of *Biochemistry* 5e). Similarly, a high-fat diet has been proposed to be beneficial as it would provide an alternative energy source; however, this remains controversial. The negative consequences of a high-fat diet may counterbalance any beneficial effects it may have in these patients.

QUESTIONS

1. Both creatine and vitamin B_6 have been tested in clinical trials for the treatment of McArdle disease. What is the rationale?

2. Why does ischemic muscle exercise fail to give rise to an increase in lactate in patients with McArdle disease?

3. Overtreatment of diabetic patients with insulin can lead to symptoms of liver disease found in some glycogen storage diseases. What might account for this?

4. The ischemic exercise test revealed a normal rise in ammonia in this patient. From where does the ammonia derive during heavy exercise, when ATP is being rapidly utilized?

5. Mutations in liver glycogen synthase cause what is known as *glycogen storage disease type 0,* with symptoms of hypoglycemia and hyperketonemia. Is this technically a glycogen storage disease?

6. Would you expect that defects in *branching enzyme,* which catalyzes the formation of α-1,6 linkages of glycogen, would result in symptoms similar to those found in McArdle disease?

FURTHER READING

1. Bartram, C., Edwards, R. H., and Beynon, R. J. McArdle's disease—muscle glycogen phosphorylase deficiency. *Biochim. Biophys. Acta* (1995) 1272(1):1–13.

2. Tarui, S. Glycolytic defects in muscle: Aspects of collaboration between basic science and clinical medicine. *Muscle Nerve* (1995) 3:S2–S9.

3. McArdle, B. Myopathy due to a defect in muscle glycogen breakdown. *Clin. Sci.* (1951) 10:13–33 (This article describes the first reported case of McArdle disease.)

For further information, see the following web sites:

National Institutes of Health, Online Mendelian Inheritance in Man, Glycogen Storage Disease V: www3.ncbi.nlm.nih.gov:80/htbin-post/Omim/dispmim?232600

Muscular Dystrophy Association: http://www.mdausa.org/disease/mpd.html

Association for Glycogen Storage Disease: www.agsdus.org/

CHAPTER 22

Fatty Acid Metabolism
Fatty Liver No Alcohol

CASE HISTORY

Obese Patient Presents for Follow-Up of Abnormal Liver Function Tests

A 43-year-old woman returns to her internist for follow-up of abnormal liver function tests discovered on routine laboratory evaluation. The patient has had a weight problem all of her life. She has been obese since childhood, and despite multiple diets has been unable to maintain any weight loss. She has a number of medical problems and complaints that are believed to be secondary to her weight. She has degenerative arthritis of both the knees and ankles, hyperlipidemia, dyspnea with mild levels of exertion, and excessive fatigue. There is no history of diabetes, viral hepatitis, gallbladder disease, or alcohol use. The only medicines she takes are a multivitamin and ibuprofen as needed for joint pain.

What are the medical complications of obesity?

What might be causing the abnormality in liver function tests?

Patient Has a Family History of Obesity

She is married with two children, both of whom are overweight, as is her husband. She does not work outside the home. Her family history is remarkable for obesity in both parents and her siblings. Her father died at the age of 61 secondary to complications of diabetes, and her mother is 64 years old and has hypertension and coronary artery disease. Her younger brother is 36 years old and is hypertensive, and her sister, age 40, has recently been started on medication for adult-onset diabetes mellitus.

Physical Examination

On physical examination her weight was 248 pounds (~113 kg) and height was 5 feet 2 inches (157 cm), with a body mass index (BMI; weight in kilograms divided by the square of the height in meters) of 45.8, consistent with severe obesity. The blood pressure was 144/88 mmHg, heart rate was 84 beats per minute, and respiratory rate was 16 per minute, and she was afebrile. The skin revealed striae (stretch marks) but no rashes or changes in pigmentation; however, there was maceration in skin folds under the breasts.

The head, eyes, ears, nose, and throat examination revealed full extraocular motion, full visual fields, and normal appearance of the fundus. The neck was supple and thick without thyromegaly. Chest examination was remarkable for distant breath sounds but no wheezes or rhonchi. The heart examination revealed a point of maximal impulse (PMI) that was difficult to appreciate because of her body habitus, and distant heart sounds with an S4 gallop sound but no murmurs or rubs. The abdomen was obese with active bowel sounds, and no fluid wave was appreciated. Although the liver was not enlarged there was tenderness to deep palpation in the right upper quadrant. She had venous varicosities of both legs and trace bilateral edema at the ankles. Neurological examination revealed no focal deficits.

What is the BMI? What range is normal?

What are the metabolic causes of obesity?

Laboratory Evaluation Reveals Mild Liver Dysfunction

She had undergone an extensive laboratory evaluation that revealed normal serum electrolytes, blood urea nitrogen (BUN), and creatinine. The fasting blood sugar was 96 mg/dl. Complete blood count was remarkable only for a mild anemia, with a hematocrit of 35%, and otherwise normal red blood cell indices. She had elevations in serum lipids, with a total cholesterol of 238 mg/dl (upper limit of normal [ULN]: 200 mg/dl), triglycerides 452mg/dl (ULN: 190 mg/dl), HDL 49 mg/dl, LDL 158 mg/dl (ULN: 160 mg/dl).

Liver function tests were abnormal on several occasions. Most recently the aspartate amino transferase (AST) was 88 IU/l (ULN: 31 IU/l), alanine amino transferase (ALT) was 112 IU/l (ULN: 31 IU/l), total bilirubin (TB) was 1.1 mg/dl (ULN: 1.2 mg/dl), and the alkaline phosphatase was 148 IU/l (ULN: 120 IU/l). Blood coagulation tests were normal. Tests for viral hepatitis A, B, and C and for heterophil antibodies were negative. Thyroid function tests were normal. Serum iron, ferritin, transferrin, and ceruloplasmin were normal. Tests for antinuclear antibodies and rheumatoid factor were unremarkable.

Why were blood coagulation tests carried out?

Why were tests for serum iron, transferrin, and ceruloplasmin carried out?

CT Scan and Biopsy Reveal "Fatty Liver"

Ultrasound examination of the abdomen revealed a normal gallbladder, bile ducts, and pancreas. She had a computed tomographic (CT) scan of the abdomen that was remarkable for hypodense regions in the liver consistent with steatosis (fatty infiltration of the liver). She had undergone a liver biopsy one week ago. The biopsy revealed fatty infiltration of the liver with sinusoidal and pericellular fibrosis with ballooning degeneration of hepatocytes and Mallory hyaline.

What are the causes of nonalcoholic fatty liver disease (NAFLD)?

Why was a liver biopsy recommended?

What is the difference between steatosis and steatohepatitis?

Diagnosis and Treatment

A diagnosis of nonalcoholic steatohepatitis (NASH) was made. It was recommended that she go on a calorie-restricted diet in an attempt to decrease her weight. The hyperlipidemia was treated with HMG-CoA reductase inhibitors ("statins") and she was started on ursodeoxycholic acid.

What are the mechanisms of liver cell injury in NASH?

What is ursodesoxycholic acid and what is the proposed mechanism of action in NASH?

How do the statins reduce cholesterol levels?

DISCUSSION

Obesity

Triacylglycerols serve as an important energy reserve

The ability to store energy in the form of fat serves an important physiological function. When food is plentiful excess energy may be stored in the form of *triacylglycerols (TAGs)*, which may later be mobilized in the form of fatty acids in times of scarce food supplies (see Section 22.1 of *Biochemistry* 5e). Adipose tissue is found throughout the body and serves as the major bodily fat repository. The storage of energy and its release is controlled via numerous endocrine and neural pathways to meet the current energy needs of the organism. For example, the hormone *epinephrine* (also called *adrenaline*), which prepares an organism for the "fight-or-flight" response, inhibits acetyl-CoA carboxylase, the enzyme that catalyzes the committed step in fatty acid synthesis, as well as activates the lipases, which catalyze the release of fatty acids from triacylglycerols. Thus, epinephrine simultaneously inhibits fatty acid synthesis and stimulates its release from adipose tissue, thus providing the energy needed to cope with an immediate threat.

BMI is the most widespread gauge of obesity

Although the ability to store fat promotes survival in times of fluctuating food supplies, in the Western industrialized world, where food supplies are now plentiful, it has led to increased prevalence of obesity, and to associated adverse health consequences. Obesity is defined as excess adipose tissue mass, and the BMI is the most widely used index of obesity (although it is not a direct measure). A BMI of >30 is generally considered to be the threshhold for obesity; however, medical problems are also associated with BMIs of 25–30. In addition, the bodily *distribution* of fat is consequential because intra-abdominal and abdominal subcutaneous fat are particularly associated with the more serious complications of obesity. Thus, the waist-to-hip ratio is helpful in assessing risk, with values > 0.9 in women and > 1.0 in men considered abnormal. Obesity is becoming increasingly prevalent in the Western world. A recent survey indicates that 22.5% of the U.S. population is obese.

Obesity is linked to a number of medical problems

Obesity is associated with a number of disorders and diseases, such as insulin resistance, diabetes, cardiovascular disease, pulmonary disease, reproductive disorders, and osteoarthritis. Obese individuals are also at increased risk for gallstones and certain types of cancer and, as will be discussed further here, are particularly susceptible to *hepatic steatosis,* which results from the accumulation of TAGs in hepatocytes. Hepatic steatosis alone is considered fairly benign; however, when it progresses and includes inflammation and tissue necrosis it can result in life-threatening cirrhosis. It has been estimated that hepatic steatosis is found in up to 75% of obese individuals and is thus very common, as well as potentially life threatening, in this subgroup.

This patient and her obese family members exhibited a number of features associated with obesity. She has suffered from arthritis in her joints as well as liver dysfunction, both her sister and father have developed diabetes, and her mother and brother suffer from cardio-vascular problems.

Nonalcoholic Steatohepatitis

Defined by hepatic histopathologic characteristics and the absence of alcoholism

NASH was first recognized in 1980 by Ludwig and colleagues, who described a condition in a group of patients who denied excessive alcohol consumption, and in which liver biopsy findings were indistinguishable from patients with alcoholic hepatitis. Most of the patients were women, mildly obese, and exhibited obesity-related diseases such as diabetes and cholelithiasis (gallstones in the gallbladder). NASH is now considered the extreme in a spectrum of NAFLD, which ranges from hepatic steatosis alone to NASH, which is characterized by hepatic steatosis, inflammation, fibrosis and necrosis, with or without Mallory hyaline (intermediate filament aggregates), or cirrhosis. The prevalence of NASH has yet to be definitively established; however, it has been reported worldwide, and from 1.2% to 9% of the general population in the United States has been estimated to be affected.

Pathogenesis

The pathogenesis of NASH is still unclear; however, it is known to occur in the setting of *steatosis*. Indeed, insulin resistance, which occurs in the large majority of patients with NAFLD, is thought to feature prominently in the development of steatosis. Resistance to insulin's normal suppression of lipolysis would result in the release of fatty acids from TAGs stored in adipocytes. The increased circulating levels of fatty acids would lead to increased influx into the liver, where TAGs are formed; this together with decreased hepatic TAG export and catabolism have been proposed to lead to steatosis. However, steatosis alone does not cause cellular injury, nor does it affect liver function, and is thus a benign condition. It has been proposed that a "second hit" is necessary for the development of NASH.

Oxidative stress has been proposed to lead to the more pathogenic forms of NAFLD. Oxidative stress is generated during peroxisomal fatty acid metabolism in the form of peroxide (see Section 22.3.4 of *Biochemistry* 5e), and evidence also implicates the microsomal (i.e., endoplasmic reticulum) enzyme, cytochrome P450 CYP2E1, in generating reactive oxygen species (ROS). CYP2E1 is an oxidase involved in microsomal metabolism of fatty acids. Fatty acids are both inducers and substrates for this enzyme, and its level is markedly elevated in patients with NASH. Oxidative stress would then trigger lipid peroxidation, which further causes cellular damage via reactive intermediates. Evidence of lipid peroxidation has been observed in animal models of the disease. The products of lipid peroxidation (such as malondialdehyde and 4-hydroxynonenol) can induce inflammation and fibrosis, which may lead to necrosis, thus accounting for all the pathological features associated with the livers of those with NASH. Although some evidence exists for this mechanism of pathogenesis, the precise sequence of events is as yet unproven. Indeed, inflammation itself generates ROS, and thus inflammation may precede the generation of oxidative stress.

Clinical features

In most cases NASH is asymptomatic, and the disease is recognized upon follow-up of mildly abnormal liver function tests, as in this patient. Patients may complain of malaise and fatigue, and of upper right quadrant abdominal pain or tenderness. Examination may reveal an enlarged liver. It has been estimated that between 65% and 100% of patients are female, 69–100% are obese, and approximately one third have diabetes (usually non-insulin dependent). Most studies have also found an association with hyperlipidaemia, occuring in 21–92% of patients. The mean age varies from 46 to 54 years. Although obesity, and the insulin resistance associated with it, are the most consistent causal factors, NASH has also been linked to protein malnutrition, jejuno-ileal bypass, prolonged parenteral nutrition, acute starvation, and various medications. No definitive studies have yet determined prognosis for NASH; however, it has been concluded that hepatic steatosis alone has a very good prognosis. One study revealed no clinical or histopathological progression over a median time period of 11 years in a group of patients with steatosis alone.

Laboratory tests usually reveal a mild elevation of the liver enzymes, AST and ALT, and the ratio of AST:ALT is generally <1, which may be used to distinguish it from alcoholic hepatitis, in which the ratio is typically >1. Serum alkaline phosphatase levels are also usually slightly elevated; however, serum bilirubin levels are usually normal, unless the disease is quite advanced.

Diagnosis

Patients are usually diagnosed upon follow-up of abnormal liver function tests. Further indication of the disease may come from ultrasound, magnetic resonance imaging, or CT scans, which may reveal features typical of fatty infiltration of the liver. However, only a liver biopsy can provide the evidence necessary to definitively diagnose NAFLD. The degree of inflammation and fibrosis can be revealed only through the biopsy, and is critical to assessing prognosis. As mentioned above, fatty liver alone is benign; however, advanced NASH can lead to cirrhosis and death from liver failure. Histological features associated with NASH are macrovesicular fat, inflammation, ballooning degeneration of hepatocytes, Mallory hyaline, hepatocyte necrosis, fibrosis, and cirrhosis in the most advanced cases. Patients are scored on these pathologic features in assessing prognosis.

Treatment

There is no established treatment for NASH, and current therapy focuses on gradual weight reduction through changes in diet and exercise patterns. Weight loss has been shown to be beneficial, improving liver function tests and histopathology, and patients also report amelioration of the fatigue and malaise associated with the disease. Weight loss has also been shown to decrease insulin resistance and hyperlipidaemia.

Data on specific therapies are preliminary; however, vitamin E (an antioxidant) supplementation has been reported to be helpful. In addition, several studies have shown ursodeoxycholic acid (USDA) to be beneficial. This naturally occurring bile acid (derived from the Chinese black bear) primarily acts to relieve cholestasis, but has also been shown to be beneficial in a number of chronic liver diseases including NASH. USDA is a particularly hydrophilic bile acid, and improves the flow of bile and appears to have hepatoprotective, immunomodulatory, and antioxidant properties.

QUESTIONS

1. Why do triacylglycerols form vesicles inside cells, and why do they not participate in the formation of cellular membranes?

2. The breakdown of triacylglycerols is induced when blood-glucose levels are low so as to generate the ATP necessary for many essential physiological processes. Identify the point at which each product of fatty acid oxidation (glycerol, acetyl CoA, NADH, $FADH_2$) enters energy-generating metabolic processes: glycolysis, the citric acid cycle, or oxidative phosphorylation.

3. Although gradual weight loss has been shown to be very effective in the treatment of NASH, very low calorie diets or fasting have been shown to worsen the condition. How might this be explained?

4. It has been proposed that nutrient deficiency, in particular, the essential amino acids, may contribute to the development of NASH as a result of reduced production of very-low-density lipoproteins (VLDL), which depend upon them for their synthesis. How might reduced levels of VLDL lead to hepatic steatosis?

5. The drug troglitazone has been used to treat patients with NASH. What is the biochemical rationale for its use?

6. The drug perhexiline is used to treat angina pectoris, and hepatic side-effects, including NASH, have been found in up to one third of patients taking the drug. The hepatic toxicity is thought to be a result of mitochondrial injury. Why would mitochondrial injury lead to NASH?

7. Genetically obese *ob/ob* mice lack *leptin,* a hormone produced by adipocytes, and are commonly found to have hepatic steatosis. What is the role of leptin in the development of obesity, and is it a factor in the development of obesity in humans?

FURTHER READING

1. McCullough, A. J. Update on non alcoholic fatty liver disease. *J. Clin. Gastroenterology* (2002) 34:255–262.

2. Yang, S. Q., et al. Obesity increases sensitivity to endotoxin liver injury: implications for the pathogenesis of steatohepatitis. *Proc. Natl. Acad. Sci.* (1997) 94:2557–2562.

3. Chitturi, S., and Farrell, G. C. Etiopathogenesis of non-alcoholic steatohepatitis. *Semin. Liv. Dis.* (2001) 21:27–41.

4. Kopelman, P. G. Obesity as a medical problem. *Nature* (2000) 404:635–643.

For further information, see the following web sites:

American Liver Foundation, NAFLD/NASH:
www.liverfoundation.org/html_filz/livheal.dir/lh_im_dox.dir/lh_pdf.dir/NAFL.pdf

Centers for Disease Control, Obesity and Overweight:
www.cdc.gov/nccdphp/dnpa/obesity/index.htm

National Heart, Lung and Blood Institute, Guidelines on Overweight and Obesity:
Electronic Textbook: www.nhlbi.nih.gov/guidelines/obesity/e_txtbk/

Protein Turnover and Amino Acid Catabolism
A Rusty Liver

CASE HISTORY

Patient Presents with Dyspnea and an Altered Mental Status

A 54-year-old woman was admitted to the hospital because of shortness of breath and an altered mental status. Three years prior to this admission she began to gain weight and experience exertional dyspnea (shortness of breath). She had no significant cardiac risk factors except for high serum cholesterol. Cardiac evaluation one year prior to admission was remarkable for an echocardiogram that revealed normal left ventricular size and function, and a normal treadmill stress test. The only significant finding at that time was a mild abnormality of liver function tests, including an alanine aminotransferase (ALT) of 42 IU/l (upper limits of normal [ULN]: 31 IU/l), an asparate amino transferase (AST) of 35 IU/l (ULN: 31 IU/l), an alkaline phosphatase of 140 IU/l (ULN: 120 IU/l), and a total bilirubin (TB) of 1.9 mg/dl (ULN: 1.2 mg/dl).

What is the link between hyperlipidemia and liver disease?

She Has a History of Gallstones and Arthritis

The patient was a cashier in a grocery store. The only remarkable past medical history was unspecified arthritis, and a cholecystectomy, performed four years ago because of gallstone-induced biliary colic. There was no personal or family history of liver disease or anemia. She did not use alcohol, illicit drugs, or tobacco. She occasionally used ibuprofen for degenerative arthritis of the knees. She had no significant occupational exposures.

Why is the history of a cholecystectomy significant?

Physical Examination

She was subsequently lost to follow-up until this admission, when she presented with increasing dyspnea, peripheral edema, and lethargy. On physical examination she was afebrile, had systolic hypertension with a blood pressure of 170/80 mmHg, moderate obesity, icteric sclera (yellow discoloration of the white part of the eye), no rash, no spider angioma, and 3+ peripheral edema. Her bowel sounds were active and the liver was of normal size and palpated at the right costal margin; the spleen was not palpable, and there was no ascites (fluid in the abdomen).

What are scleral icterus and spider angiomata, and what causes these physical findings?

What are the causes of ascites?

Laboratory Tests, Radiographs, Echocardiogram, and Sonogram

Laboratory tests revealed further increases in the TB, at 6.8 mg/dl, conjugated bilirubin, at 2.4 mg/dl (ULN: 0.3 mg/dl), ALT, at 126 IU/l, AST at 298 IU/l, alkaline phosphatase at 175 IU/l, and a reduced serum albumin concentration with a normal total protein concentration due to an increase in serum globulins. Serological tests for hepatitis B and C were negative. She had mild anemia with a hematocrit of 32% with macrocytosis, a white blood cell count of 14,000 per cubic mm, with a normal differential count and a prolonged prothrombin time. A toxicology screening test of serum revealed no substances of abuse.

The serum ferritin level was markedly elevated at 1200 µg/l (ULN: 300 µg/l), with a mild increase in the serum iron (serum iron is transferrin-associated) at 169 µg/dl (ULN: 150 µg/dl), and normal total iron binding capacity, 295 µg/dl (normal: 250–400 µg/dl). The urine was positive for urobilinogen.

Radiographs of the chest showed evidence of pulmonary congestion and a small left-pleural effusion. An echocardiogram revealed left-atrial enlargement and normal left-ventricular function. An ultrasonographic examination of the abdomen showed diffuse, heterogeneous hyperechogenicity of the liver, consistent with fatty change, with no dilation of the biliary ducts.

What are the causes of macrocytic anemia?

What is the difference between total bilirubin and conjugated bilirubin, and what is the significance with respect to the type of liver dysfunction?

What causes an elevation in serum iron and ferritin, and what is the relationship to liver disease?

Why does the patient have evidence of heart failure (dyspnea and pulmonary vascular congestion) with normal left-ventricular function?

What is the significance of hyperechogenicity in the liver on the ultrasound examination?

She Is Treated for Edema, but Becomes More Lethargic and Vomits "Coffee Grounds"

Furosemide was given intravenously, with resolution of the edema. She was treated with conjugated estrogens, furosemide, potassium chloride, and propoxyphene as needed for pain. On the third hospital day she vomited bilious liquid mixed with "coffee grounds" on two occasions, and became even more lethargic. She remained afebrile and her physical examination was not significantly changed except that she responded to her name. However, she did not follow commands, and she had asterixis, but without focal neurological deficits.

The ammonia level was elevated. Specimens of blood and urine were sent for culture and were negative.

What is furosemide and why was it prescribed?

What is the significance of the "coffee grounds" in her emesis? Did it have anything to do with the deterioration in her mental status?

What is asterixis?

What is the significance of an elevated ammonia level?

She Receives Treatment for Complications, and Her Condition Improves

Vitamin K, lactulose, thiamine, and multivitamins were given. Intravenous somatostatin/octreotide, Pantoloc, ampicillin and ceftriaxone were administered. A lumbar puncture was performed after the intravenous administration of fresh frozen plasma. The opening pressure was normal and there were no abnormalities of the cerebrospinal fluid. During the next several days, the patient became more alert. By the tenth hospital day, she was fully alert and oriented, and the asterixis had resolved.

What is vitamin K and why was it given in this case?

Why was Pantoloc administered? What is one of the precipitants of hepatic encephalopathy?

Why was fresh frozen plasma administered prior to lumbar puncture?

Diagnosis

The patient underwent a liver biopsy. Microscopical examination revealed micronodular cirrhosis. Staining for iron showed fine granules within many hepatocytes. The liver contained 226 μmol of iron per gram, yielding a hepatic iron index (HII) of 4.2, consistent with hemochromatosis. Genetic testing revealed homozygosity of the C282Y mutation in the *HFE* gene, confirming the diagnosis. She was prescribed a restricted protein diet (60g/day), with avoidance of iron supplements and vitamin C, and referred for periodic phlebotomy.

What is hemochromatosis and how does it mechanistically link the patient's signs and symptoms?

Why was she instructed to avoid vitamin C?

Can you list the causes of cirrhosis?

DISCUSSION

Cirrhosis

The liver has many vital functions

The liver, which weighs about 3 pounds, is the largest visceral organ in the body and one of the most complex. It has numerous functions, including metabolism of toxic substances absorbed from the intestine and maintenance of steady blood-glucose levels. It generates the bile salts used in the digestion of fats and blood-clotting factors used in the healing of wounds, and manufactures approximately 50% of bodily cholesterol, an essential component of cell membranes. It also stores a number of lipid-soluble vitamins such as vitamins A, B_{12}, D, E, K, and folate, and minerals such as copper and iron. In addition, the liver generates most of the proteins suspended in blood plasma, including albumin, and is an important site for amino acid metabolism. Notably, the urea cycle, in which ammonium ion derived from amino acids is converted into urea, takes place in the liver (see Section 23.4 of *Biochemistry* 5e). The liver generates bile, which contains cholesterol, bilirubin (a by-product of heme breakdown), various excretory products, and the bile salts, and secretes it into the duodenum via the bile duct.

Cirrhosis is characterized by impaired liver function

Cirrhosis results from injury to the liver and the subsequent response, with the end result being extensive fibrosis within the organ, and the formation of regenerative nodules. The abnormal structure of the organ and reduced blood flow through it result in severely compromised liver function, which leads to a characteristic set of clinical features, including jaundice, hepatomegaly, portal hypertension, cholestasis (blockage of the flow of bile), ascites, hepatic encephalopathy, and in severe cases, liver failure. Not all of these features are manifested in all patients, and the onset of the disorder is usually slow and gradual. Consequently, it is an underdiagnosed condition, especially in its early stages. Cirrhosis has many possible causes, but the pathology of the disorder is similar in all cases.

Etiology

Cirrhosis has numerous causes

Chronic alcoholism, chronic viral hepatitis (types B, C, and D), and autoimmune hepatitis are common causes of cirrhosis. In addition, cirrhosis can occur indirectly, from chronic inflammation of the bile ducts, which lead from the liver to the duodenum. Although much less common, some genetic diseases also predispose one to cirrhosis, including hemochromatosis, with which this patient was diagnosed. Other genetic diseases associated with cirrhosis include cystic fibrosis and Wilson's disease.

Molecular genetics of hemochromatosis

Hemochromatosis (also referred to as *hereditary hemochromatosis*, or *HHC*) is an autosomal recessive disorder, characterized by excessive iron absorption and its deposition in bodily organs, particularly the liver. In most cases it is caused by mutation of the *HFE* gene, with the most common mutation—and the one carried by this patient—resulting in a cysteine to tyrosine substitution at amino acid 282 (C282Y). A second mutation, causing a histidine-to-aspartate substitution at amino acid 63 (H63D), has also been associated with disease, but is far less common and is less severe. The penetrance of the C282Y mutation is not complete, and preliminary data indicate that between 40% and 70% of homozygotes will develop iron overload. Thus, additional genetic and/or environmental factors play a role in the development of the disorder.

The HFE protein is a 343 amino acid transmembrane protein found in the crypt cells of the duodenum, where iron uptake occurs. There it interacts with the *transferrin receptor* and mediates iron uptake by the intestine. The C282Y mutation has been found to inhibit

HFE transport to the cell surface, as well as to inhibit its interaction with the transferrin receptor. Although further research is necessary, the increased iron uptake in hemochromatosis patients has been hypothesized to be the result of the *DMT-1* (*divalent metal transporter-1*) gene.

Epidemiology

Approximately 25,000 people die of cirrhosis each year in the United States, with approximately 50% of those deaths being alcohol related. Chronic hepatitis is the second most common cause of cirrhosis in the United States; however, in many parts of Asia and Africa, chronic hepatitis is the most common cause of cirrhosis.

Hemochromatosis is most common among those of northern European (especially Irish) extraction, and C282Y homozygosity accounts for the majority of cases. In Europe and in North America, approximately 1:200 individuals is homozygous for C282Y. In Europe, it has been estimated that between 52% and 96% of cases are caused by C282Y homozygosity, while in North America C282Y homozygosity accounts for 88% of cases. The C282Y mutation is very rare among Asian, Indian, African, Middle Eastern, and Australian populations.

Pathophysiology

Pathophysiology stems from excess iron absorption

Normally, the 3–4 grams of iron in the body are kept at a constant level by the loss and gain of approximately 1 mg/day in men, and 1.5 mg/day in women. However, in those with hemochromatosis, the gain well outstrips the loss, at approximately 4 mg/day or more; iron progressively accumulates and is deposited in the organs, especially the liver, a major bodily iron repository, but also the pancreas, heart, joints, and pituitary gland. Damage to cells presumably occurs because the amount of iron absorbed exceeds the body's capacity to complex it with *ferritin*, the cellular iron-storage protein (see Figure 31.37 of *Biochemistry* 5e). Free iron is very reactive and leads to the formation of oxygen radicals, which damage cellular proteins and may lead to cell death. Iron accumulation occurs over a period of years, and the disease is not usually manifested until between the ages of 40 and 60.

Liver, heart, and joints are affected in this patient

This patient presented with a number of signs of cirrhosis, and also exhibited features of cardiomyopathy (pulmonary vascular congestion and dyspnea), suggesting that both her liver and heart were affected. In addition, she had a history of arthritis, suggesting that her joints had also been affected by the disease.

Laboratory results indicate liver dysfunction

The myriad functions carried out by the liver are reflected in the variety and number of abnormal test results and symptoms experienced by this patient. The elevated levels of hepatic enzymes are typical, as these are released into the bloodstream upon necrosis of hepatic tissue, and the elevated bilirubin and cholesterol levels are also characteristic. Both substances are normally eliminated by secretion into the intestine as components of bile. In the absence of normal liver function, they accumulate in the bloodstream.

Similarly of the elevated ammonia level reflects a reduced capacity of the urea cycle (see Section 23.4 of *Biochemistry* 5e), which occurs in the liver. Approximately 40% of the ammonia in the body is derived from the bacterial breakdown of nitrogenous compounds in the intestine, while the remaining 60% is derived from amino acid metabolism (see Section 23.3 of *Biochemistry* 5e). In the urea cycle, ammonia derived from either source is used to make urea, a less toxic substance, which is subsequently excreted from the body via the kidneys. The elevated ammonia level likely contributes to the altered mental status of this patient that was noted upon admission, and to the hepatic encephalopathy she subsequently developed (see below).

This patient also is deficient in biological molecules normally synthesized by the liver, such as albumin and the blood coagulation factors: her serum albumin level was low, and the extended prothrombin time indicated a shortage of coagulation factors. The macrocytic anemia is also typical and results from vitamin B_{12} and folate deficiencies, cofactors normally stored by the liver and released when needed. Both of these play a role in nucleic acid synthesis, and are thus necessary for the process of erythropoiesis (red-blood-cell production).

Hepatic encephalopathy and variceal hemorrhage are complications of cirrhosis

This patient presented with an altered mental status, which is a sign of *hepatic encephalopathy*, a complication of cirrhosis. As mentioned above, the liver has a purifying function, metabolizing harmful substances absorbed from the intestine, as well as harmful metabolites produced by the body itself. When liver function is impaired these toxins may build up and affect brain functioning. Diversion of the portal system, a result of *portal hypertension* (see below), such that blood flow from the intestine completely bypasses the liver, may contribute to the disorder. Although the particular substance responsible for the effect is not known, ammonia is a likely candidate. Most patients with hepatic encephalopathy have elevated ammonia levels, and symptoms improve as levels decrease. In addition, children with elevated ammonia levels as a result of genetic abnormalities in the urea cycle have neuropsychiatric abnormalities similar to those manifested in hepatic encephalopathy. Lethargy, disorientation, and confusion are typical features of the disorder, as is asterixis or "flapping tremor," which is demonstrated as flapping of the arms when held outstretched. Hepatic encephalopathy can lead to coma, and thus must be treated promptly. The lactulose she was given to treat the condition serves to reduce protein absorption by the intestine, and thereby ammonium production.

On the third hospital day this patient vomited what looked like coffee grounds, a sign of *variceal hemorrhage,* another complication of cirrhosis. Blood flows from the intestine to the liver through the portal vein, and exits through the hepatic vein to join the general venous circulation (Figure 23.1). In cirrhosis scarring of the liver causes resistance to blood flow, and as a consequence, collateral vessels develop between the portal vein and the general circulation such that bloodflow from the intestine completely bypasses the liver. These collateral vessels commonly form at the lower end of the esophagus, near the entrance to the stomach. These vessels are often engorged due to elevated pressure, and susceptible to bleeding, causing blood seepage into the intestinal tract. Variceal hemorrhage is life-threatening and must be treated promptly. The vasoconstrictors, somatostatin and octreotide, were infused to reduce bleeding at the sites, and vitamin K, essential in the synthesis of clotting factors, was given to promote healing of the damaged vessels.

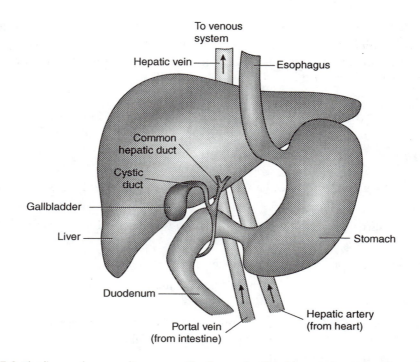

FIGURE 23.1 The liver and surrounding organs. The liver and gallbladder are shown, together with the cystic duct and the common hepatic duct, through which bile flows into the duodenum. The portal vein carries blood directly from the capillaries of the stomach, intestines, pancreas, and spleen to the liver; the hepatic artery carries blood from the heart to the liver, which feeds the organ itself. The capillaries from the portal vein and the hepatic artery interconnect within the organ, and blood leaves the liver via the hepatic vein, which leads to the venous system of blood vessels.

Diagnosis

Early diagnosis of hemochromatosis is important because early treatment is effective in preventing iron accumulation and organ damage. If diagnosed and treated early, patients may lead normal, healthy lives. Symptoms of liver dysfunction, diabetes mellitus, heart disease, arthritis, and hypogonadism suggest the diagnosis; however, evidence of iron overload is needed to confirm it. Transferrin (an iron-transport protein) levels, iron saturation, and ferritin (an iron-storage protein) levels have been used as a measure of iron stores, however liver biopsy is more definitive. The hepatic iron index of 4.2 measured in this patient was well above normal values. Normal individuals have HHIs <1, while 85–90% of those with hemochromatosis have HHIs >1.9. In addition, genetic testing is becoming an increasingly important diagnostic tool because a single mutation is responsible in most cases.

Treatment

Treatment for cirrhosis depends upon the cause and consists of halting any further injury to the liver, and treating any complications. In this case, excess absorption of iron is the cause, and as the body does not possess metabolic pathways for clearing the metal, it is physically removed by drawing blood on a regular basis. Most of the iron in the body is found in hemoglobin, and regular phlebotomy has proven to be safe and effective in treating the disease. A restricted protein diet is recommended to limit the production of ammonium ion, which can build up to dangerous levels in patients with liver dysfunction. Many symptoms ameliorate with a return to normal iron levels. However, because cirrhosis is irreversible, the patient's liver function will not return to normal, although it may improve.

QUESTIONS

1. What is cryptogenic cirrhosis and approximately what fraction of cases can be attributed to this type?

2. Patients with cirrhosis are often very sensitive to the effects of medications. Why might this be?

3. Why are cystic fibrosis patients at increased risk for cirrhosis?

4. Hemochromatosis is more often manifested in men than in women. What might be the reason?

5. The transferrin saturation (TS) is a common screening test for hemochromatosis, with levels greater than 45% suggesting disease. From the laboratory results provided in the case, calculate the TS in this patient.

6. Hepatic encephalopathy is often triggered by an event that causes liver damage, such as an alcoholic binge or an infection, or by consumption of a large amount of protein. Why might a large amount of protein cause hepatic encephalopathy?

7. How does administration of lactulose in the treatment of hepatic encephalopathy inhibit protein absorption by the intestine?

8. What causes edema in patients with cirrhosis?

9. The liver-function tests performed on this patient included measurement of levels of ALT, ASP, and alkaline phosphatase, as well as total and conjugated bilirubin. Why were so many different tests performed?

FURTHER READING

1. Feder, J. N., Gnirke, A., Thomas, W., et. al. A novel MHC class I-like gene is mutated in patients with hereditrary haemochromatosis. *Nature Genetics* (1996) 13:399–408. (This article describes the identification of the HFE gene as a factor in hereditary hemochromatosis.)

2. Gerber, T., and Schomerus, H. Hepatic encephalopathy in liver cirrhosis: pathogenesis, diagnosis and management. *Drugs* (2000) 60(6):1353–1370.

3. Hanson, E. H., Imperatore, G., and Burke, W. HFE gene and hereditary hemochromatosis. *American Journal of Epidemiology* (2001) 154(3):193–206.

4. Gochee, P. A., and Powell, L. W. What's new in hemochromatosis. *Curr. Opin. Hematol.* (2001) 8(2):98–104.

For further information, see the following web sites:

National Digestive Diseases Information Clearing House:
www.niddk.nih.gov/health/digest/pubs/hemochrom/hemochromatosis.htm

The American Gastroenterological Association, "Cirrhosis of the Liver":
www.gastro.org/public/cirrhosis.html

Online Mendelian Inheritance in Man, Hemochromatosis:
www3.ncbi.nlm.nih.gov/htbin-post/Omim/dispmim?235200

American Hemochromatosis Society: www.americanhs.org/

Centers for Disease Control, Iron Overload and Hemochromatosis:
www.cdc.gov/nccdphp/dnpa/hemochromatosis/index.htm

CHAPTER 24

Amino Acid Biosynthesis
A Vitamin a Day Keeps the Cardiologist at Bay

CASE HISTORY

Man Experiences Chest Pain and Summons Emergency Medical Service

A 39-year-old man with a history of hypertension and hypercholesterolemia was admitted to the emergency department complaining of chest pain. Two days prior to admission he experienced similar discomfort that was intense for approximately 10 minutes and then subsided over about an hour. On the day of admission, he complained of chest pain while drinking coffee and watching television at 6:30 a.m. The pain was described as intense squeezing pressure in his anterior chest, radiating to left axilla and shoulder and accompanied by diaphoresis and nausea but no vomiting, and shortness of breath but no palpitations. He waited for 15 minutes before summoning the emergency medical services, and he transferred to an outside hospital.

Does knowing the time of day help in making the diagnosis?

What is the significance of the character of the pain?

On arrival in the emergency department he was given a crushed aspirin and an electrocardiogram (ECG) was performed, which revealed inferior and anterior ST elevation. He was treated with nitroglycerin, heparin, eptifibatide, and supplemental oxygen by nasal cannulae, with resolution of the pain and ST-segment elevations after a total of 40 minutes.

What is the significance of ST segment elevation on the ECG?

Why the crushed aspirin?

What are heparin and eptifibatide and why were they administered?

He Had Several Risk Factors for Coronary Artery Disease

He is a married insurance adjuster and describes his job as stressful. He does not smoke, drinks alcohol only socially and denies use of illicit drugs. He has one younger and one older brother; the older brother had a myocardial infarction (heart attack) at age 45, and the younger brother had coronary angioplasty at age 42. His father died at age 56 of congestive heart failure after several myocardial infarctions. He has one paternal uncle and two paternal aunts with a history of angina pectoris. His mother is alive and has hypertension; she had two brothers, both smokers, who died in their mid-60s of "heart attacks."

What are the familial causes of premature atherosclerosis and coronary artery disease?

He denied any history of chest pain prior to the two days before this admission. He was sedentary but was able to engage in mild exercise (walking) without limitation. There has been no recent change in weight and he denied having orthopnea, paroxysmal nocturnal dyspnea, or pedal edema. There was no history of claudication, stroke, transient ischemic attack, and no personal or family history of diabetes mellitus. There was no history of headaches or cold or discolored extremities. His only medication was atenolol, 50 mg daily, for high blood pressure.

What are orthopnea and paroxysmal noctural dyspnea, and what do they imply?

Why was he asked about headaches and discoloration of the extremities?

Physical Examination

The physical examination revealed a well-appearing man, looking his stated age, and in no acute distress. The blood pressure was 129/81 mmHg, and pulse was 67 beats per minute (bpm). The head, eyes, ears, nose, and throat examination revealed mild arteriolar narrowing with no arteriolar-venous crossing changes and no other abnormalities. The lungs were clear to auscultation and percussion. His cardiac examination revealed a normally sized and positioned point of maximal impulse (PMI), normal first and second heart sounds, and a fourth heart sound with no pathological murmurs or pericardial friction rub. The abdomen was soft and nontender with no masses, and there were active bowel sounds. He had no peripheral edema, and the pulses in the legs were symmetrical and of normal volume.

Why is examination of the fundus of the eye useful in patients with suspected atherosclerotic cardiovascular disease?

What is the significance of a pericardial friction rub in a patient with a myocardial infarction?

What are the physical signs of heart failure?

Laboratory Evaluation

The laboratory examination revealed normal serum electrolytes, a blood urea nitrogen (BUN) of 30 mg/dl, and a creatinine of 0.9 mg/dl. The complete blood count was normal. The initial creatine kinase (CK) was 330 IU/l with 13% MB (myocardial specific) fraction, and it peaked at 1073 IU/l with 23% MB fraction. The initial serum troponin was 0.27 ng/ml (upper limit of normal [ULN]: .03 ng/ml), the total cholesterol was 242 mg/dl (ULN: 200 mg/dl), triglycerides 252mg/dl (ULN: 190 mg/dl), HDL 46 mg/dl, LDL 166 mg/dl (ULN: 160 mg/dl). A serum homocysteine level was 52.4 µmol/L (ULN: 15 µmol/L). Chest radiogram was normal, and the ECG at this hospital revealed nonspecific ST- and T-wave abnormalities.

What is CK and what does an increase in the serum signify?

What is the importance of the elevation in the level of homocysteine?

What are HDL ("good") and LDL ("bad") cholesterol, and what do the levels in this patient suggest?

Coronary Angiography Localizes Site of Injury

The patient underwent a cardiac catheterization, and coronary angiography revealed a normal left main coronary artery, a large caliber left anterior descending (LAD) artery with a large caliber first LAD diagonal branch with a mid-60% stenosis. The LAD had a 90% stenosis after the takeoff of the diagonal branch (Figure 24.1). The more distal LAD had mild irregularities, but no significant stenoses. The left circumflex artery gave rise to a single circumflex marginal branch with no significant disease. The right coronary artery was a large-caliber dominant vessel with a 50% stenosis in its mid portion. The ventriculogram demonstrated anteroapical akinesis (no contraction) with an overall ejection fraction of 45% (normal ≥ 55%). A percutaneous coronary intervention (PCI) was performed with deployment of a 3.0 × 8.0–mm stent, eliminating the 90% stenosis of the LAD (see Figure 24.1). The patient then was given aspirin, eptifibatide, clopidogrel, and transferred back to the CCU for monitoring.

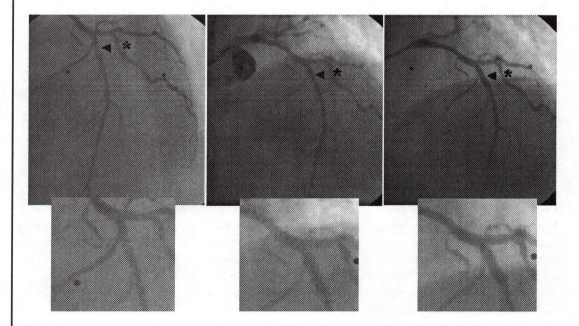

FIGURE 24.1 Stenosis of the left anterior descending (LAD) artery (arrowhead) and a diagonal branch (asterisk), as visualized by coronary angiography (left panel). In the far right panel a stent has been deployed in the LAD, eliminating the stenosis in this artery. In the middle panel the guide wire used to deploy the stent has not yet been removed and may be seen. The insets show magnified views of the area of stenosis and stent development.

What is PCI? What is a stent, and what does it do?

What degree of stenosis of a coronary artery is flow limiting?

Discharge and Treatment

The patient was discharged four days later on the following medicines: aspirin, 325 mg daily; clopidogrel, 75 mg daily for 30 days; metoprolol, 75 mg twice daily; atorvastatin, 20 mg daily; and valsartan, 160 mg daily. In view of the elevation in the serum homocysteine, he was treated with vitamins B_6, B_{12}, and folate supplements.

What is the rationale behind treatment with vitamins?

DISCUSSION

Coronary Obstruction and Myocardial Infarction

Interruption of blood flow to the myocardium by obstruction of one or more of the coronary arteries may result in necrosis of the tissue supplied by that artery, or *myocardial infarction* (MI). Despite some debate over the past few decades, it is now generally agreed that thrombosis at the site of rupture of an unstable atherosclerotic plaque obstructing a coronary artery is the cause of MI. If the obstruction is not relieved promptly, tissue necrosis results; the longer the obstruction exists, the less myocardium will be salvaged. This notion that "time is tissue" has prompted extraordinary efforts to initiate therapy of myocardial infarction as soon as the diagnosis is made.

Incidence of Coronary Artery Disease

Although unadjusted mortality rates have declined in the past four decades, myocardial infarction remains the single most common cause of death in Western industrialized countries. It is estimated that over 12 million Americans have coronary heart disease and that there are over one million coronary events annually, of which 20% are fatal. It is clear that the economic burden of coronary artery disease (CAD) is enormous.

Signs, Symptoms, and Laboratory Findings in MI

Chest pain in early morning typical

Pressure-like chest discomfort is a classical symptom of compromise in the blood flow to the heart; it is either reversible and called angina pectoris, or is persistent and results in tissue necrosis or MI. The pain with myocardial infarction typically does not occur with exertion, but there is a circadian pattern to the incidence of MI, with a peak in the early morning coincident with peaks in serum adrenocorticotropic hormone (ACTH) and catecholamine levels. In some patients, especially diabetics, chest pain may not occur, perhaps because of autonomic and sensory neuropathy.

Other symptoms associated with chest pain may include diaphoresis, nausea, dizziness, and shortness of breath, and may result from a redistribution of blood flow or compromise in the performance of the heart as a result of the infarction. In as many as 25% of myocardial infarctions the initial presentation may be sudden death, most likely due to the development of a malignant ventricular arrhythmia such as ventricular fibrillation (rapid abnormal electrical activity).

Laboratory results and imaging studies

The diagnosis of myocardial infarction is typically made on the basis of changes in the electrocardiogram and the presence of enzymes such as CK from myocardial cells in the serum, indicating myocardial necrosis. CK levels begin to rise four hours after cardiac injury, peak after 18–24 hours, and return to normal within a week. There are differing isoforms of CK, and thus the myocardial specific fraction (%MB) is used to verify that the source of elevated CK is the heart. Troponin T or I (cardiac muscle proteins) levels are more sensitive indicators of myocardial necrosis than CK and are used to support the diagnosis of MI. Imaging studies such as the coronary angiography carried out on this patient define the anatomical extent of CAD and can pinpoint the location of the obstruction(s) in the coronary circulation and determine what type of therapy (medicines, PCI, or coronary bypass surgery) is most appropriate.

Management of MI and Other Acute Coronary Syndromes (ACS)

Pharmacologic therapies

The key in the management of MI is restoration of myocardial blood flow as soon as possible. The means for reestablishing blood flow is either pharmacological or mechanical. Thrombolytics or "clot busting" drugs are a mainstay of therapy in acute MI.

These drugs (such as tissue plasminogen activator and anistreplase) activate the natural fibrinolytic system by converting the proenzyme plasminogen to its active form, *plasmin*, with subsequent dissolution of the clot. Other pharmacologic agents target platelet function (aspirin, glycoprotein IIb-IIIa inhibitors) or inhibit thrombin (heparin), and are used to hasten and/or sustain coronary patency.

Mechanical interventions

The other major method of reestablishing coronary blood flow is by mechanical opening of the blood vessel by percutaneous coronary angioplasty (PTCA), atherectomy, and/or deployment of a stent. Collectively these methods are referred to as PCI. In centers with trained coronary interventionalists and teams that can be quickly mobilized, this may be the best method of quickly opening coronary blood vessels.

The Association of Elevated Homocysteine with Vascular Disease Is Well Established

Homocysteine is a sulfur-containing amino acid that is not incorporated into protein but is an important intermediary in the metabolism of methionine and cysteine (see Sections 24.2.7–24.2.9 of *Biochemistry* 5e). The association of vascular disease with elevated plasma levels of homocysteine is well documented in children with homocystinuria due to cystathionine β-synthase deficiency. In addition, a role for homocysteine in the pathogenesis of more common atherosclerosis was suspected on the basis of greater elevations in serum homocysteine levels after a challenge with oral methionine in patients with CAD compared to controls. Since the initial observation, extensive epidemiological data in over 20 published studies has confirmed the association of hyperhomocyteinemia and atherosclerosis. A meta-analysis of over 4000 patients showed a graded risk for atherosclerosis with increasing plasma homocysteine levels.

Etiology of Hyperhomocysteinemias

Hyperhomocysteinemia is a condition with distinct causes and clinical features that predispose patients to atherosclerosis and thrombosis. Hyperhomocysteinemia can result from an enzyme deficiency, the most common being cystathionine β-synthase deficiency; however, a number of other enzyme deficiencies, as well as nutritional deficiencies, various drug therapies, and diseases are also associated with the disorder. Homocysteine lies at the intersection of two metabolic pathways (Figure 24.2), and inhibition of either one can result in the build-up of this amino acid to dangerous levels. Mild hyperhomocysteinemia occurs in 5–7% of the population and is asymptomatic until the third to fourth decade of life, when premature CAD and arterial and venous thrombosis occur.

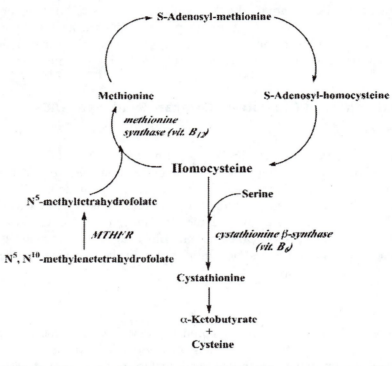

FIGURE 24.2 Metabolic pathways involving homocysteine. Homocysteine is a component of the activated methyl cycle that produces methionine, and is a substrate in the condensation reaction that generates cystathionine, a cysteine precursor. MTHFR: N^5,N^{10}-methylenetetrahydrofolate reductase.

Cystathionine β-synthase deficiency

Cystathionine β-synthase deficiency is the most common genetic abnormality but is nevertheless extremely rare, with the incidence of homozygous deficiency being ~1/200,000. The consequences of homozygous deficiency are protean and are associated with defects in the eye, skeleton, central nervous system, and, most prominently, thromboembolism and premature atherosclerosis. Heterozygotes are less severely affected.

N^5,N^{10}-methylenetetrahydrofolate reductase deficiency

Deficiencies of N^5,N^{10}-methylenetetrahydrofolate reductase (MTHFR) produce a range of symptoms including severe atherogenic predisposition and neurological dysfunction. MTHFR catalyzes the conversion of N^5,N^{10}-methylenetetrahydrofolate to N^5-methyltetrahydrofolate, a cofactor in the generation of methionine from homocysteine (see Section 24.2.7 of *Biochemistry* 5e). Those with this deficiency have hyperhomocysteinemia and methionine deficiency, and both these abnormalities are thought to contribute to the pathogenesis of the disorder.

Nutritional deficiencies and drug therapies

Nutritional deficiencies such as deficiencies in folate, vitamin B_6 (pyridoxal phosphate), and vitamin B_{12} (cobalamin), are a more frequent cause of hyperhomocysteinemia. Vitamin B_{12} and folate are cofactors in the reaction catalyzed by methionine synthase, whereas pyridoxal phosphate is necessary for the activity of cystathionine β-synthase (see Figure 24.2). Vitamin supplementation can normalize homocysteine levels, but it is not clear whether it reduces the risk of clinically significant CAD. Drugs such as methotrexate, likely through its depletion of folate, and the anticonvulsant diphenylhydatoin (dilantin), through interference with folate metabolism, increase homocysteine levels. Theophylline and other phosphodiesterase inhibitors interfere with pyridoxal phosphate synthesis, and smoking lowers its level; both are associated with hyperhomocysteinemia.

Diseases and disorders associated with hyperhomocysteinemia

Kidney disease may be associated with increased plasma homocysteine levels; hypothyroidism, pernicious anemia, and carcinoma of the breast, pancreas, and ovary have also been associated with hyperhomocysteinemia.

Formation of the Atheromatous Plaque: Role of Homocysteine

The development of an atherosclerotic plaque

The development of an atherosclerotic plaque is the result of a complex sequence of events involving the vascular endothelium, smooth muscle, and formed elements of the blood, and it is orchestrated by a number of signaling molecules. A prominent feature in the birth of an atheroma is endothelial dysfunction, perhaps mediated by lipid and lipid oxidation products, and followed by the recruitment and accumulation of white blood cells. Maturation of the atherosclerotic plaque involves migration, and perhaps dedifferentiation, of vascular smooth muscle cells.

The majority of the volume of an atherosclerotic plaque consists of extracellular matrix components, such as collagens (types I and III), proteoglycans, and elastin, rather than cellular elements, and molecules such as bone morphogenic protein may contribute to the well-known calcification, or mineralization, of the plaque. The process of evolution of an atherosclerotic plaque occurs over decades, during which time the patient is asymptomatic. With extensive remodeling and increasing size of the plaque, the lumen of the blood vessel becomes compromised, limiting blood flow. Stenoses of the coronary arteries of 60–70% are flow-limiting under circumstances of increased demand.

Risk factors for atherosclerosis

A number of factors are associated with increased risk of atherosclerosis and clinically significant CAD. The traditional risk factors include a familial predisposition, increase in serum lipids (cholesterol and triglycerides), hypertension, diabetes, and smoking. More recently, homocysteine has been found to be associated with increased risk of cardiovascular disease and hence is considered a nontraditional risk factor. This patient, with elevated cholesterol and triglycerides, a familial predisposition, and elevated homocysteine levels, thus had a number of risk factors for cardiovascular disease.

Mechanism of homocysteine-induced atherosclerosis

Homocysteine has a number of pro-atherogenic actions (Table 24.1), including the promotion of endothelial dysfunction, increasing thrombogenicity and pro-elastase activity, the promotion of collagen deposition, and the inhibition of nitric oxide production and action. It has been suggested that these effects all derive from a common pathway: the production of deleterious *reactive oxygen species (ROS)*. Homocysteine is autooxidized in the blood to homocystine, homocysteine thiolactone, and a group of mixed disulfides, with the elaboration of ROS, including superoxide, hydrogen peroxide, and hydroxyl radicals. ROS may directly injure the endothelium, exposing the underlying matrix and smooth muscle cells, with the consequent activation of platelets and white blood cells. Homocysteine alters the antithrombotic character of the endothelium in a number of other ways, including alteration in the expression of coagulation factors and antithrombin molecules. It may also initiate an inflammatory response, perhaps by activation of the redox-sensitive transcription factor, NF-κB.

Management of Elevated Homocysteine Levels

Modification of homocysteine levels is straightforward in many cases: administration of vitamin cofactors (B_6, B_{12}, folate) involved in the metabolism of this amino acid lower the level in many patients. However, although studies are currently ongoing, a reduced risk of cardiovascular disease as a result of vitamin supplementation has yet to be definitively established. Nevertheless, homocysteine is an attractive target for risk factor modification in patients predisposed to CAD.

TABLE 24.1 Atherogenic effects of homocysteine or its oxidation by-products

Effect	Possible mediators
Endothelial injury	ROS (H_2O_2, O_2^-, OH^-) impaired production of endothelial nitric oxide
Intimal injury	ROS (H_2O_2, O_2^-, OH^-), lipid oxidation
Derangement of matrix	elastase activation, deposition of sulfated glycosaminoglycans (homocysteine thiolactone)
Lipid peroxidation	O_2^-, OH^-
Foam cell production	homocysteine, thiolactone-induced aggregation of LDL
Increased thrombogenicity	increased Factor XII and V activity, inhibition of thrombomodulin and heparan sulfate expression by endothelial cells, increased tissue factor expression
VSMC mitogen production	increased cyclins D1 and A, increased NOS through activation and nuclear translocalization of NFκB
Inflammation	initiated by activation and nuclear translocalization of NFkB, with increased TNF-α; increased expression of receptor for advanced glycation products (RAGE) and its ligand (EN-RAGE); increased VCAM-1, tissue factor, and matrix metalloprotein-9 expression

ROS: reactive oxygen species; VSMC: vascular smooth muscle cell; VCAM-1: vascular cell-adhesion molecule 1; NOS: nitric oxide synthase; TNF-α: tumor necrosis factor α; LDL: low-density lipoprotein

QUESTIONS

1. How might vitamin B_6 and B_{12} supplementation combat atherosclerotic cardiovascular disease?

2. Homocysteine is a substrate in the biosynthesis of both cysteine and methionine. Which of these pathways is referred to as the *transsulfuration* pathway, and which is referred to as the *remethylation* pathway?

3. Betaine, a product of choline metabolism, has been used to reduce homocysteine levels in patients with hyperhomocysteinemia. What is the biochemical basis for the effect?

4. Folic acid supplementation in pregnancy has been found to reduce the risk of neural tube defects in infants, and experiments on avian embryos have shown that exposure to elevated homocysteine levels results in increased incidence of neural tube defects. Given these two observations, what might be the role of folic acid in decreasing the risk of neural tube defects?

5. The *methionine challenge test* may be administered to detect defects in homocysteine metabolism. Describe this test and its biochemical basis.

6. This patient had an elevated BUN of 30 mg/dl (ULN: 20 mg/dl). What does this indicate, and is it related to hyperhomocysteinemia?

FURTHER READING

1. Boushey, et al. A quantitative assessment of plasma homocysteine as a risk factor for vascular disease. *JAMA* (1995) 274:1049–1057. (This article describes an epidemiological analysis of the link between homocysteine and atherosclerosis.)

2. Ross, R. Atherosclerosis as an inflammatory disease. *New Engl. J. Med.* (1999) 340:115–126.

3. Hajjar, K. Homocysteine: a sulph'rous fire. *J. Clin. Invest.* (2001) 107:663–664.

4. Welch, G. N., and Loscalzo, J. Homocysteine and atherothrombosis. *New Engl. J. Med.* (1998) 338:1042–1050.

For further information, see the following web sites:

Center for Disease Control Morbidity and Mortality Weekly Reports
http://www.cdc.gov/mmwr/

American Heart Association http://www.americanheart.org/
presenter.jhtml?identifier=1200000

National Institutes of Health: National Heart Lung and Blood Institute
http://www.nhlbi.nih.gov/health/public/heart/other/homocyst.txt

CHAPTER 25

Nucleotide Biosynthesis
A Target for the Treatment of "Evil Air"

CASE HISTORY

A Congolese Man Is Admitted for Fever and Abdominal Pain

A 28-year-old man, a native of Kisangani in the Democratic Republic of the Congo, was admitted to the hospital because of fever and chills. Two years before admission he immigrated to Great Britain, where he was employed as a waiter. He reported a history of *Giardia lamblia* and *Entamoeba histolytica*, for which he had been treated with metronidazole, and malaria, for which he had been treated with chloroquine.

He subsequently immigrated to the United States to attend college. Upon immigration he had a positive tuberculin skin test, and isoniazid was administered for six months.

What is the importance of his country of origin?

What is chloroquine, and what is the basis for its anti-malarial action?

He was admitted to this hospital for fever and pain in the left upper-abdominal quadrant. The pain had been increasing in severity over the past two weeks. He denied nausea, vomiting, or diarrhea, but had not eaten much because of abdominal discomfort. There was no history of headache, seizures, hematemesis, melena, hematochezia, receipt of blood products, or extramarital sex. He had never been tested for hepatitis viruses or for antibodies to human immunodeficiency virus (HIV). The patient was married, with one child, and both his wife and son were healthy. He denied use of tobacco or alcohol, and did not consume pork. There was no prior history of occupational exposures to toxins.

Physical Examination

On admission he was acutely, but did not appear chronically, ill, and his temperature was 38.9°C, blood pressure 118/60 mm Hg, heart rate 108 beats per minute (bpm), and respiratory rate 26 per minute. He had no rash, and shotty anterior cervical lymph nodes were palpated. His neck was supple, the lungs were clear, and his heart was normal. The remainder of the physical examination was remarkable for tender enlargement of the spleen (splenomegaly), which was palpated 4 cm below the left costal margin. The liver was not palpated. The bowel sounds were active and he had diffuse abdominal tenderness without guarding or rebound, and no costovertebral-angle tenderness. There was no digital clubbing, cyanosis, or edema. Rectal examination revealed no masses, and a stool specimen was negative for occult blood. Neurological examination revealed no focal deficits.

What are possible causes of splenomegaly?

Laboratory Results

The results of blood tests were: hematocrit 36%; mean corpuscular volume (MCV) 89 fl; reticulocyte count 1.9%; erthyrocyte sedimentation rate (ESR) 18 mm/hr; white blood cell count (WBC) 5600 per mm^3, with a differential count of 72% polymorphonuclear cells, 6% band forms, 10% lymphocytes, and 12% monocytes, and a platelet count of 98,000 per mm^3. The prothrombin time (PT) and partial thromboplastin time (PTT) were normal. Blood chemistry studies revealed normal blood urea nitrogen, creatinine, amylase aspartate transaminase, and lactate dehydrogenase concentrations, but an elevated total bilirubin of 4.2 mg/dl with a conjugated bilirubin of 1.3 mg/dl. A test for glucose-6-phosphate dehydrogenase deficiency was negative. The urine was yellow with a pH of 7.0 and a specific gravity of 1.006, and was +++ positive for urobilinogen; the sediment contained two white cells and two red cells per high-power field. Blood cultures were negative and a test for heterophil agglutinins was negative.

What are the causes of an elevated serum bilirubin? Does the level of conjugated bilirubin help in the differential diagnosis?

What does the slightly elevated reticulocyte count indicate?

Do the blood tests suggest an acute bacterial infection? Why might an acute bacterial infection be suspected?

Thin and thick Giemsa-stained smears of the blood were examined for malarial parasites and none were found, and the red blood cells were morphologically normal. Cultures of the stool and examination for ova and parasites were negative. Examination of stained specimens of urine and sputum for acid-fast bacilli was negative and cultures for mycobacteria were sent. The ratio of CD4+ to CD8+ T lymphocytes was increased, and a test for HIV antibodies was negative. Serum protein electrophoresis revealed an increase in IgG 1600 mg/dL (upper limits of normal 1350 mg/dl) and IgM 360 mg/dl (upper limits of normal 260 mg/dl), but normal IgA levels.

Does the absence of malaria parasites in these blood specimens definitively exclude the diagnosis of malaria? Why are both thin and thick smears examined?

CT Scan Confirms Splenomegaly and Reveals Slight Hepatomegaly

Radiographs of the chest showed scarring in the right upper lobe without other significant abnormalities. A computed tomographic (CT) scan of the abdomen after oral and intravenous contrast revealed enlargement of the spleen extending to the lower pole of the left kidney, without focal defects or calcification, slight hepatomegaly, and no intrabdominal or pelvic lymphadenopathy. There was no evidence of free fluid within the abdomen or pelvis. There was no evidence for a focal fluid collection suggestive of an abscess.

Diagnosis and Treatment

On the third hospital day a Giemsa-stained thick smear of blood revealed mildly enlarged red blood cells with late-stage asexual malarial parasites (Figure 25.1). A diagnosis of falciparum malaria with tropical splenomegaly, the result of chronic infection, was made. The diagnosis of *P. falciparum* infection was confirmed with an enzyme-linked immunosorbent assay (ELISA) for detection of a *P. falciparum* antigen (histidine-rich protein 2). He was treated for presumed chloroquine resistant falciparum malaria and received Malarone (atovaquone 250 mg, proguanil hydrochloride 100 mg), four tablets daily for three days. Therapy was associated with resolution of his fevers and dramatic improvement in the abdominal pain.

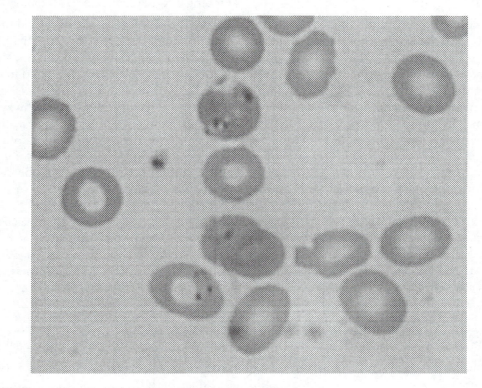

FIGURE 25.1 Thick smear of the blood containing malarial parasites.

What is malaria and how is it typically diagnosed?

What is the treatment for malaria and what are the therapeutic targets?

DISCUSSION

Malaria Is an Age-Old Disease

Malaria, a name that is derived from the Italian for *evil air,* is an ancient, protozoan disease transmitted by the *Anopheles* mosquito. References to what is presumed to have been malaria date back to China in the *Nei Ching Canon of Medicine* in 1700 BC, and to Egypt in the *Ebers Papyrus* in 1570 BC. Hippocrates, a Greek physician living in the fourth and fifth centuries BC, clearly recognized the syndrome of malaria and its relationship to marshes. The long co-evolution of parasite and host has led to the development of unique biochemical pathways in the parasite and, conversely, to the loss of pathways in the parasite when a similar pathway in the host could be exploited. These differences in biochemical pathways between host and parasite have been exploited in treating the disease: pathways unique to the parasite may be targeted without harming the host. Even pathways common to host and parasite may be targeted when the host possesses an alternative pathway, or when there is sufficient specificity for the parasite pathway.

Many Drugs Target Differences Between Host and Parasite Biochemical Pathways

To the biochemical chemotherapist, it is not only a matter of faith, but an obvious fact, that every cell type must have a characteristic biochemical pattern, and therefore be susceptible to attack at some locus or loci critical for its survival and replication.
—G. H. Hitchings, Nobel Laureate in Medicine, 1988

Quinine and its derivatives (including chloroquine) are the oldest class of drugs used to treat malaria, and are thought to inhibit pathways unique to the parasite for the detoxification of heme, a byproduct of parasite hemoglobin digestion. Differences between the nucleotide biosynthetic pathways utilized by the parasite and the host have also been exploited in treating the disease (see Table 25.1). For example, unlike the mammalian host, the parasite cannot synthesize pyrimidines via a salvage pathway, but only via the de novo pathway (see Section 25.1 of *Biochemistry* 5e), and effective drug treatments targeting de novo synthesis have been developed. Similarly, inhibition of de novo synthesis of folate, a pathway not found in humans, has proven an effective treatment strategy. Although the quinine derivatives remain a major class of antimalarial therapeutic agents, inhibitors of folate metabolism also constitute one of the major classes of antimalarial drugs.

TABLE 25.1 Comparison of nucleic acid biosynthetic pathways in *Plasmodia* and mammals. While mammals can synthesize pyrimidines and purines via de novo or salvage pathways, *Plasmodia* can do only one or the other. Mammals cannot synthesize folate and thus rely upon dietary sources, whereas *Plasmodia* synthesizes its own folate. Antifolate drugs, especially, have been used extensively to combat malaria, but the usefulness of many is now limited because of the emergence of drug-resistant strains.

Biosynthetic pathway	*Plasmodia*	Mammals
Pyrimidines	de novo, no salvage	both de novo and salvage
Purines	salvage, no de novo	both de novo and salvage
Folate	both de novo and salvage	no biosynthetic pathway

Epidemiology

Up to 500 million cases of malaria are estimated to occur per year worldwide, and the disease is responsible for between 1.5 and 3.5 million deaths annually. It is most common in tropical and subtropical regions of Sub-Saharan Africa, South and Central America, the Caribbean, the Middle East, the Indian subcontinent, southeast Asia, and Oceania. Within these regions the disease is most often found in low-lying areas with high annual rainfall, where the *Anopheles* mosquito breeds. Approximately 90% of fatalities occur in Sub-Saharan Africa, and victims are most often children under the age of five and pregnant women. Although sporadic cases still occur occasionally in North America, the disease has largely been suppressed thanks to programs implemented in the mid-1900s.

Drug resistance is on the rise

Drug-resistant strains have become extremely widespread, with resistance to all commonly used drugs having been reported. Resistance to chloroquine, one of the previously most effective and commonly used drugs, has reduced its efficacy in some of the most seriously affected areas, including the Democratic Republic of the Congo, from where this patient emigrated.

Malaria Life Cycle

The malaria parasite is of the genus *Plasmodium,* and four species infect humans—*P. vivax, P. malariae, P. ovale,* and *P. falciparum*; however, *P. falciparum* is the most virulent. The parasite is transmitted to the human host by the *Anopheles* mosquito via injection of infected saliva into the blood stream (Figure 25.2). The parasite first invades and replicates within a liver cell, which eventually lyses, releasing multitudes of parasites into the bloodstream. This first stage is clinically silent and lasts about two weeks.

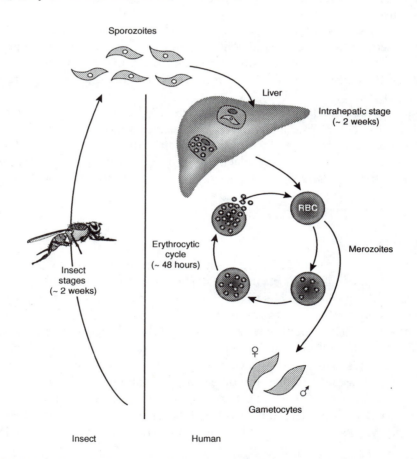

FIGURE 25.2 Malaria life cycle. *Sporozoites* (thread shaped) are injected into the human bloodstream by a feeding mosquito. They invade the liver and replicate there, forming thousands of *merozoites* (round shaped), which are released into the bloodstream when the cells lyse. The merozoites then invade red blood cells (RBC), forming *trophozoites,* which will then begin to replicate, forming multitudes of merozoites. Upon cell lysis these go on to infect other red blood cells, and this erythrocytic cycle is repeated, accounting for the cyclic nature of the fevers associated with the disease. In an alternative developmental pathway, merozoites form *gametocytes,* which upon ingestion by a feeding *Anopheles* mosquito, leads to the sexual stage of the cycle, which ultimately results in the colonization of the insect's salivary glands by sporozoites and a return to the beginning of the cycle.

The parasite then enters and replicates within red blood cells (RBCs or *erythrocytes*), feeding upon the hemoglobin found there. The cells eventually lyse, releasing parasites into the bloodstream, which are then free to infect other RBCs. This erythrocytic cycle occurs repeatedly, and is often synchronized, with the RBCs lysing simultaneously in a periodic fashion. This accounts for the characteristic cyclic nature of the fevers that occur during this stage. Ingestion of blood from an infected individual by a mosquito leads to the completion of the cycle. The importance of the *Anopheles* mosquito in the maintenance of the cycle has been clearly recognized by public health organizations, and programs to eradicate it have been prominent in attempts to control the disease.

Diagnosis

The standard for diagnosis of malaria is direct, microscopic observation of the parasites in a blood sample. Both thick and thin smears are usually examined; the greater number of cells in a thick smear increases the sensitivity of detection, but the *Plasmodium* species is more easily discerned in a thin smear. In many cases malaria parasites are not found in peripheral blood smears of patients with tropical splenomegaly, with which this patient was diagnosed. In this case, despite failure to find parasites in the first samples taken from the patient, a subsequent thick smear revealed the presence of the pathogen. The diagnosis was confirmed by the ELISA test for an antigen, histidine-rich protein-2, specific for *P. falciparum*.

In most areas where the disease occurs this standard of diagnosis cannot be achieved due to the lack of medical equipment and trained personnel, and medical workers must instead rely upon the symptoms experienced by the patient to make the diagnosis. Characteristic signs are fever, muscle and joint pain, and headache. In practice, in many regions where the disease is endemic, patients presenting with a fever with no other apparent cause are treated for malaria.

Pathophysiology

During the erythrocytic stage of the life cycle the parasite multiplies very rapidly, sometimes occupying up to 30% of the host's erythrocytes. The lysis of infected erythrocytes releases toxic substances and parasite antigens that induce inflammation, accounting for the fevers. The extensive hemolysis that occurs results in anemia and elevated levels of bilirubin, a metabolite of hemoglobin. Splenomegaly, enlargement of the spleen, is common as this organ plays a critical role in the cell-mediated immune response to infections, and in the clearance of damaged erythrocytes. This patient, having presented with fever, anemia, elevated bilirubin levels, and splenomegaly, thus had multiple symptoms and signs of malaria.

In most cases malaria is not fatal and patients recover after an illness lasting between 10 and 20 days; however, 1–3% of infected individuals die of the disease. Life-threatening complications of malaria include hypoglycemia, lactic acidosis, vital organ dysfunction, and cerebral malaria.

Tropical Splenomegaly (or Hyperreactive Malarial Splenomegaly)

Tropical splenomegaly results from an abnormal response to chronic malaria infection. It occurs in areas where malaria is endemic, especially tropical Africa. Via mechanisms still unclear, repeated malaria infection induces the production of cytotoxic IgM antibodies to suppressor T cells (CD8+). This results in an increase in the ratio of CD4+ to CD8+ cells, and to B cell overproduction of IgM antibodies, which have a tendency to aggregate into *cryoglobulins*. The mobilization of the reticuloendothelial system to clear these aggregates results in severe splenomegaly.

The sharp pains experienced by this patient are typical of the syndrome and are probably due to perisplenitis, splenic infarcts, or splenic distension. Elevation of antibodies directed toward malarial antigens, reflected in the increased IgG and IgM levels in this patient, are also typical. The increased reticulocyte count is the result of increased turnover of red blood cells due to malarial parasite induced hemolysis. Some level of pancytopenia (overall reduction in blood cell levels) is also usually observed as a result of hypersplenism (overactive spleen).

Treatment

Choice of drugs depends upon several factors

Drugs may be taken to *prevent* malaria, or to *treat* either severe or uncomplicated malaria. Severe malaria, when the percentage of infected erythrocytes is high and vital organ dysfunction may occur, is life-threatening and is treated with intravenous administration of quinine or quinidine. These drugs may cause disturbances of the heart rhythm and are thus primarily used when treating severe malaria. A variety of drugs exist to treat uncomplicated malaria, or to prevent the disease altogether in travelers to endemic areas; the choice depends upon drug availability and the drug-resistance pattern in the area where the parasite was, or may be, acquired.

Malarone inhibits parasite nucleotide biosynthesis

The drug used to treat this patient, Malarone, is a combination of two very effective antimalarials, *atovaquone* and *proguanil hydrochloride* (proguanil HCl) (Figure 25.3). Given the patient's country of origin, chloroquine-resistance was suspected, and as his condition was not life-threatening, quinine and quinidine were avoided. Atovaquone and proguanil HCl act synergistically to inhibit pyrimidine biosynthesis, and the combination has proven very effective, curing >98% of cases with little evidence of the emergence of resistance.

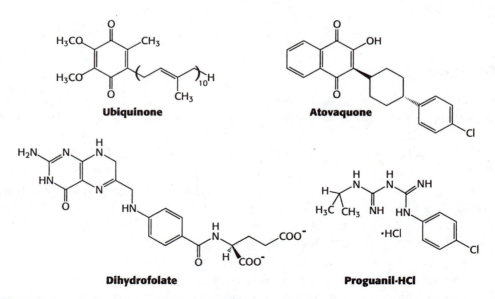

FIGURE 25.3 Chemical structures of the antimalarial drugs atovaquone and proguanil HCl. The inhibitory mechanisms are based on structural similarities between the drugs and their natural counterparts—ubiquinone and dihydrofolate.

Mechanism of atovaquone

Although the exact mechanism for antimalarial action has not yet been elucidated, atovaquone is known to inhibit the mitochondrial electron transport chain, thereby inhibiting de novo pyrimidine synthesis. The parasite mitochondrial enzyme, dihydroorotate dehydrogenase (DHODase), oxidizes dihydroorotate to orotate, a step in the biosynthetic pathway (see Section 25.1.4 of *Biochemistry* 5e). This oxidation reaction is a major source of electrons for the electron-transport chain, and thus inhibiting transport results in reduced production of orotate, and hence reduced de novo pyrimidine biosynthesis. Atovaquone structurally resembles ubiquinone and inhibits the electron transport chain at complex III (see Section 18.3 of *Biochemistry* 5e). Structural differences between mammalian and parasite cytochrome *b*, the component of complex III bound by the drug, may account for the antimalarial selectivity of the drug. Unlike mammalian cells, the malaria parasite does not depend upon the electron-transport chain for energy production, but instead derives most of its ATP from glycolysis (see Chapter 16 of *Biochemistry* 5e).

Mechanism of proguanil hydrochloride

Proguanil hydrochloride—via its metabolite, cycloguanil—selectively inhibits the parasite *dihydrofolate reductase (DHFR)*, and thus interferes with thymidylate synthesis (see Section 25.3.2 of *Biochemistry* 5e), as well as the biosynthesis of purine nucleotides and certain amino acids. Thus, atovaquone and cycloguanil act via independent mechanisms to inhibit pyrimidine biosynthesis. The rapidly dividing parasites of the erythrocytic stage, which depend upon large quantities of nucleotides for DNA replication and RNA synthesis, are thus highly susceptible to Malarone.

Interestingly, some evidence suggests that the unmetabolized proguanil HCl may have antimalarial activity distinct from that of its metabolite, cycloguanil. Proguanil HCl is only partially metabolized to cycloguanil by liver cytochrome P450 isozymes, and thus both forms of the drug are found in the patient. Proguanil HCl may contribute to the collapse of the proton gradient across the mitochondrial inner membrane, thus enhancing the effect of atovaquone. Many metabolic processes depend upon the gradient, and thus its collapse would be expected to have multiple deleterious consequences for the parasite.

QUESTIONS

1. Individuals from countries where malaria has been suppressed or eradicated are especially vulnerable to the disease when visiting areas where malaria is endemic. For example, malaria was the most common cause for hospitalization among American troops sent to Somalia in 1993, and the second most common cause (after combat-related injuries) among American troops in Vietnam. How can this be explained?

2. Antisense nucleic acids to parasite hypoxanthine-guanine phosphoribosyltransferase (HGPRT) mRNA, an enzyme in the purine salvage pathway (see Section 25.2.1 of *Biochemistry* 5e), have been shown to kill parasites in vitro. What explains the sensitivity of the parasite to suppression of this enzyme?

3. The drug sulfadoxine has been used to treat malaria in combination with pyrimethamine. Sulfadoxine competitively inhibits dihydropteroate synthase, an enzyme in the parasite's folate biosynthetic pathway. Why does this drug *not* affect human folate metabolism?

4. This patient was tested for glucose-6-phosphate dehydrogenase deficiency. What is the significance of this?

5. A vaccine would be an invaluable tool in the fight for global eradication of malaria, especially in the face of increasing drug-resistance. Although significant effort has been invested in this cause, it has not yet yielded an effective vaccine. What are some of the properties of the malaria parasite that have most hampered the development of an effective vaccine?

6. The malaria parasite can synthesize isopentenyl phosphate, a precursor of cholesterol, via the 1-deoxy-D-xylulose 5-phosphate (DOXP) pathway, whereas animal cells possess another pathway, the mevanolate pathway (see Section 26.2.1 of *Biochemistry* 5e), for the synthesis of this intermediate. The drugs, fosmidomycin and FR900098, which inhibit DOXP reductoisomerase, an enzyme in the DOXP pathway, have been shown to inhibit the growth of *P. falciparum* in culture, and to cure mice infected with a related parasite. What can you infer from these data regarding the pathways that exist in parasite and host for isopentenyl phosphate biosynthesis?

FURTHER READING

1. Olliaro, P. L., and Yuthavong, Y. An overview of chemotherapeutic targets for antimalarial drug discovery. *Pharmacology and Therapeutics* (1999) 81(2):91–110.

2. Korsinczky, M., Chen, N., Kotecka, B., Saul, A., Rieckmann, K., and Cheng, Q. Mutations in *Plasmodium falciparum* cytochrome *b* that are associated with atovaquone resistance are located at a putative drug-binding site. *Antimicrob. Agents Chemother.* (2000) 44(8):100–108.

3. Rosenblatt, J. E. Antiparasitic agents. *Mayo Clin. Proc.* (1999) 74(11):1161–1175.

4. Looareesuwan, S., Chulay, J. D., Canfield, C. J., and Hutchinson, D. B. Malarone (atovaquone and proguanil hydrochloride): A review of its clinical development for treatment of malaria. Malarone Clinical Trials Study Group. *Am. J. Trop. Med. Hyg.* (1999) 60(4): 533–541.

5. Ryan, E. T. Malaria: Epidemiology, pathogenesis, diagnosis, prevention and treatment—an update. *Curr. Clin. Top. Infect. Dis.* (2001) 21:83–113.

For further information, see the following web sites:

Malaria Foundation International: www.malaria.org

Centers for Disease Control, Parasitic Diseases Division: www.cdc.gov/ncidod/dpd/parasites/malaria/default.htm

World Health Organization, Malaria information page: www.who.int/health-topics/malaria.htm

Malaria Parasite Metabolic Pathways: sites.huji.ac.il/malaria/FramIntroduction.html

CHAPTER 26

The Biosynthesis of Membrane Lipids and Steroids
Premature Menopause and Facial Hair

Patient History

A 44-year-old woman was evaluated in the endocrinology clinic because of male-pattern baldness and hirsutism (excessive body and facial hair in a male pattern). Two years earlier she had noticed thinning of the hair on the vertex of her scalp and an increase in facial hair. Laboratory tests performed at the time revealed elevated total and free testosterone of 220 and 6 ng/dl respectively (upper limit of normal [ULN] 110 and 3.2 ng/dl), normal androstenedione and dehydroepiandrosterone sulfate (DHEAS) levels, and normal thyroid function tests.

Does the age of onset help in the differential diagnosis in this case? What diagnoses would you consider if she had hirsutism or virilization with an onset at birth?

What are the normal sources of androgens in women?

What is the significance of an elevated testosterone but normal androstenedione and dehydroepiandrosterone levels?

The patient was 13 years old at the time of menarche (onset of menses) and had been sexually active during her married life without using contraceptives, but never became pregnant. She had regular menses without menstrual molimina (premenstrual syndrome) or dysmenorrhea. Eight years ago her menses became less frequent and ultimately ceased without other symptoms, but at approximately the same time mild hypertension and hyperlipidemia were diagnosed. There was no evidence for diabetes mellitus. She denied any history of muscle weakness, skin rash, thinning of skin, hoarseness, major changes in weight, change in libido, visual disturbance, hot or cold intolerance, postmenopausal bleeding, or family history of hirsutism or infertility.

The patient experienced apparently normal ovulatory menstrual cycles and developed typical female secondary sexual characteristics. What is the significance of these two observations?

What is virilization, how does it differ from hirsutism, and what are the implications regarding the underlying causes?

She had a history of anxiety and depression, which became exacerbated at the time of life stresses such as the death of her parents and an older sibling. She had taken antidepressant medications intermittently in the past, and currently was taking paroxetine and desipramine. The remainder of her medicines were verapamil, atorvastatin, and Xanax as needed for anxiety.

What is the significance of concomitant depression in this patient?

She was employed as a receptionist in a medical office and had done clerical work all of her adult life without any history of occupational exposures. She was a former smoker of one pack per day, but quit approximately ten years ago. She denied any use of alcohol or illicit drugs. She did not weight train, and denied the use of anabolic steroids or other supplements.

Physical Examination

She was well-appearing but mildly obese on physical examination. Her blood pressure was 146/92 mmHg, with a pulse rate of 88 beats per minute, and a normal respiratory rate. The general physical examination was remarkable for coarse, dark hairs on the chin, upper lip, and sideburn areas, with thinning of scalp hair in a male-baldness pattern. She had male-pattern distribution of hair on the chest and low back with pubic hair that was widely distributed over the lower abdomen and extended upward in the midline to the umbilicus. There was no axillary hyperpigmentation, striae (stretch marks), or buffalo hump, suggestive of hypercortisolism. The thyroid gland was not enlarged, and the heart and lung examinations were normal. She had no abdominal tenderness or masses, and bowel sounds were active. The pelvic examination was remarkable for clitoral enlargement, a normal uterus, and ovaries that were difficult to palpate. Neurologic examination revealed full extraocular movements and full visual fields. She had no focal motor or sensory deficits.

What is the significance of the absence of hyperpigmentation and striae?

What would you be concerned about if she had deficits in visual field testing?

Laboratory Studies and Scans

Additional laboratory studies were performed and revealed normal serum chemistries and normal blood sugar. The complete blood count was normal. Measurement of the reproductive hormone levels revealed, as previously, elevated total (235 ng/dl) and free (6.4 ng/dl) testosterone levels with normal prolactin, DHEA, DHEAS, estradiol, luteinizing hormone (LH), follicle-stimulating hormone (FSH), and human chorionic gonadotropin (HCG) levels. A repeat thyroid stimulating hormone (TSH) level was normal. Suprapubic and transvaginal ultrasound examination of the pelvis revealed a normal uterus for the patient's age. Both ovaries were slightly enlarged, and the right ovary showed a single simple cyst two centimeters in diameter. A CT scan of the abdomen revealed that the adrenal glands were of normal shape and size.

Is the pelvic ultrasound consistent with polycystic ovarian syndrome? What other findings argue against this diagnosis?

Is the presentation consistent with any of the adrenal hyperplastic syndromes such as 21-hydroxylase deficiency, 11β-hydroxylase deficiency, or 3β-hydroxysteroid dehydrogenase deficiency?

Treatment and Diagnosis

The patient underwent an exploratory laparatomy, hysterectomy, and bilateral salpingo-oophorectomy (removal of the uterus and ovaries). The ovaries were enlarged for a woman of her age, without multiple follicles or cysts. The most prominent change on low-power microscopic examination was nodular and diffuse hyperplasia of the ovarian stroma, especially prominent in the medullary region, warranting the diagnosis of stromal hyperthecosis. Further microscopic examination revealed no evidence for an ovarian tumor.

What hormone(s) do stromal cells secrete? What normally controls their hormone secretion?

In retrospect, what was the first manifestation of the ovarian-induced androgen excess?

One month after surgery the patient returned for follow-up. The testosterone level had normalized, and there was a dramatic improvement in her mood that coincided with a decrease in the amount of facial hair and scalp hair thinning.

DISCUSSION

Hair Growth

Growth cycle

Hair grows from a bulb-shaped structure in the skin called a hair follicle. The number of follicles does not change over an individual's lifetime; however, the growth rate and type of hair can vary, and is affected by numerous factors, especially *androgens*. There are two phases in the cycle of hair growth: (1) *anagen*, the growth phase; and (2) *telogen,* a rest phase, which culminates in shedding of the hair and the start of a new cycle. Variation in the length of time spent in anagen accounts for differences in hair length at different sites of the body. Scalp hair can grow for 2–5 years, while eyelashes only grow for a few months before entering telogen.

The hair on the body is catagorized into two types: the fine, unpigmented *vellus* hair, and the longer, coarser *terminal* hair. Most of the body is covered in vellus hair until puberty, when androgens produced at that time cause the conversion of vellus hair into terminal hair at various body sites. The extent to which this occurs depends upon both the sensitivity of hair follicles to androgens and on the amount of androgens produced. Thus, women, with lower levels of circulating androgens, develop axillary and pubic hair, whereas men, with higher androgen levels, typically develop terminal hair on their faces and chests as well. Some follicles—such as those on the eyebrows, eyelashes, and scalp—are not dependent upon androgens for the generation of terminal hair, and are thus of this type even before puberty.

Hormonal control of hair growth

Testosterone (via its metabolite, dihydrotestosterone [DHT], see below) is the most important mediator of androgen-dependent hair growth. In women, testosterone is secreted in approximately equal quantities by the ovaries and the adrenal glands, and ultimately mediates hair growth through the *androgen receptor* found in cells of the hair follicle. In addition, testosterone is also produced in peripheral tissues from precursors such as *androstenedione* and DHEA (see Figure 26.1). The androgen receptor–androgen complex binds gene regulatory sequences and stimulates the synthesis of *cytokines*, which either stimulate or inhibit hair growth, depending upon intrinsic properties of the follicle.

Testosterone is transported in the blood by sex-hormone binding globulin (SHBG), and is converted into DHT by *5α-reductase* at the hair follicle. DHT is far more potent than testosterone, binding the androgen receptor with a higher affinity, and dissociating more slowly. DHT is thus the primary mediator of androgen-dependent hair growth and, interestingly, is also the major culprit in male-pattern baldness. Although DHT stimulates the formation of terminal hair at some body sites, it has the opposite effect on the follicles at the top of the scalp, causing them to shrink and cease producing hairs. Androstenedione and DHEA (most of which is in the sulfated form, DHEAS) are adrenal androgens which have also been associated with androgen-dependent hair growth. However, they have little direct stimulatory effect on the androgen receptor and exert their effects primarily by conversion to testosterone and DHT in peripheral tissues.

Hirsutism: Etiology

Androgen excess is the most common etiology

Hirsutism is defined as the excessive growth of thick, pigmented hair in areas where women typically have little hair, such as the face, chest, and areolae. The follicles in these areas are androgen-dependent, and the most common cause of hirsutism is androgen overproduction. However, as outlined above, there are several steps in the pathway between testosterone production and stimulation of the androgen receptor, any of which may lie at the root of a case of hirsutism. Decreased SHBG levels, enhanced activity of 5α-reductase, or increased level or sensitivity of the androgen receptor, although rare, may also cause excessive hair growth. When androgen excess is the cause, it may be due to either adrenal or ovarian abnormalities, or to exogenous administration of anabolic steroids. This patient denied use of steroids and lab results corroborated this, thus indicating a likely adrenal or ovarian abnormality as the cause of hirsutism.

FIGURE 26.1 Androgen biosynthetic pathways. The starting material for the synthesis of androgens is cholesterol, most of which is absorbed from the bloodstream by androgen-producing cells. There are two pathways, diverging at pregnenolone, for the synthesis of androstenedione, and the pathway that predominates depends upon cell type. The steps leading to the production of testosterone occur in the adrenal glands and the ovaries; however, the principal androgen secreted by the adrenal is DHEA and its sulfated form (DHEAS). Conversion of testosterone to 5α-dihydrotestosterone occurs in peripheral tissues. The mitochondrial enzyme, P450-SSC, converts cholesterol into pregnenolone. Deficiencies in 21-hydroxylase, 11β-hydroxylase (in the glucocorticoid pathway), or 3β hydroxysteroid dehydrogenase cause congenital adrenal hyperplasia (CAH). Note that the early steps in androgen biosynthesis (up to the production of progesterone) are shared with the mineralocorticoid and glucocorticoid biosynthetic pathways.

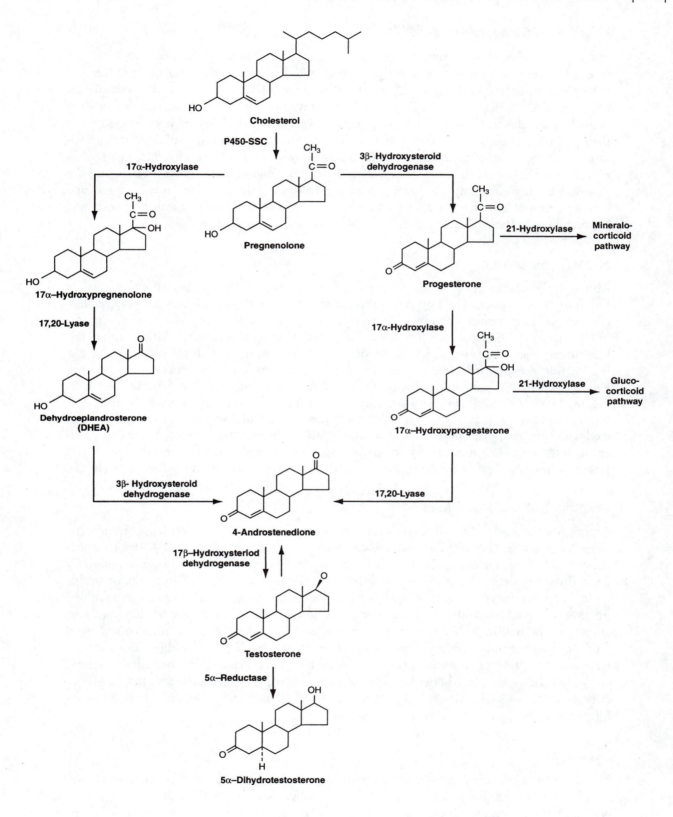

Polycystic ovarian syndrome

There was reason to suspect *polycystic ovarian syndrome (PCOS)* in this patient because this is the most common condition associated with hirsutism. Although the fundamental cause of the syndrome is unknown, it is characterized by the enlargement of the ovaries and the accumulation of undeveloped follicles (where the ova are formed) within them. The ultrasound examination argued against this diagnosis—although the ovaries were enlarged, they were not polycystic, and the histological studies carried out on the ovaries after the oophorectomy confirmed this. In addition, the age of onset of hirsutism in PCOS is usually soon after menarche, and amenorrhea typically occurs shortly thereafter. This patient had regular menses until the age of 36, and she first noticed hirsutism only two years ago, at age 42. Finally, the syndrome is often associated with an excess of luteinizing hormone and a normal to low level of follicle stimulating hormone; abnormalities in these hormone levels were not observed in this patient.

Stromal hyperthecosis

Although the diagnosis of PCOS was excluded in this patient, the ovaries were nevertheless found to be the source of androgen overproduction. She was diagnosed with *stromal hyperthecosis*, a condition characterized by a hypercellular stroma (a supportive tissue) containing foci of luteinized cells. Stromal cells produce androgens (including testosterone) during normal follicle development; however, they are functionally aberrant in those with the condition, resulting in androgen overproduction. This caused the increase in facial hair and thinning of scalp hair noticed by this patient two years ago. The abnormal physiology of the ovaries results in failure of follicle development, and amenorrhea, which occurred in this patient eight years ago. Stromal hyperthecosis is most often observed in post-menopausal women, and may be associated with very high levels of circulating androgens and, consequently, signs of virilization, deepening of the voice or clitoromegaly. In addition to the excessive terminal hair observed on this patient's trunk and face, she exhibited clitoral enlargement, a reflection of high androgen levels.

Congenital adrenal hyperplasia

Androgen excess as a result of overproduction by the adrenal glands is a far less common etiology for hirsutism; however, a condition called *congenital adrenal hyperplasia (CAH)*, in which the adrenal glands are enlarged from birth, can cause the disorder. CAH is usually caused by 21-hydroxylase deficiency (95% of cases), but may also be caused by 11β-hydroxylase or 3β-hydroxysteroid dehydrogenase deficiencies (see Figure 26.1). All three enzymes are involved in adrenal steroid biosynthesis, and a shortage of any one results in the accumulation of precursors of the reactions they catalyze. These precursors are then diverted into the androgen pathway, and hence the androgen excess characteristic of the disorder. Additionally, reduced feedback inhibition of adrenocorticotropic hormone (ACTH) caused by the reduced glucocorticoid levels results in adrenal hyperplasia, and a further increase in the production of precursors. CAH is usually manifested at birth or in early childhood; however, a delayed-onset form may cause hirsutism in adult women.

Neoplasms

The fairly late onset of symptoms in this patient made the diagnosis of an ovarian or adrenal neoplasm a distinct possibility. Neoplasms are usually associated with high levels of androgens and often produce male characteristics, such as deepening of the voice, breast atrophy, increased muscle bulk, clitoromegaly, and increased libido, in addition to hirsutism. A tumor may be suspected when signs of virilization appear, especially with a fairly sudden onset. Fortunately, adrenal and ovarian tumors are rare, and together account for fewer than 2% of cases of hirsutism. They were ruled out in this case on the basis of the pathology results.

Diagnosis

In most cases hirsutism is benign and is primarily a cosmetic concern; however, occasionally it indicates a serious underlying condition such as a tumor. Identifying the cause of hirsutism is necessary because it allows for the distinction between benign and serious conditions, and guides treatment decisions.

History and physical examination

The history and physical examination can yield relevant information as to the cause of hirsutism. Medication-induced hirsutism can be identified, and in some cases, especially when disease is quite advanced, adrenal or ovarian tumors may be palpated. The rate of progression of hirsutism may suggest a neoplastic versus benign origin to the disorder, and the pattern of the menstrual cycle may indicate an ovarian versus an adrenal abnormality. Excessive weight of the patient is significant, as obesity is associated with hirsutism. Various cutaneous signs may also suggest an etiology. For example, a form of PCOS is associated with acanthosis nigricans (and insulin resistance), and hypercortisolism is often accompanied by axillary hyperpigmentation and striae.

Quantitation of hirsutism

A clinical assessment of the degree of hirsutism is among the first steps in the evaluation. Assessment of the distribution, quantity, and type of hair can distinguish between hirsutism and hypertrichosis (non-androgen-dependent hair growth), and in cases of hirsutism, gives a rough estimate of the degree of androgen excess and provides a baseline by which to gauge the effects of treatment. Quantitation is usually done using the Ferriman-Gallwey model, in which the amount of hair is scored (0 to 4) in nine anatomical areas.

Laboratory evaluation, scans, and pathology

A laboratory evaluation of hormone levels can be used to locate the predominant source of androgen excess and may give an indication of the probability of a neoplasm. Elevated testosterone levels may be associated with either an ovarian or adrenal abnormality; however, elevated DHEAS levels are associated with adrenal dysfunction, such as in CAH. The hormonal evaluation of this patient, with elevated testosterone levels and normal DHEA and DHEAS levels, thus suggested an ovarian abnormality. The testosterone level, in the 200 ng/dl range, was high enough to warrant the possibility of a tumor; however, the scans and pathology report following the oophorectomy did not bear this out. The ultrasound and CT scans were consistent with an ovarian source of androgen excess—the ovaries were enlarged, whereas the adrenal glands were of normal size and shape. Finally, microscopic examination of ovarian tissue, which revealed nodular hyperplasia of the stroma, elicited the diagnosis of stromal hyperthecosis.

Treatment

The treatment for hirsutism depends upon the cause. Pharmacological treatment includes suppression of adrenal or ovarian androgen secretion; however, agents that block androgen action at the hair follicle appear to be the most effective (e.g., spironolactone, which binds the androgen receptor). Mechanical hair removal (such as shaving, waxing, and depilatory creams) is also an option, especially for milder cases. In this case a bilateral salpingo-oophorectomy was performed because the patient was post-menopausal, and oophorectomy has been shown to halt the virilization and some of the associated characteristics, such as hypertension, in those with stromal hyperthecosis. A hysterectomy was also performed because there was a concern about the possibility of a pelvic tumor originating in the ovaries, and perhaps also involving the uterus.

QUESTIONS

1. Testosterone is secreted into the bloodstream and exerts its effects via the androgen receptor found in target cells. Unlike many receptors, the androgen receptor is found within the cell, rather than in the cell membrane. Considering that receptor and ligand are separated by the cell membrane, how do they interact to effect the changes mediated by the hormone?

2. The *dexamethasone-androgen suppression test* is often used to broadly distinguish between ovarian and adrenal androgen overproduction in cases of hirsutism. What is dexamethasone, and what is the mechanistic basis for the test?

3. What is the difference between *hirsutism* and *hypertrichosis*?

4. Propecia (finasteride) has been used to inhibit male-pattern baldness in men and is also an option in the treatment of hirsutism. What is the mechanism of drug action?

5. Given the reactions they catalyze (see Figure 26.1), which enzyme substrates would you expect to accumulate if the gene encoding 17,20 lyase were deleted?

6. What is steroidogenic acute regulatory protein (StAR) and what is its role in steroid biosynthesis?

FURTHER READING

1. Ayuk, P., Stringfellow, H., Donnai, P., et al. Hirsutism of recent onset with marked hyperandrogenaemia and ovarian hyperthecosis after the menopause. *Ann. Clin. Biochem.* (1998) 35(Pt.1):145–148.

2. Miller, W. L. Early steps in androgen biosynthesis: from cholesterol to DHEA. *Baillieres Clin. Endocrinol. Metab.* (1998) 12(1):67–81.

3. Stocco, D. M. StAR protein and the regulation of steroid hormone biosynthesis. *Annu. Rev. Physiol.* (2001) 63:193–213.

4. Lobo, R. A. Ovarian hyperandrogenism and androgen-producing tumors. *Endocrinol. Metab. Clin. North. Am.* (1991) 20(4):773–805.

For further information, see the following web sites:

National Library of Medicine, Medlineplus health information, ovarian overproduction of androgens: www.nlm.nih.gov/medlineplus/ency/article/001165.htm

American Medical Association, male pattern baldness: www.medem.com/MedLB/article_detaillb.cfm?article_ID=ZZZ6Q72SWAC&sub_cat=300

DNA Replication and Repair
Punch Drunk or Something Else?

CASE HISTORY

Patient History

A 47-year-old, right-handed man was admitted to the hospital because of dementia, inappropriate behavior, and a movement disorder. Two years ago an obvious decline in the patient's cognitive function was noted by his wife and son, manifesting as difficulty in concentrating, forgetfulness, and problems naming objects. About eight months ago he began to develop involuntary movements of his arms and head.

What are the types of movement disorders?

Does the presence of cognitive dysfunction help in making the diagnosis?

The patient was employed as a sales representative for a fastener company and was an avid woodworker and reader in his spare time. He had completely stopped reading, complaining that nothing captured his interest. He noted increasing difficulty with furniture making because of his movement disorder, and more important, difficulty remembering how to perform the tasks required. He initially denied that he had a problem, and was defensive and hostile when his cognitive difficulties were brought to his attention. At other times he was very passive and withdrawn.

What is the significance of his depressed mood?

The patient had boxed professionally in his early twenties, having been in over 50 fights. He had a losing record, but claimed that he had never been knocked out, although he had been knocked down on a number of occasions. He recalls headaches and dizziness while he was boxing but did not regularly suffer from these symptoms in recent years. He drank alcohol excessively while he was boxing and for a few years thereafter, but had not had a drink in 20 years. He did not take any medicines regularly and did not use illicit drugs. He smoked as a teenager but quit when he began boxing. He has never worked in an occupation in which he was exposed to industrial toxins or chemicals.

The remainder of his past medical history was unremarkable except for traumatic injuries including a broken nose, jaw, and ribs, which he suffered in the ring. He had no other significant medical illness. He had not had any surgery other than closed reduction of fractures, and he never received a blood transfusion. He was adopted, so details of the family history were unavailable.

Six months ago, at the request of his employer, he was seen by a neurologist and diagnosed with progressive dementia pugilistica. As a result, he was placed on a permanent leave of absence. Initially, no specific treatment was recommended. The movement abnormalities became more noticeable over the past two months, as did his mood swings. He has had more difficulty walking and has nearly fallen a number of times over the past several months. In the past month he got lost driving home from the store.

> *What is dementia pugilistica, and what are its features?*

Physical Examination

On physical examination he was a well-appearing, well-kempt man. Occasional involuntary movements of the head and arms were noted while he was at rest. The temperature was 37°C, the pulse was 68 per minute and regular, and the respirations were 14 per minute. His blood pressure was 145/80 mmHg, without orthostatic changes. The eyes, ears, nose, and throat examination revealed no evidence for stare, lid-lag, Kayser-Fleischer rings, and saccades at the extremes of movement of the eye. The retinal examination was normal. His neck revealed no thyromegaly. The remainder of the general physical examination was normal.

> *What is the significance of a stare and lid-lag?*
> *What are Kayser-Fleischer rings?*

Neurologic Examination

On neurologic examination, the patient was alert and oriented to person, time, and situation, but he often seemed uninterested. He knew the name of the sitting president and his predecessor. His speech and language were normal without dysarthria, hypophonia (decreased amplitude of speech), aprosody (lack of inflection), palilalia (repetition of the first syllables of words), tachyphemia (very rapid speech), or coprolalia (uttering obscenities). He could not perform serial or abstract tasks. He did not recall any of three objects after five minutes, even with cues. Remote memories were preserved. Cranial-nerve functions were intact.

Muscle Examination

The muscle examination was remarkable for normal tone and strength, without atrophy of the major muscle groups. He had full range of motion without cogwheel rigidity. There were no tremors, fasciculations, or myoclonus; however, he did exhibit periodic choreiform involuntary movements of both arms, and when asked to grip a hand he involuntarily tightened and loosened the grip (milk maid's grip). The patient's posture was normal. His normal, spontaneous gait was interrupted by occasional involuntary movements of the left leg, producing a slapping, dancing step. He could not perform tandem walking.

Sensation of light touch, proprioception (sense of joint position), and perception of vibration were intact. The deep-tendon reflexes were ++ and symmetric, and the Babinski sign was absent.

What kind of movement disorder does the muscle examination suggest?

What is chorea?

What are muscle fasciculations and what do they signify?

Diagnosis and Treatment

Routine blood chemistries and a complete blood count were normal. Thyroid function tests and vitamin B_{12} levels were normal. The chest radiogram and electrocardiogram were also normal. A magnetic resonance imaging (MRI) examination of the brain showed atrophy of the caudate nucleus with enlarged lateral ventricles, and there was mild cerebral atrophy. Examination of the cerebrospinal fluid revealed a normal cell count, total protein, and immunoelectrophoresis. Cultures of the spinal fluid for bacteria and fungus were negative. Analysis of the gene for the huntingtin protein (*HD*) revealed 68 CAG repeats in the first exon, consistent with the diagnosis of Huntington's disease (HD).

What is HD, and what is the significance of the trinucleotide repeats in huntingtin?

What is his son's risk of development of HD?

His movement disorder was not disabling so dopamine antagonists were not prescribed so as to avoid serious side-effects, such as tardive dyskinesia (produced by neuroleptics, such as haloperidol and fluphenazine) and worsening depression (produced by dopamine-depleting drugs, such as reserpine and tetrabenazine). He was prescribed Prozac for depression.

Disease Worsens

His disease progressed with worsening chorea and frequent falling. His impulse control worsened, and he became prone to violent outbursts directed toward his wife and son. His confusion worsened to the point where he could not be left alone. Despite their best efforts, his wife and son could no longer care for him, and he was admitted to a long-term care facility.

Patient's Son Undergoes Genetic Testing

The patient's son was 22 years old and healthy, without evidence for any neurological disease. He was unmarried but had plans to have a family in the future. He underwent testing for the defect in the huntingtin gene and was found to have eight CAG repeats in the first exon. The median number of repeats in the normal population is 11; thus he was not considered to be at risk for the development of HD.

There is no treatment for HD, so would you have advised the son to be tested at this time?

DISCUSSION

Triplet Repeat Diseases

Caused by expansion of trinucleotide repeats

HD, an inherited, autosomal dominant disorder, is one of more than a dozen known triplet repeat diseases, a class that also includes fragile X syndrome, a common cause of mental retardation, and Kennedy's disease, a rare neurodegenerative disorder (Table 27.1). Although the gene affected in each disease differs, they are all caused by the same type of mutation—*triplet repeat expansion*—in which a tract of trinucleotide repeats is expanded during replication, resulting in increasing numbers of repeats with each subsequent generation (see Section 27.6.6 of *Biochemistry* 5e). The repeats show both *somatic* and *germ-line* instability, meaning that the gene is subject to expansion in both somatic and germ-line cells. The mechanism(s) underlying the instability in the repeats is unknown; however, some evidence indicates that strand slippage during DNA replication may be partially to blame. In addition, the formation of nonstandard structures, such as hairpins, during replication could lead to the activation of DNA repair mechanisms that inadvertently expand the repeat. Another possibility is that unequal strand exchange between chromatids during crossing over in meiosis could result in repeat expansion. Although both repeat expansion and contraction are possible, for reasons not yet clear, there is a strong bias for expansion.

Anticipation is a feature of triplet repeat diseases

An earlier onset and increased severity of symptoms generally correlate with larger numbers of repeats, and are associated with successive generations within an affected family. This phenomenon is referred to as *genetic anticipation*. Thus, an affected child of an HD patient will typically become ill at a younger age and will have more severe symptoms than his parent. Interestingly, the parental origin of the mutant allele also appears to influence the severity of disease, with paternal origin (especially with older fathers) generally causing more serious disease with earlier onset. This may be explained by the ongoing spermatogenesis throughout the lives of men, thus increasing the possibility for repeat expansion.

Pathogenesis

Impaired gene expression in one subclass of triplet repeat diseases

Triplet repeat diseases are categorized into two subclasses, one in which the repeats are found in noncoding regions (untranslated regions), and the other in which the repeats are found within coding sequences. Defective gene regulation, resulting in little or no gene expression, is thought to account for the pathogenesis in diseases of the first class, which includes fragile X syndrome.

The polyglutamine disease subclass

Those in the second class, which includes HD, all result from the expansion of the same triplet of nucleotides, CAG, which encodes the amino acid glutamine, and this subclass is therefore referred to as the *polyglutamine diseases*. Expanded polyglutamine stretches appear to alter the conformation of a protein, rendering it more susceptible to aggregation, which lies at the heart of the pathogenesis. The polyglutamine diseases are caused by mutations in different genes; however, they are thought to share the same pathogenic mechanism by virtue of the expanded polyglutamine tract the mutant proteins all possess.

TABLE 27.1 Diseases caused by trinucleotide repeat expansion. The first six diseases are caused by expansion within noncoding regions, and the last eight are polyglutamine diseases, caused by expansion of the CAG glutamine codon within coding sequence.

Disease	Gene	Protein	Trinucleotide repeat	Normal repeat number	Disease repeat number
Fragile X syndrome	FMR1	FMR-1 protein	CGG	6–53	>230
Fragile XE syndrome	FMR2	FMR-2 protein	GCC	6–35	>200
Friedreich's ataxia	X25	Frataxin	GAA	7–34	>100
Myotonic dystrophy	DMPK1	Myotonic dystrophy protein kinase	CTG	5–37	>50
Spinocerebellar ataxia type 8	SCA8	None	CTG	18–37	110–250
Spinocerebellar ataxia type 12	SCA12	PP2A-PR55β	CAG	7–28	66–76
Kennedy's disease	AR	Androgen receptor	CAG	9–36	38–62
Huntington's disease	HD	Huntingtin	CAG	5–35	36–121
Haw-River syndrome	DRPLA1	Atrophin-1	CAG	6–35	49–88
Spinocerebellar ataxia type 1	SCA1	Ataxin-1	CAG	6–44	39–82
Spinocerebellar ataxia type 2	SCA2	Ataxin-2	CAG	15–31	36–63
Spinocerebellar ataxia type 3	SCA3	Ataxin-3	CAG	12–40	55–84
Spinocerebellar ataxia type 6	SCA6	α_{1A} voltage-dependent calcium channel subunit	CAG	4–18	21–33
Spinocerebellar ataxia type 7	SCA7	Ataxin-7	CAG	4–35	37–306

Interestingly, although the proteins involved in each disease vary in their normal subcellular localization, in most cases when mutant they form *intranuclear* aggregates, and thus some type of nuclear dysfunction has been hypothesized to be the basis for pathogenesis. Possibilities include disruption of transcription or RNA processing, or sequestration of an essential nuclear protein. Many transcription factors contain polyglutamine stretches, and in fact, the CREB-binding protein (CBP), a transcriptional coactivator, has been found to be associated with the intranuclear aggregates in several of the polyglutamine diseases, including HD. Evidence indicates that the apoptotic pathway (programmed cell death, see Section 18.6.6 of *Biochemistry* 5e) is triggered in the afflicted cells; however, further research is needed to determine if this is merely secondary to the toxicity caused by the protein aggregates.

Polyglutamine diseases primarily affect neuronal cells

All polyglutamine diseases are progressive neurodegenerative diseases; although the genes involved are generally widely distributed throughout the body, for reasons unknown neuronal cells are predominantly affected. In addition, different subpopulations of neurons are principally affected in the various diseases. The protein "context" in which the polyglutamine tract is found is thought to account for the particular subpopulation affected in each case. Although isolated, expanded polyglutamine tracts are generally toxic to cells, animal studies have shown that selective neuronal toxicity is mediated by the protein in which the tract is found. In HD, neuronal loss occurs in the neostriatum, particularly the caudate nucleus. These regions lie within the basal ganglia, a brain structure found at the base of the cerebrum, which is involved in controlling movements.

Molecular Genetics of HD

HD is caused by a mutation in exon 1 of *HD* (also called *IT15*), a gene on chromosome 4 that encodes a protein called *huntingtin*. The gene encodes an unstable stretch of glutamine residues, beginning at residue 18. Normal individuals have from 10 to 35 CAG repeats, while patients with HD have between 36 and 121 repeats. Those with 27–35 CAG repeats are not at increased risk for HD; however, they are at increased risk of having a child with HD. Those with 36–41 CAG repeats may not develop symptoms within their lifetime. Generally, there is an inverse correlation between the number of repeats and the age of onset of disease, with the median age of onset being 66 in those with 39 repeats, and 27 in those with 50 repeats. This patient, with 68 repeats and the obvious decline in his condition not having occurred until age 45, was spared the early onset of the disease. However, symptoms often arise slowly and gradually, and he may have had minor manifestations of the disease prior to the prominent decline at age 45.

The normal physiological role of huntingtin is unknown. It is a 348 kDa protein and possesses 10 HEAT domains, which are thought to be involved in protein–protein interactions. It is widely expressed in both neurons and other cell types, and the reason for the selective destruction of cells in the neostriatum is unknown; however, it appears that apoptosis is the predominant mode of cell death.

Clinical Features of HD

HD is characterized by progressive chorea (derived from the Greek word for "dancing") and dementia, and the juvenile form may be associated with seizures. It was first described by George Huntington in 1872 in a family on Long Island. The disease usually strikes in middle age (typically between 35 and 44), and lasts for between 10 and 20 years.

The onset of the disease is typically characterized by the development of cognitive and behavioral symptoms that may precede the onset of the movement disorder by a decade. Cognitive disturbances include forgetfulness, poor judgment, slowed thought processes, and visuospatial impairment; however, memory is usually preserved until the late stages of disease. Psychiatric disturbances, such as depression, irritability, and apathy are also common, and delusions may occur. The chorea usually begins fairly harmlessly as subtle fidgeting or clumsiness; however, it gradually progresses to jerking of the limbs, face, and trunk, and eventually becomes disabling. In the late stages motor impairment predominates over behavioral disturbances, and the patient becomes completely dependent. Death usually occurs on average between 13 and 15 years after symptoms first appear, and is often precipitated by pneumonia or a fatal injury caused by a fall.

Epidemiology

The estimated prevalence of HD is 5 to 10 per 100,000 in those of western European descent, and is particularly high in South Wales and parts of Scotland and Northern Ireland. It is less common among African and Asian populations. A prevalence of less than 1 per 100,000 was found among South Carolinians of African-American descent. The most common *HD* alleles worldwide have between 15 and 20 CAG repeats; however, the distribution among western European populations is typically in the higher part of this range, while Africans and Asians are more often in the lower portion of the range.

Diagnosis

Diagnosis of HD may be made on the basis of the presence of progressive motor disability and cognitive and behavioral disturbances. A positive family history can also indicate a diagnosis of HD; however, in this case the patient was adopted and had no knowledge of his family's medical history. Imaging studies, such as the MRI performed here, can give further support to the diagnosis or, conversely, may indicate other conditions. In this case, the atrophy of the caudate nucleus revealed by the MRI supported the diagnosis of HD. Finally, DNA testing can reveal the number of CAG repeats in the *HD* gene, and can be used to make the diagnosis definitively.

The availability of a DNA-based test for HD makes possible the evaluation of individuals in affected families even before any clinical manifestations arise; however, psychological factors must be carefully considered in these cases. As no treatment has yet been found to prevent or slow the onset of HD, the near certainty of a long and debilitating illness in one's future may have devastating psychological effects.

Treatment

To date there is no treatment that can prevent or even slow the onset of HD; thus treatment is limited to symptomatic therapies. Neuroleptics may be used to treat the chorea, and a variety of drugs may be helpful in treating the psychiatric disturbances. Although the identification of the mutation underlying HD has indeed been a substantial step toward understanding the disease, medical professionals are currently in the unfortunate position of being able to accurately diagnose HD, while being unable to provide any relief from its relentless assault. The intense pace of research in the area of triplet repeat diseases will likely soon remedy this situation.

QUESTIONS

1. Given that HD is fatal, why has selective pressure not caused the mutation to die out?

2. Triplet repeat diseases do not always behave according to the laws of Mendelian genetics. For example, monozygotic twins are not always equally affected by the disease, disease may disappear from a branch of an affected family, and sporadic cases occasionally arise. How might these anomalies be explained?

3. Deletion of a single copy of the *HD* gene does not result in the disease. What does this tell you about the nature of the HD mutation?

4. Many of the fatal afflictions of our day are the result of DNA mutations. Do you think that through natural selection the accuracy of DNA replication and the effectiveness of DNA repair mechanisms will increase to the point where DNA mutations will no longer occur?

5. Two unrelated patients come to you, a genetic counselor, and ask to be tested by the DNA test for the *HD* mutation. When your receive the results from the laboratory, you see that patient #1 has the sequence $5'$ ...GGG AAT $(CAG)_{35}$ GGT TTC ATA... $3'$ in the first exon of *HD*, while patient #2 has the sequence $5'$ ...GGG AAT $(CAG)_{33}$ CTG $(CAG)_8$... $3'$. What can you say about the relative risk of the two patients, or of their children, developing the disease?

6. You have just received the DNA test results for an individual who has all the classic features of HD. You believed that the DNA test would merely confirm your presumed diagnosis of HD; however, to your surprise the test reveals that both alleles of the patient's *HD* gene have normal numbers of CAG repeats. How could this be explained (laboratory errors excluded!)?

FURTHER READING

1. Martin, J. B. Molecular basis of neurodegenerative disorders. *New England Journal of Medicine* (1999) 340:1970–1980.

2. The Huntington's Disease Collaborative Research Group. A novel gene containing a trinucleotide repeat that is expanded and unstable on Huntington's disease chromosomes. *Cell* (1993) 72:971–983. (This article describes the cloning of the HD gene.)

3. Pearson, C. E., and Sinden, R. R. Trinucleotide repeat DNA structures: Dynamic mutations from dynamic DNA. *Current Opinion in Structural Biology* (1998) 8:321–330.

4. Cummings, C. J., and Zoghbi, H. Y. Fourteen and counting: Unraveling trinucleotide repeat diseases. *Human Molecular Genetics* (2000) 9(6):909–916.

For further information, see the following web sites:

Huntington's Disease Society of America: http://www.hdsa.org/

Huntington's Disease Association of the UK: http://www.hda.org.uk/

Huntington's Disease Advocacy Center: http://www.hdac.org/

NINDS Huntington's Disease Information Page:
http://www.ninds.nih.gov/health_and_medical/disorders/huntington.htm

OMIM: http://www.ncbi.nlm.nih.gov/entrez/query.fcgi?db=OMIM

CHAPTER 28

RNA Synthesis and Splicing
You Don't Have to be Liberal to Have a "Bleeding Heart or Blood Vessels"

A Young Man Is Admitted with Chest Pain and Dyspnea

A 19-year-old man was admitted to the hospital because of chest pain and dyspnea after blunt trauma to the chest. He had been well until one week earlier, when he was pushed into a wall while playing basketball. He had tenderness and bruising over the sternum (breastbone), but no other symptoms. On the following day he developed mild dyspnea, and the chest discomfort increased in intensity. He was evaluated in the emergency room where a chest radiogram was performed and revealed an enlarged cardiac silhouette and a widened superior mediastinum (central part of the chest cavity). He was subsequently admitted to the hospital.

What is the significance of the radiographic findings?

What are the causes of cardiac enlargement?

Patient History

The patient was a college student who had always enjoyed excellent health. He was physically active and played intercollegiate basketball. There was no known personal or family history of cardiovascular disease and he was not hypertensive. He had no prior history of chest pain or exercise intolerance, and no history of heart murmurs or rheumatic fever. He did not have arthritis or joint problems, ligamentous laxity, visual problems, or neurological abnormalities. He had no rashes or other abnormalities of the skin. He had myopia (nearsightedness) for which he wore contact lenses.

Why is the absence of hypertension, joint, neurological, and skin disease relevant in this patient?

Physical Examination

The patient was anxious and in mild respiratory distress on presentation. His height was 6 feet 5 inches and weight was 200 pounds. He had arachnodactyly (long fingers). The temperature was 37.5°C orally, the pulse was 106 per minute and regular, and the respirations were 24 per minute. His blood pressure was 108/66 mmHg, with a pulsus paradoxus (a decrease of blood pressure with inspiration) of 22 mmHg.

> *What is a paradoxical pulse and what is its significance?*
>
> *Does the patient's stature give you any clue as to what is going on here?*

The head, eyes, ears, nose, and throat examination was normal—specifically, there were no abnormalities of the lens of the eye or the palate. He had distended neck veins that rose with inspiration (Kussmaul's sign) and the carotid pulses were normal. The chest examination was remarkable for mild pectus excavatum (inward deformity of the breastbone), with no kyphosis (outward curvature of the spine) or scoliosis (lateral deviation of the spine). The lungs were clear, but there were decreased breath sounds over the left lower-lung field. The heart exam revealed a laterally displaced point of maximal impulse, a two-component pericardial friction rub was heard, and a grade-II/VI holosystolic murmur was heard over the apex of the heart, radiating to the axilla. An S^4, but no S^3, gallop was detected. The peripheral pulses were of low volume but symmetrical, and there was no peripheral edema. The neurological examination was normal.

> *What is the significance of symmetrical peripheral pulses?*
>
> *What causes a pericardial friction rub?*
>
> *Is the heart murmur significant? What is the etiology?*
>
> *What would you be concerned about if his neurological examination was abnormal?*

Laboratory Evaluation

Routine laboratory evaluation revealed normal serum electrolytes, normal liver function tests, and other blood chemistries except for an elevated creatine kinase (CK) of 388 IU/liter (normal 24–170 IU/l) with an MB fraction of 15 micrograms/liter for a CK-MB/CK-Total index of 3.8% (upper limits of normal [ULN]: 3%), and a troponin I of 0.4 ng/ml (ULN: 0.15 ng/ml). The prothrombin (PT) and partial-thromboplastin times (PTT) were normal. Complete blood count revealed a mild anemia with a hematocrit of 37%, a normal white blood cell count, and a normal platelet count of 230,000.

> *What are CK (and the MB isoform) and troponin I, and what does their elevation in the blood suggest?*

Electrocardiogram, Echocardiogram, and Doppler Flow Studies

The electrocardiogram was remarkable for sinus tachycardia at a rate of 110 beats per minute, with normal intervals and nonspecific ST-segment and T-wave changes in leads

I, II, and aV$_F$. Macroscopic QRS alternans was apparent. An emergently obtained echocardiogram revealed a small but circumferential pericardial effusion, with diastolic right atrial collapse. The right- and left-ventricular function was normal. The aortic root was dilated with a diameter of 5.8 cm at the level of the sinuses of Valsalva (ULN: 4.0 cm), there was no obvious defect in the wall of the aorta (Figure 28.1), but the mitral valve leaflets were redundant and prolapsed into the left atrium during early systole. On Doppler flow study there was moderate aortic regurgitation (backward flow of blood through the aortic valve) and mild mitral regurgitation.

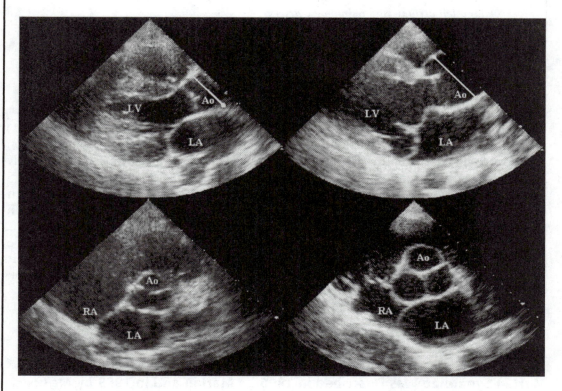

FIGURE 28.1 Echocardiographic view of the aortic root in a normal subject (left) and a patient with the Marfan syndrome (MFS) (right). The yellow line is the dimension of the aortic (Ao) root at the level of the Sinuses of Valsalva in a long-axis view of the heart (top panel). The left ventricle and left atrium are enlarged in the patient with MFS. The lower panel shows a cross-sectional view through the aortic root with the three valve leaflets seen coapting in the center of the aorta (normal left and MFS right). Images courtesy of Ellen Flynn and Dr. João Lima.

What is QRS alternans and what is its significance in this patient?

What are the sinuses of Valsalva, and what is the significance of the dilation of the aortic root at the sinuses?

What does the presence of pericardial fluid signify?

Cardiac Catheterization, Coronary Angiography, and Aortogram

The patient was started on intravenous β-adrenergic antagonists to lower the heart rate and the rate of rise of pressure in the aorta. Cardiac catheterization was performed and the right atrial, mean pulmonary artery, and pulmonary capillary wedge pressures (an index of left atrial pressure) were nearly equal at 19, 22, 20 mmHg, respectively. Coronary angiography revealed a normal origin of the right and left coronary arteries in the right and left sinuses of Valsalva respectively, and there were no coronary occlusions. An aortogram (injection of radiographic contrast material into the proximal aorta) was performed and revealed a false channel in the proximal aortic wall with minor contrast staining of the pericardial space.

What is the significance of normalization of the pressures in the right atrium, left atrium, and pulmonary artery?

Why was coronary angiography performed?

What does the aortogram tell you?

Diagnosis and Treatment

A magnetic resonance scan of the chest was performed. The study revealed a dissection (seepage of blood through a blood vessel wall) of the proximal aorta involving the non-coronary sinus of Valsalva, extending proximally in the ascending aorta but not involving the origin of the right innominate artery. A diagnosis of Marfan syndrome (MFS) complicated by a proximal aortic dissection with hemopericardium (blood in the pericardial space) was made, and he was scheduled for surgical correction. The patient underwent repair of the aortic dissection and valvular regurgitation by placement of a composite graft (mechanical aortic valve connected to a Dacron graft) in the aortic position. He subsequently did well, but was advised to avoid contact and competitive sports and isometric exercise. He was treated with β-adrenergic blocking drugs and was anticoagulated with warfarin.

DISCUSSION

Marfan Syndrome Was Described in 1896 by Marfan and in 1955 by McKusick

What do Abraham Lincoln and the violin virtuoso of the 19th century, Nicolo Paganini, have in common? Perhaps both were afflicted with the same genetic disorder, MFS. The skeletal features of MFS were first described by Marfan in 1896, a condition that he referred to as dolichostenomelia (long, thin limbs). The first comprehensive description of the natural history and complications in a large number of patients was carried out by McKusick in 1955. His work established the mode of inheritance, the range of pleiotropism, and variability and focused on the life-threatening cardiovascular complications of the disease. MFS is a rare disorder, with an incidence of approximately 1 in 5000 cases across racial and ethnic groups.

Pathophysiology

Cardiac manifestations are the most serious

MFS is an autosomal dominant, inherited systemic disorder of connective tissue caused by mutations in the fibrillin-1 (*FBN*-1) gene. The syndrome is pleiotropic and exhibits variable expression, even among those bearing the same mutation. The hallmark features involve the heart, skeleton, and the eye; however, the most serious and life-threatening manifestations of

MFS are those that involve the heart and major blood vessels. Dilatation of the aortic root at the level of the Sinuses of Valsalva, with aortic dissection, is the most common cause of death in patients with MFS. Rupture of the aorta, through which oxygenated blood flows to all parts of the body, is a particular threat, especially during exercise, and can rapidly lead to death. Other cardiac abnormalities include redundancy, prolapse, and incompetence of the atrioventricular (mitral and tricuspid) and aortic valves. This patient, with dilatation of the aortic root (at 5.8 cm, as revealed by the echocardiogram) and aortic dissection (observed in the aortogram and MRI), thus exhibited some of the most serious cardiac manifestations of the disease. The compromised heart function caused the dyspnea he presented with at the emergency room.

Skeletal and visual disturbances are common

Skeletal manifestations are the most immediately obvious and include tall stature, abnormal sternal structure (pectus excavatum or carinatum), arachnodactyly, scoliosis, and reduced thoracic kyphosis. Erosion of the lumbosacral vertebrae may occur, secondary to dural ectasia (distension of the covering of the spinal cord). MFS patients may be nearsighted, like this patient, and exhibit a displacement of the lens of the eye, known as ectopia lentis. Craniofacial features include a high-arched palate, maxillary overjet (overbite), and crowding of the teeth. Hernias and stretch marks (striae atrophicae) are common skin findings in these patients. Spontaneous pneumothoraces also occur due to rupture of apical pulmonary blebs.

Diagnosis

The diagnosis of MFS is not always straightforward because of the multiplicity of clinical features, and the variable expression of these features. If a first-degree relative has MFS, the presence of one major criterion (aortic root dilatation or dissection, characteristic skeletal abnormalities, ectopia lentis, dural ectasia) and involvement of one other system is sufficient to make the diagnosis. In the absence of a family history (about 30% of cases are the result of sporadic mutations), one major criterion and involvement of two other systems, as in this case, is required to make the diagnosis. New DNA tests for some of the known disease-causing mutations may also assist in the diagnosis.

Treatment

There is no treatment for MFS; however, some physicians recommend the administration of β-adrenergic blockers to lower the blood pressure and the rate of rise of pressure in the aorta, thus slowing the rate of aortic dilatation. In this case, with aortic dilatation already having occurred, surgical replacement of a portion of the aorta and the aortic valve was necessary to correct the defect. Surgical replacement of the mitral valve is also carried out in cases where there is hemodynamically significant incompetence of the valve.

Defects in the Fibrillin Protein Cause the Clinical Manifestations of MFS

Fibrillin is a component of connective tissue

Fibrillin is a widely distributed protein and a major constituent of the elastin-associated microfibrils, first isolated from the medium of cultured human fibroblasts. A well-established histological finding in MFS is fragmentation and disorganization of elastic fibers in the aortic media. Elastic fibers are an essential component of all structures affected by MFS, that is, the wall of the aorta, the leaflets and rings of heart valves, the periosteum, and the ciliary zonule of the eye. A fibrillin-induced defect in microfibrils explains the pleiotropic manifestations of MFS.

Structure of the fibrillin protein

Mature fibrillin-1 undergoes cleavage of a signal peptide and has a molecular weight of 347 kDa. Profibrillin-1 is a secreted, cysteine-rich glycoprotein comprised of 2871 amino acids arranged into five structurally distinct regions (Figure 28.2). The protein contains 47 epidermal growth factor (EGF)-like repeats, 43 of which are of the Ca^{2+}-binding type (cbEGF), and seven 8-cysteine motifs that have homology to a motif first recognized in transforming growth factor β-1-binding protein. Calcium binding appears to orient contiguous cbEGFs into a linearly rigid structure, facilitating fibrillin monomer interaction. The binding of Ca^{2+} is essential for the formation of the macromolecular microfibril structure, and for protection against proteolysis. The most frequently observed mutations that cause MFS are amino acid substitutions in the cbEGF domains that disrupt calcium binding.

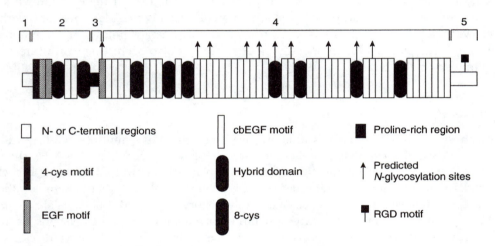

FIGURE 28.2 Structure of fibrillin-1. The protein structure can be divided into five distinct regions: unique N- and C-terminal regions and two cysteine-rich regions separated by a proline-rich region. The 4-Cys, 8-Cys, and hybrid domains are cysteine-rich, and the RGD motif is an integrin-recognition motif. Both the calcium-binding (cbEGF) and non-calcium-binding (EGF) repeats are shown.

Structure of microfibrils

Microfibrils are composed of two types of fibrillin (1 and 2) with very similar structures. The details of the process involved in the formation of elastic fibers is unknown; however, polymerized fibrillin monomers are generally believed to constitute the building blocks of the microfibrillar networks. The composition of these networks with respect to the types of fibrillin is not known, nor has the contribution of each type to the formation and function of microfibrils been distinguished. However, the role of each of the fibrillins in the microfibril is believed to be different, despite their similar structures. The strongest piece of indirect evidence for this conjecture is that mutations in fibrillins-1 and -2 give rise to clinically distinct phenotypes, MFS, and congenital contractural arachondactyly, respectively.

The Fibrillin Gene Is Complex

Most of the EGF-like repeats are encoded by single exons; thus the fibrillin gene is large (~200 Kb), with a highly fragmented (65 exons) coding sequence. Although most of the nearly 200 fibrillin mutations that have been described in MFS patients are missense mutations, a significant number alter RNA splicing. Mutations in *FBN-1* have been found to have effects on synthesis, secretion, and incorporation of fibrillin-1 monomers into the extracellular matrix. The mechanism by which a defect in a single *FBN-1* allele produces MFS is a matter of debate, and may depend upon the particular mutant. A *dominant negative* suppression mechanism, by which mutant fibrillin would sequester normal fibrillin into abnormal elastic fibers, or fibers that have been targeted for degradation, has been suggested.

RNA Splicing Defects

Over 20 mutations in splice junctions of *FBN-1* have been described. These mutations alter the donor (also known as the left or 5′) or acceptor (right or 3′) splice junctions, and many have profound effects on the protein structure. Most of the mutations that disrupt splicing of the fibrillin-1 pre-mRNA produce exon skipping, causing deletion of functionally important regions of the protein. Interestingly, even point mutations outside of either splice junction can cause exon skipping: in some cases introduction of a termination codon into an exon results in skipping of the faulty exon, suggesting that nonsense mutations may affect splice site selection.

Splice site mutations may also result in the use of cryptic splice sites located elsewhere in the intron, causing insertion of a portion of the intron into the coding sequence, with unpredictable results. For example, a G to A transversion at the +1 position of the 46th intron of *FBN-1* was isolated from a patient with MFS. Sequencing revealed the inclusion of the first 33 nucleotides of the intron in the mRNA, as a result of the use of a cryptic splice site found 33 nucleotides downstream of the normal site. Although the mRNA remained in frame, the protein produced was aberrant and resulted in disease symptoms.

Other insertions introduce termination codons into the mRNA that result in the production of truncated fibrillin-1. Premature termination of the polypeptide generally results in a nonfunctional protein; however, if the truncated protein is secreted and retains interaction domains, it may bind normal fibrillin-1 and sequester it from microfibrils. Alternatively, the mutant fibrillin-1 may be incorporated into microfibrils, likely altering their function. In any case, the loss of normal fibrillin-1 compromises the mechanical integrity of elastic structures such as the aortic wall, leading to the phenotypic features of MFS.

QUESTIONS

1. Which step in the generation of an mRNA molecule has the most stringent requirement for precision: transcriptional initiation, splicing, or transcriptional termination?

2. What are some of the limitations for genetic testing in MFS?

3. It has been suggested that mutated fibrillin-1 may produce the phenotype of MFS by a dominant negative effect. Can you speculate as to the mechanism of the dominant negative effect of mutant fibrillin-1?

4. What are the advantages of post-transcriptional processing and splicing of pre-mRNAs?

5. What step(s) of mRNA splicing are splice junction mutations likely to affect?

6. A G + 5 → T transversion in intron 37 was found in an MFS patient, which ultimately resulted in deletion of 16 amino acids, and the substitution of one amino acid, from exon 37. How do you think this occurred?

Exon 37 ←——→ Intron 37

CTG **GGT AAA GCC TGG GGT ACT CCT TGT GAG ATG TGT CCT GCT GTG AAC ACA** T   **gtaattggacat**

Lau Gly Lys Ala Trp Gly Thr Pro Cys Glu Met Cys Pro Ala Val Asn Thr

FIGURE 28.3 Nucleotide sequence at the junction of exon and intron 37 of mutant *FBN-1* gene. Uppercase nucleotides represent exon sequence, and the lowercase nucleotides represent intron sequence. Boldface nucleotides in the exon represent nucleotides deleted as a result of the splice site mutation, and the mutated base at the splice junction is indicated.

FURTHER READING

1. Maron, B. J., et al. Impact of laboratory molecular diagnosis on contemporary diagnostic criteria for genetically transmitted cardiovascular diseases: hypertrophic cardiomyopathy, long-QT syndrome, and Marfan syndrome. A statement for healthcare professionals from the Councils on Clinical Cardiology, Cardiovascular Disease in the Young, and Basic Science, American Heart Association. *Circulation* (1998) 98:1460–1471.

2. Robinson, P. N., and Godfrey, M. The molecular genetics of Marfan syndrome and related microfibrillopathies. *J. Med. Genet.* (2000) 37:9–25.

3. Dietz, H. C., et al. Marfan syndrome caused by a recurrent de novo missense mutation in the fibrillin gene. *Nature* (1991) 352:337–339.

4. De Paepe, A., Devereux, R. B., Dietz, H. C., Hennekam, R. C. M., and Pyeritz, R. E. Revised diagnostic criteria for the Marfan syndrome. *Am. J. Med. Genet.* (1996) 62: 417–426.

5. McKusick, V. A. The cardiovascular aspects of Marfan's syndrome. *Circulation* (1955) 11:321–342.

For further information, see the following web sites:

National Marfan Foundation: http://www.marfan.org/

Online Mendelian Inheritance in Man: http://www.ncbi.nlm.nih.gov/omim/

CHAPTER 29

Protein Synthesis
Acute Then Persistent Otitis Media

CASE HISTORY

Child with Signs of an Ear Infection Visits His Pediatrician

A three-year-old boy is brought to his pediatrician with fever, irritability, and anorexia. He was well until three days earlier when he developed a cough, rhinorrhea, and a low-grade fever of 38.6°C. There was no rash, vomiting, or diarrhea, and he did not complain of ear pain, but was tugging on his right ear lobe. He was given acetaminophen alternating with ibuprofen every 6–8 hours, and Dimetapp in the evenings. Despite this his fever persisted, he was less active and was eating less over the last day, but he was drinking fruit juices. He attended preschool daily, but had been kept home for the past three days. Several of his classmates were out of school with symptoms of a "viral infection."

There Are No Abnormalities in His Medical History

He was born by normal spontaneous vaginal delivery after a normal gestation, with Apgar scores of 8 and 9. His size at birth was appropriate for gestational age. He had been a healthy child without significant medical illness who reached all developmental milestones on time. He had a history of a middle-ear infection about one year ago that was treated with a 10-day course of antibiotics, and several colds since beginning preschool at age 18 months. He has one older brother who had several ear infections as a child and seasonal allergies but no other significant medical illness. His parents were well without significant medical illnesses; his mother was 29 years old at the time of his birth. All four grandparents are alive and well in their mid-60s, and his paternal grandfather has well-controlled hypertension. His father smokes cigars occasionally, but never in the home.

What are the risk factors for the development of acute otitis media (AOM)?

Which bacterial species most frequently cause acute otitis media?

Physical Examination Reveals Inflammation of Right Ear

On physical examination he was alert with fits of crying. There were no skin rashes. The blood pressure was 94/52 mmHg, the heart rate was 118 beats per minute (bpm), and the respiratory rate was 22 per minute and not labored. The head, eyes, ears, nose, and throat examination was remarkable for mild conjunctival injection (redness) and pharyngeal redness without exudates. Otoscopy revealed an erythmatous external auditory canal and tympanic membrane in the right ear. There was an air-fluid level behind the right tympanic membrane and decreased movement of the membrane with pneumatic otoscopy on the right compared to the left ear. There was no evidence for rupture of the tympanic membrane. The left ear revealed only mild injection of the tympanic membrane. His neck was supple without a Kernig's or Brudzinski's sign, and there was shotty cervical adenopathy. The chest was clear and resonant to auscultation and percussion, and there were no intercostal muscle contractions. The heart examination was normal. The abdomen was soft and nontender with normally active bowel sounds. There were no focal deficits on neurological examination.

> *What are the physical signs of AOM?*
>
> *What are the complications of AOM?*
>
> *What is otoscopy?*

He Is Treated with Amoxicillin, Then Clarithromycin When Symptoms Persist

He was begun on a course of 40 mg/kg daily of amoxicillin, divided in doses given every eight hours. On the fourth day he continued to have intermittent fevers and developed an urticarial rash (a rash with hives) on his trunk without lesions of the mucous membranes or difficulty breathing. The physical examination revealed no change in the appearance of the right middle ear. The antibiotics were switched from amoxicillin to clarithromycin, 15 mg/kg/day, divided in doses given every 12 hours for persistent otitis media.

> *What is the time course of response to appropriate antibiotic therapy in AOM?*
>
> *What is the significance of the urticarial rash in the setting of antibiotic treatment?*
>
> *What is persistent versus recurrent otitis media?*

He defervesced in 24 hours and the rash disappeared over the next two days. Clarithromycin was continued for a total of 10 days. A follow-up visit two weeks after the start of antibiotic treatment revealed resolution of the abnormalities of the right middle ear, with no air-fluid level and a normal appearance and movement of both the right and left tympanic membranes.

> *What is the bactericidal mechanism(s) of action of the antimicrobial agents used to treat this child?*
>
> *Avoidance of the use of antibiotics in the management of uncomplicated AOM has been recommended. Why?*

DISCUSSION

Many Antibiotic Drugs Inhibit Translation

Cellular targets are varied

The cellular targets of the antibiotic drugs are many and varied, including cell wall synthesis (e.g., the penicillins, such as the amoxicillin prescribed here), nucleic acid synthesis, cell membranes, metabolic pathways, as well as a large contingent that target protein synthesis (see Table 29.4 of *Biochemistry* 5e). In most cases, the antibiotic protein synthesis inhibitors target the ribosome, and in fact, much of what is known of the various functions of the ribosome was learned from studies of the mechanisms of inhibition of this drug class. Most of the protein synthesis inhibitors target the rRNA components of the ribosome, which play the most active role in ribosomal function.

Spectinomycin inhibits translocation

A number of antibiotics target the 30S subunit, which binds the mRNA and serves as the starting point for the formation of the 70S initiation complex. The 30S subunit, together with the 50S subunit, also participates in the translocation and termination steps. *Spectinomycin* (Figure 29.1), for example, has been known for some time to inhibit the elongation factor G (EF-G)-catalyzed translocation of the tRNAs after each round of peptide-bond formation. Recent crystallographic data have provided the basis for this mode of inhibition: the binding site was localized to helix 34 of the 16S rRNA, a region likely to be involved in translocation.

Streptomycin affects fidelity

Like spectinomycin, *streptomycin* (see Figure 29.1) binds the 16S rRNA, but it affects the fidelity of decoding rather than translational elongation. The ribosome possesses alternative conformations that affect the fidelity of anticodon:codon pairing; the low-fidelity conformation is known as *ram* (ribosomal *ambiguity*), whereas the high-fidelity conformation is called the *restrictive* conformation. A variety of evidence indicates that the function of the conformational change is in discriminating between cognate and noncognate codon–anticodon pairings, with the *ram* form being *less* discriminating than the restrictive form. Thus, a tRNA would bind a ribosome in the *ram* form, and subsequently the switch to the restrictive conformation would increase the prominence of the codon–anticodon interaction in overall binding such that noncognate tRNAs would dissociate. The *ram* form has a higher affinity for tRNA than the restrictive conformation, and thus its function would be to recruit the tRNA; the restrictive form would then play a role in ousting any noncognate tRNAs before the incorrect amino acid is incorporated into the polypeptide.

Crystallographic data have shown that streptomycin interacts with a central portion of 16S rRNA, stabilizing the *ram* conformation, and hence reducing the fidelity of decoding. Streptomycin also makes contacts with the S12 protein, which is also believed to be involved in the conformational change between the *ram* and restrictive forms.

Clarithromycin blocks tunnel

Clarithromycin (a macrolide antibiotic; see Figure 29.1), one of the drugs prescribed for this child, is similar in structure and acts via the same mechanism as *erythromycin*, the prototype of this class of antibiotics. Although the mode of binding of erythromycin and clarithromycin to the ribosome appears to be identical, clarithromycin has a longer serum

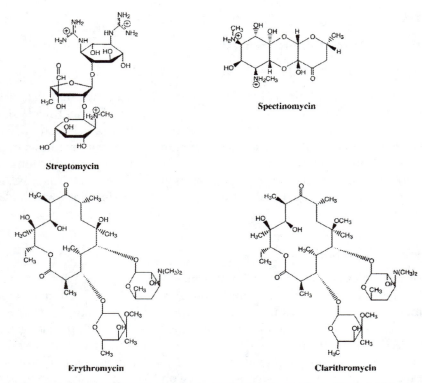

Streptomycin

Spectinomycin

Erythromycin

Clarithromycin

FIGURE 29.1 Chemical structures of some commonly used antibiotic protein synthesis inhibitors.

half-life and better tissue penetration, and is thus clinically more useful. Crystallographic data have shown that clarithromycin binds the 23S rRNA in the *peptidyl transferase center* (see Section 29.3.7 of *Biochemistry* 5e), at a site near the entrance of the tunnel through which the polypeptide chain is threaded during protein synthesis. The nascent protein can reach a length of 6–8 amino acids before the entrance to the tunnel is reached, which is consistent with previous observations indicating that erythromycin-bound ribosomes can synthesize peptides of eight or fewer amino acids. Thus, once the nascent polypeptide reaches the tunnel entrance, all further steps are blocked and protein synthesis comes to a halt.

The variety of antibiotics reflects natural selection among soil microorganisms

The variety of inhibitory mechanisms exhibited by the antibiotic drugs is a reflection of the long-standing contest for survival among microorganisms—most antibiotics used today are synthetic derivatives of natural substances first isolated from soil microorganisms. Their remarkable potency and wide range of activities is a result of the long co-evolution of multiple competing strains. Equally important, this long co-evolution has also allowed for the development of equally intricate mechanisms for evading the toxic effects of antibiotics, and hence the effectiveness of virtually every clinically useful antibiotic is now limited by the spread of drug resistance (see Chapter 2). Thus, many research efforts are now directed toward the development of entirely synthetic drugs against novel targets in the hopes of limiting the potential for development of resistance among microorganisms.

Acute Otitis Media (AOM)

Physiology of hearing

The ear is divided into the external ear, the middle ear, and the inner ear (Figure 29.2). A sound is heard when a sound wave passes through the auditory canal of the outer ear, strikes the tympanic membrane (eardrum), which in turn strikes the ossicular chain (malleus, incus and stapes) of the middle ear. The last ossicle of the chain, the stapes, then strikes the oval window, a flexible membrane that transmits a wave through the fluid-filled cochlea of the inner ear. Stimulation of the hair cells of the cochlea then triggers the transmission of a nerve impulse through the auditory nerve to the brain, where the sound is perceived (see Section 32.4 of *Biochemistry* 5e). The middle ear is connected to the nasopharynx through the eustachian tube, and is thus filled with air. The "popping" in the ears that occurs upon change of altitude results from opening of the eustachian tube and equalization of the pressure on both sides of the tympanic membrane.

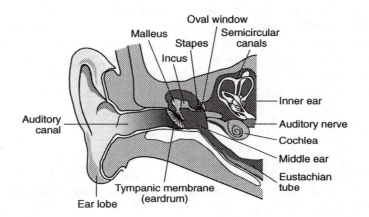

FIGURE 29.2 Structure of the ear, showing the components of the outer, middle, and inner ear.

Cause of AOM

AOM is an inflammation of the middle ear. It usually occurs when the eustachian tube is blocked, frequently as a result of an upper respiratory tract infection that causes inflammation and swelling of the tube. Under these conditions fluid can build up in the middle ear, providing a medium in which bacteria can grow. Otitis media may be caused by viral or bacterial infection; however, fewer than 10% of cases may be attributed to viral infection alone. The bacterial strains most commonly associated with otitis media are *S. pneumoniae, H. influenzae,* and *M. catarrhalis.* Early recurrences of AOM are usually the result of infection with a different organism from the one that caused the first episode. This child, however, suffered from *persistent* otitis media rather than a recurrence. His symptoms did not resolve with the initial amoxicillin treatment, and a second antibiotic had to be prescribed before his symptoms improved.

Treatment of AOM

AOM is the most common infection for which antibiotics are prescribed in the United States, and high doses are typically used to attain sufficient serum concentrations to combat the infection in the closed-space of the middle ear. In addition, high doses are more effective when treating strains that have acquired partial antibiotic resistance. Antibiotic resistance among all the strains that commonly cause AOM is on the rise, and thus treatment is empirical. Amoxicillin is usually the first drug prescribed as it is effective in most cases; however, not only was it ineffectual in this case, but it caused an acute allergic reaction. Other drug treatments, such as amoxicillin/clavulanate, erythromycin/sulfisoxazole, cephalosporin, cephuroxime, or clarithromycin may be used when amoxicillin treatment fails. Subsequent administration of clarithromycin led to the resolution of this child's symptoms. Pain usually subsides within days of treatment; however, fluid may remain in the middle ear for 3–6 weeks, during which time hearing will be less sensitive than normal.

Studies have shown that over 80% of cases of AOM resolve spontaneously, without antibiotic treatment. Other studies have shown that the use of antibiotics simply results in relief from pain two days earlier than without treatment. These findings, together with the growing threat of antibiotic resistance (which is promoted by antibiotic overuse) has led some physicians to advocate withholding antibiotics in uncomplicated cases.

When AOM recurs frequently a surgical procedure called a myringotomy may be performed. This involves making a hole in the tympanic membrane and inserting a small tube, such that the middle ear is open to the air. This will allow any fluid to drain, and allow the child to hear. Without treatment a child with recurrent episodes of AOM may be delayed in learning to speak because of the hearing loss that accompanies each episode.

Symptoms and complications

Ear pain, fever (sometimes as high as 105°F), loss of appetite, and decreased hearing acuity are typical signs of AOM. The fever and irritability noticed by this child's parents are thus typical, and tugging on the ear is a common sign of otitis media in young children. Upon examination, the tympanic membrane is usually red and may bulge. In addition, it does not move as well with insufflation, as measured by pneumatic otoscopy, because of the fluid behind it. Spontaneous perforation of the tympanic membrane may occur; however, it usually heals spontaneously and does not require surgical intervention.

If left untreated the infection may spread to nearby areas of the head or the brain. Complications may be intracranial or extracranial. *Extracranial* complications include hearing loss, facial paralysis, or labyrinthine (inner ear) infection, which may cause vertigo. The most common *intracranial* complication is meningitis; however, epidural abscess, brain abscess, or dural venous thrombophlebitis may also occur. The onset of a severe headache, nausea, or vertigo may signal a serious complication. Complications are rare, even in untreated cases.

Incidence

Otitis media is most common among children between the ages of three months and three years. Seventy-five percent of children under three years of age have experienced an episode of AOM, and nearly half of these experience three or more episodes before age three. Children who attend day care centers or who routinely attend preschool, like this child, are particularly susceptible because of the frequent upper respiratory tract infections they experience. The immature immune system in children, together with the shorter and more horizontal eustachian tube, renders children particularly prone to AOM. With a prevalence of 0.25%, it is rare in adults.

QUESTIONS

1. Why is puromycin, which prematurely terminates protein synthesis, not used for antibiotic drug therapy?

2. What is the difference between a bacteriostatic antibiotic and a bacteriocidal antibiotic?

3. Genetic studies have revealed that mutations in the 50S subunit proteins, L4 and L22, are associated with resistance to erythromycin; however, the crystal structure of erythromycin bound to the 50S subunit revealed that neither of these proteins was close enough to the drug to interact with it. What might account for the results of the genetic studies?

4. Why is otitis media more common in young children than in adults?

5. Treatment with therapeutic levels of aminoglycoside antibiotics (such as gentamycin and streptomycin) can have the catastrophic consequence of inducing irreversible deafness in certain individuals. Susceptibility to the ototoxic effects of these aminoglycosides was found to correlate with a polymorphism in the mitochondrial 12S rRNA (functionally equivalent to bacterial 16S rRNA) gene, at the predicted aminoglycoside binding site. What does this suggest about the mechanism of toxicity?

6. Many of the bacterial strains that cause AOM have become resistant to the drug most commonly used to treat it: penicillin and its related compounds (including amoxicillin). What is the most common resistance mechanism to penicillin?

7. The antibiotic *mupirocin* (pseudomonic acid) inhibits protein translation by inhibiting prokaryotic isoleucine tRNA synthetase. Among the antibiotic protein synthesis inhibitors, what is unique about the mechanism of inhibition employed by mupirocin?

FURTHER READING

1. Schlünzen, F., Zarivach, R., Harms, J., Bashan, A., Tocilj, A., Albrecht, R., Yonath, A., and Franceschi, F. Structural basis for the interaction of antibiotics with the peptidyl transferase centre in eubacteria. *Nature* (2001) 413:814–821.

2. Pichichero, M. E. Acute otitis media: part II. Treatment in an era of increasing antibiotic resistance. *Am. Fam. Physician* (2000) 61:2410–2416.

3. Carter, A. P., Clemons, W. M., Brodersen, D. E., Morgan-Warren, R. J., Wimberly, B. T., and Ramakrishnan, V. Functional insights from the structure of the 30S ribosomal subunit and its interactions with antibiotics. *Nature* (2000) 407:340–348.

4. Spahn, C. M., and Prescott, C. D. Throwing a spanner in the works: antibiotics and the translation apparatus. *J. Mol. Med.* (1996) 74(8):423–439.

5. Green, R., and Noller, H. F. Ribosomes and translation. *Annu. Rev. Biochem.* (1997) 66:679–716.

For further information, see the following web sites:

National Institute on Deafness and Other Communication Disorders:
http://www.nidcd.nih.gov/health/parents/otitismedia.htm

Center for Disease Control—Antimicrobial Resistance Page:
http://www.cdc.gov/drugresistance/

American Academy of Otolarygolomy—Head and Neck Surgery, Ears:
www.entnet.org/ENTNet/healthinfo/ears/index.cfm

CHAPTER 30

Integration of Metabolism
The Curse of the Supermodels

CASE HISTORY

Patient with Eating Disorder Is Admitted to the Hospital

A 17-year-old woman was admitted to the hospital for bradycardia and abnormalities in serum electrolytes. She had a history of an eating disorder, with food restriction and excessive exercising, and currently has a body mass index (BMI) of 14. She was evaluated by her primary physician for a two-week history of decreased oral intake, decreased energy, blurred vision, and increasing bilateral lower-extremity edema. The patient was agreeable to hospitalization because of these symptoms.

> *What is the cause of the edema and blurred vision in this patient?*
>
> *What is anorexia nervosa? What is bulimia nervosa?*

Patient Had a History of Anorexia Nervosa

She first began dieting at age 15 at a weight of 127 lbs. Despite dieting she felt as if she was gaining weight so she began an intensive exercise program approximately 18 months ago, with even more severe restriction of food. She began consistently skipping meals and limiting intake. She lost about 3–4 lbs. per week to a weight of 98 lbs. Nine months prior to this admission she was taken by her parents to her primary care physician, who referred her for inpatient treatment. She initially refused, but ultimately was admitted to an outside hospital for management of anorexia. Her discharge weight at that time was 120 lbs. Since then she had been seen regularly by a psychiatrist and counselor. She had been on paroxetine hydrochloride (Paxil), 30 mg daily, for about four months for obsessive-compulsive disorder. She continued to lose weight and experienced increasing fatigue and lower-extremity edema. She was brought to the emergency room by her parents.

What is Paxil?

What is obsessive-compulsive disorder, and what is its relationship to anorexia?

How could an eating disorder produce edema?

She had no history of being overweight and had not been the object of critical comments about her weight by family or peers. She first became preoccupied with fears of being fat at age 14; her perceived weight at that time was 140 lbs., whereas her actual weight was 118 lbs. She described herself as introverted but had a close circle of friends. She is a senior in high school and is an above average student whose grades have not changed significantly over the past two years. She denied any alcohol or drug use. She did not take diuretics or laxatives. There was no family history of psychiatric disease or eating disorders. She has two younger siblings and both are healthy. Her mother teaches aerobics and her father is an attorney.

Why was she asked about laxative and diuretic use?

The review of systems was remarkable for menarche at age 13, but she has been amenorrheic for the last two years. She had been sexually active in the past but was not currently in a relationship. Other medical problems included thrombocytopenia, diplopia, and rhabdomyolysis, all felt to be secondary to her eating disorder.

Physical Examination

Her physical examination was remarkable for a weight of 94 lbs. and height of 59.5 inches. She was afebrile and the blood pressure was 102/74 mmHg, the heart rate was 48 beats per minute (bpm), and the respiratory rate was 16 per minute. She was cachectic but in no acute distress. Her head, eyes, ears, nose, and throat examination was remarkable for thinning hair and temporal muscle wasting. The pupils were equal, round, reactive to light and accommodation, with full extraocular motion. The neck was supple without thyromegaly or adenopathy.

The lungs were resonant to percussion and clear to auscultation. Heart examination revealed a slow but regular rate with a grade II/VI systolic ejection murmur heard best along the left sternal border. The pulses were 1+ and symmetrical in the distal lower extremities, and full and symmetrical elsewhere. She had 3+ pitting edema in both legs to the knees, without clubbing or cyanosis. The abdomen was extremely scaphoid, with active bowel sounds and no hepatosplenomegaly. She had profound wasting in all muscle groups, with weakness commensurate with the degree of wasting.

She was alert and fully oriented, with full and prompt speech. There were no sensory defects. The gait was slow but steady. Cerebellar examination was normal. She stated that her mood was good, and she denied suicidal/homicidal ideation, phobias, obsessions, compulsions, false perceptions, or panic attacks. The score on a mini-mental examination was 29/30, but her judgment and insight were poor.

Laboratory Evaluation

On admission to the hospital her serum electrolytes were remarkable for a low potassium of 2.8 mEq/l. Her creatinine was low at 0.4 mg/dl, with a blood urea nitrogen (BUN) of 20 mg/dl. Blood chemistries revealed a low phosphate of 1.1mg/dl, total calcium of 7.9

mg/dl, ionized calcium of 1.21 mMol/l (normal range: 1.3–1.32 mMol/l), total protein of 5.5 g/dl, albumin of 3.0 g/dl, and a glucose of 55 mg/dl. Liver function tests (LFTs) were abnormal, with an elevated alanine amino transferase (ALT) of 1060 IU/l and asparate amino transferase (AST) of 1350 IU/l, and a normal alkaline phosphatase. The pro-thrombin time (PT) was elevated at 19.5 seconds, with an international normalized ratio (INR) of 2.7, but the partial thromboplastin time (PTT) was normal.

Why are the serum potassium and phosphate low?

The ionized calcium in this patient was normal despite a low total calcium. Can you explain this finding?

Why are the PT and INR abnormal?

The white blood cell count was high at 15.8/mm³; she had a normal hemoglobin concentration and normal hematocrit. The platelet count was depressed at 79,000/mm³ (lower limits of normal: 150,000/mm³). Thyroid function tests were normal. The chest radiogram was normal and the electrocardiogram revealed sinus bradycardia, but was oth-erwise unremarkable.

Treatment

The patient was hospitalized on Eating Disorders Service. She was started on a 1500 kcal per day diet and maintained on intravenous fluids to correct the dehydration, hypo-glycemia, and electrolyte abnormalities. She was initially unable to take in enough calo-ries to meet her nutritional needs, so Ensure plus supplements were added to her diet. Her electrolytes were monitored daily, and she developed worsening hypophosphatemia that required intravenous supplementation.

What are the clinical manifestations of hypophosphatemia?

With improvement in her caloric intake, her liver function abnormalities resolved and total protein and albumin increased with resolution of the peripheral edema. The bradycardia improved with her heart rate increasing to the 60–70 bpm range. An echocardiogram revealed normal heart chamber sizes and function. The hematological problems, abnormal clotting, and low platelet count improved with refeeding. A noncontrast head and orbital computed tomo-graphic scan revealed no abnormalities, and her blurred vision improved over time.

She and her family participated in group and family counseling, which continued after hospital discharge.

DISCUSSION

Metabolic Effects of Starvation

Evolution has favored the ability to survive starvation

In the parts of the world where food is plentiful most healthy individuals alternate between mild starved and fed states, and rarely endure a period of starvation that lasts for more than a few hours. However, humanity evolved under less benign circumstances, when food supplies fluctuated unpredictably. We thus developed the ability to store extra calories in the form of triacylglycerols when food was plentiful, and the ability to mobilize them when it was scarce (Figure 30.1). Indeed, individuals of normal weight can survive for 1–3 months without food, and overweight individuals may survive for many months longer.

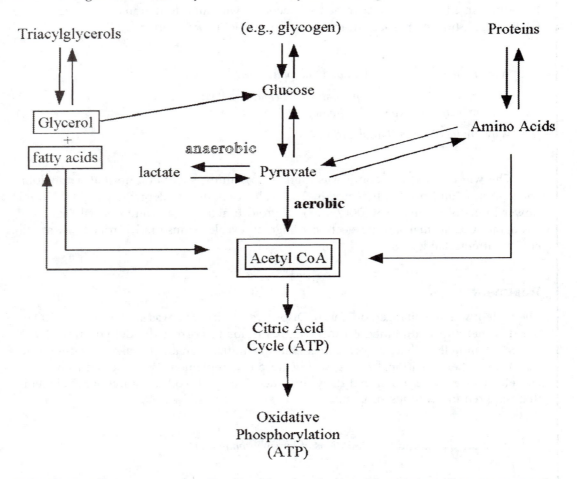

FIGURE 30.1 Overview of metabolism. Triacylglycerols, polysaccharides, and proteins may all serve as sources of energy. Acetyl CoA, which may be generated from all three types of macromolecules, is a central focus of metabolism. Glycogen is an important short-term energy reserve, providing energy for ~1 day; however, triacylglycerols are the body's main energy reserve, with sufficient stores to sustain life for several weeks or months. Acetyl CoA derived from fatty acid catabolism cannot be used to generate glucose because acetyl CoA cannot be converted into pyruvate, and thus, early in the course of starvation amino acids are important substrates in the generation of glucose via gluconeogenesis. The brain depends upon glucose as a source of energy in the first few days of starvation. Later in the course of starvation the brain adapts to the use of ketone bodies (acetoacetate and 3-hydroxybutyrate) derived from acetyl CoA, thus reducing the dependence on protein catabolism.

First glycogen, then triacylglycerols, as primary energy sources during starvation

Although protein may be used as a source of energy during starvation, overdependence on it would rapidly become detrimental to the organism because proteins carry out many vital functions throughout the body. Excessive breakdown of protein would result in the destruction of bodily tissues and eventually lead to death from organ failure. Thus, although protein is utilized early in the course of starvation to generate glucose via *gluconeogenesis*, after about three days gluconeogenesis slows, thereby sparing bodily proteins. At this point, the body becomes increasingly dependent upon *triacylglycerols* as a source of energy. The brain, which previously depended on glucose as its primary energy source, adapts to the use of ketones, derived from the breakdown of triacylglycerols (see Section 30.3.1 of *Biochemistry* 5e). Along with adaptive changes in substrate utilization, the metabolic rate gradually decreases with starvation, and after one month of fasting the resting metabolic rate is reduced by as much as 30%.

When stores of triacylglycerols become depleted, no other energy stores remain (glycogen having been utilized early in the course of starvation), and proteins become the primary source of ATP. Serious clinical manifestations soon appear, leading to death unless an outside source of food is obtained.

Anorexia Nervosa Is an Eating Disorder

Anorexia nervosa is a curious disorder characterized by *voluntary* self-starvation, or the refusal to maintain normal body weight. Like *bulimia nervosa* and *binge-eating,* it is an *eating disorder.* There are two types, the *restricting* type and the *binge-eating/purging* type, with the former being characterized by restriction of food intake and excessive exercise. In the binge-eating/purging type, binges are compensated for by self-induced vomiting or abuse of laxatives, diuretics, or enemas. What distinguishes bulimia nervosa from this type of anorexia is the weight of the patient: those with anorexia are by definition underweight, whereas those with bulimia are of normal or above-normal weight. Binge-eating disorder is characterized by bingeing without compensatory purging behavior. These patients are typically obese, but may be distinguished from other obese patients by increased frequency of depression and anxiety.

Etiology

Cultural factors

The etiology of anorexia is unknown; however, it is likely to be multifactorial. Cultural, psychological, and physiological factors are believed to be involved. Ninety-five percent of patients are adolescent women, and most cases occur in Western industrialized countries, where a slender physique is particularly valued. Even children are aware of these attitudes, and two-thirds of adolescent girls have dieted or taken other measures to control their weight. However, only a small fraction become anorexic and thus other factors contribute to the disorder. Although anorexia nervosa is becoming increasingly common, it has been recognized for centuries, well before the notion of thinness as the ideal body type became pervasive.

Psychological factors

Those who develop anorexia are frequently perfectionists, and hence the concurrent obsessive-compulsive disorder in this patient is not unusual. In addition, they are typically competitive and high achievers, placing high demands upon themselves in work or school. The disorder usually begins as a diet, and patients feel a sense of accomplishment in losing weight; conversely, weight gain is viewed as a personal failure. A low self-image is also associated with the illness, as is a history of depression and anxiety. Weight loss gives patients a sense of control and enhances self-esteem. A family history of mental disorders, as well as traumatic experiences (such as sexual abuse), also seems to increase susceptibility.

Physiological factors

A number of physiological changes have been found to occur in those with anorexia; however, the role these play in the etiology is unclear. Most of the changes vanish with a return to normal eating habits, suggesting that these are more likely to be secondary to the eating disorder. Although as yet undefined, genetic factors have been implicated from the high concordance rates among monozygotic twins compared to dizygotic twins.

Incidence

Women in the United States have between 0.5% and 1% risk of developing anorexia nervosa in their lifetimes. The disorder also occurs in men, but much less frequently, with only approximately 5% of cases occurring in men. The incidence is similar in other industrialized countries where food is prevalent and thinness is valued; however, it is rare in developing nations where food is scarce. Dancers, models, athletes, actors, and others for whom thinness is particularly emphasized are at increased risk for the disorder. Anorexia is most common among middle- and upper-socioeconomic strata. The disorder typically strikes in adolescence (~75% of cases) but may occur at any age.

Diagnosis

The diagnosis of anorexia nervosa rests upon four main criteria. The first is a lower than normal weight, defined as <85% of normal, which corresponds to a BMI of roughly 18.5 kg/m^2. Second, patients have an irrational fear of becoming fat and resist gaining weight for fear of weight gain getting out of control. Third, they have a distorted body image. Although even healthy women without eating disorders typically have distorted body images, this is much more pronounced in those with eating disorders. Despite being emaciated, women with anorexia perceive themselves (or certain parts of themselves) as being fat. The final criterion is amenorrhea, the absence of at least three menstrual cycles in postmenarchal females. This reflects diminished release of gonadotropin-releasing hormone (GnRH) from the hypothalamus. The release of GnRH is highly sensitive to body weight, exercise, and stress, all of which are likely to be factors in patients with anorexia nervosa.

Clinical Features

Begins with a diet, which increasingly becomes the focus of mental life

The disorder typically begins in adolescence with a diet, and as weight loss progresses, patients become increasingly preoccupied with it. They obsessively fantasize about food, but rarely indulge in any of the items over which they obsess. They develop idiosyncrasies in their eating behavior, and often attempt to conceal their eating habits and purging behaviors. Excessive exercise is also a common feature of the disorder. Patients typically deny the disorder and are usually brought for treatment by friends or family.

Damage to the organs and tissues

On examination patients are thin and may complain of constipation and intolerance to cold. All bodily organs are affected by starvation, but the greatest loss of organ mass occurs in the liver and intestines, followed by the heart and kidneys. The absorptive surface of the gastrointestinal tract shrinks, thereby exacerbating malnutrition. The heart weakens and reduced cardiac output and bradycardia are typical findings. The blood pressure decreases, as does the respiratory rate. Muscle mass decreases, which is manifested as weakness, as in this patient. The elevated prothrombin time and abnormal LFTs reflect damage to the liver.

Electrolyte disturbances and nutritional deficiencies

Electrolyte disturbances are common among those with anorexia nervosa, and are particularly dangerous as they can cause cardiac disturbances and lead to sudden death. Dehydration, reduced nutrient intake, and the use of diuretics all contribute to the problem. The hypokalemia found in this patient is particularly dangerous, and may lead to life-threatening cardiac rhythm disturbances. Dehydration, such as that caused by diuretics, often triggers a feedback response that causes water retention, or "rebound" edema, and patients frequently notice swelling in their extremities and faces. Low serum protein levels (as a result of malnutrition) contribute to edema as well: serum proteins, especially albumin, are important mediators of serum osmolarity, and when their concentration decreases, water tends to pass through the vessel walls and into the tissues, causing swelling. Dehydration often causes slight elevations in BUN; the other serum marker of renal function, the creatinine, is often low because its source is skeletal muscle, which is greatly reduced in mass in these patients.

Reduced calcium intake causes hypocalcemia, which leads to bone resorption and the loss of bone mass, which may be irreversible. However, in this patient the ionized calcium level was normal despite low total calcium levels. Albumin is a major serum calcium-binding protein, and when albumin levels are low, the free or ionized calcium concentration increases. The ionized calcium level is a better indicator of what is available at the cellular level. Low blood sugar (hypoglycemia) is also common, and reflects low intake and the low rate of gluconeogenesis during prolonged starvation.

Prognosis

Mortality is 10–20% among those with anorexia nervosa, with most deaths resulting from starvation (usually the result of cardiac abnormalities) or suicide. Fifty percent of patients recover completely; however, approximately 40% continue to have residual signs of the disorder and remain thin or emaciated. Few long-term adverse consequences are associated with those that make a full recovery, with the notable exception of osteopenia, which may not normalize. Peak bone mass is attained during the adolescent years, and thus adequate nutrition is crucial during this period.

Treatment

The primary goal of treatment is the restoration of 90% normal weight. The initial treatment depends upon the extent of starvation, and hospitalization is recommended for patients <75% of normal body weight. Acute problems, such as the dehydration, electrolyte disturbances, and hypoglycemia found in this patient, are addressed, and restoration of nutrition is initiated. Feeding can almost always be accomplished orally, under the close supervision of hospital staff.

Psychiatric treatment is important to long-term treatment. Patients with anorexia channel much of their energy into losing weight, and this "accomplishment" becomes their primary source of self-esteem. The aim of therapy is to redirect the patient's efforts into more meaningful endeavors and to enhance the self-esteem they derive from these. This, of course, is easier said than done, especially because most patients deny having the disorder and resist treatment. Patients require extensive emotional support during the recovery phase, and the involvement of family members is crucial, especially among younger patients.

QUESTIONS

1. Why is the name *anorexia nervosa* a misnomer?

2. During fasting the focus of metabolism can be said to be on providing fuel to the brain. Once glycogen stores in the liver have been exhausted, there is first a short-term and then a long-term solution to providing the brain with energy. What are these?

3. Describe the role of triacylglycerols in providing fuel to the brain during fasting.

4. Upon *refeeding* of patients with anorexia nervosa, the serum phosphate levels must be closely monitored to avoid a precipitate and dangerous drop. What accounts for the hypophosphatemia that often occurs upon refeeding (and that occurred in this patient)?

5. What is the role of glucagon in mobilizing energy sources during starvation?

6. A 22-year-old woman presents to clinic distressed about her bingeing and purging behavior. She admits to bingeing 7–8 times per week and purging using laxatives and diuretics. Her height is 162 cm (5 ft., 3 in.) and her weight is 57 kg (125.4 lbs.). Would you be inclined to diagnose this patient with anorexia nervosa?

7. Early in the course of a fast the brain relies heavily upon liver glycogen stores, which last for approximately one day, as a source of glucose. Why are the muscle glycogen stores, which account for 75% of total bodily glycogen, not utilized once liver stores are depleted, thus sparing valuable protein?

FURTHER READING

1. Melchior, J. C. From malnutrition to refeeding during anorexia nervosa. *Curr. Opin. Clin. Nutr. Metab. Care.* (1998) 1(6):481–485.

2. Walsh, B. T., and Devlin, M. J. Eating disorders: Progress and problems. *Science* (1998) 280(5368):1387–1390.

3. Powers, P. S., and Santana, C. A. Eating disorders: A guide for the primary care physician. *Prim. Care* (2002) 29(1):81–98, vii.

4. Herzog, W., Deter, H. C., Fiehn, W., and Petzold, E. Medical findings and predictors of long-term physical outcome in anorexia nervosa: a prospective, 12-year follow-up study. *Psychol. Med.* (1997) 27(2):269–279.

6. Haglin, L. Hypophosphataemia in anorexia nervosa. *Postgrad. Med. J.* (2001) 77(907):305–311.

For further information, see the following web sites:

National Eating Disorders Association: http://www.nationaleatingdisorders.org/

Anorexia Nervosa and Related Eating Disorders: http://www.anred.com/

Academy for Eating Disorders: http://www.aedweb.org/

CHAPTER 31

Control of Gene Expression
The Death Cap and RNA Polymerase

Amateur Mycologist Taken to the Hospital with Abdominal Pain

A 22-year-old girl was transferred to this hospital for management of acute liver failure. She was transferred from a community hospital one day after ingesting wild mushrooms she had collected from a local park. She had consumed about 500 g of sautéed mushrooms. Approximately 10 hours later she developed nausea, vomiting, abdominal pain, and diarrhea. She was taken by her roommate (who had not consumed the mushrooms) to a local hospital, where she was found to be alert and oriented, but in moderate distress because of the gastrointestinal symptoms. She was not jaundiced and had a low-normal blood pressure of 92/60 mmHg and a heart rate of 114 beats per minute (bpm). The laboratory evaluation was remarkable for a low serum potassium and mild anemia, but the liver function tests (LFT) were normal. She was given intravenous fluids and underwent gastric lavage and administration of activated charcoal by nasogastric tube. The regional poison control center was notified and recommended the transfer of the patient to a tertiary care hospital.

She had no significant past medical history, did not smoke, drank alcohol socially, and denied using illicit drugs. She was an amateur botanist and mycologist and had picked and eaten wild mushrooms in the past without incident. She was not taking any medicines regularly.

What is the significance of the consumption of wild mushrooms to her clinical presentation?

Can you speculate as to the etiology of the gastrointestinal symptoms?

Why was the serum potassium low?

What is activated charcoal and why was it given?

She Is Transferred to a Tertiary Care Hospital, and Her Condition Improves Somewhat

On arrival at this hospital, about 30 hours after the initial ingestion, she was somnolent but easily arousable and oriented to person, place, and situation. Her gastrointestinal symptoms had subsided. Her blood pressure and heart rate were unchanged. Her skin and sclerae were non-icteric (not yellow). There were no stigmata of chronic liver disease. The chest and heart examinations were normal except for tachycardia. The abdomen was soft with audible bowel sounds and diffuse tenderness; there were no masses or hepatosplenomegaly. The neurological examination revealed no focal deficits.

Laboratory Results Reveal Liver and Kidney Dysfunction

The laboratory evaluation revealed anemia with a hematocrit of 31%; she had abnormal blood coagulation with a prothrombin time (PT) of 24 seconds (upper limit of normal [ULN]: 13 seconds), with an international normalized ratio (INR) of 6.0. LFT were abnormal with an aspartate amino transferase (AST) of 240 IU/l (ULN: 31 IU/l) and an alanine amino transferase (ALT) of 190 IU/l (ULN: 31 IU/l), total bilirubin (TB) of 1.2 mg/dl (ULN: 1.2 mg/dl), and an alkaline phosphatase of 196 IU/l (ULN: 120 IU/l). Tests for viral hepatitis A, B, and C and for heterophil antibodies were negative. Renal function tests were abnormal with a blood urea nitrogen (BUN) of 36 mg/dl (ULN: 20 mg/dl) and a creatinine of 2.0 mg/dl (ULN: 1.4 mg/dl). She was treated with charcoal hemoperfusion for presumed mushroom poisoning. Her roommate retrieved some uneaten mushrooms that were found to be *Amanita phalloides* (Death cap).

What is the mechanism of liver damage in mushroom poisoning?

How does charcoal hemoperfusion work?

What are heterophil antibodies?

Her Condition Deteriorates

Despite treatment the patient continued to worsen, developing stage 2 hepatic encephalopathy with increasing somnolence, disorientation and asterixis, a continued rise in liver enzymes (AST, ALT, alkaline phosphatase) and TB, a worsening coagulopathy (INR rising to 12), and progressive abnormality of the renal function tests. Vitamin K was administered subcutaneously and she was begun on penicillin G, ascorbic acid, and cimetidine. She continued to worsen, with deterioration of her mental state and descent into stage 3 hepatic encephalopathy. The coagulopathy also worsened despite treatment with vitamin K.

What is hepatic encephalopathy and what are its stages?

How do penicillin G, ascorbic acid, and cimetidine work in liver failure?

Vitamin K was given to treat the bleeding disorder. How does this vitamin work to reverse abnormal blood clotting?

Liver Transplantation Performed in the Face of Liver Failure

On the sixth hospital day she underwent orthotopic liver transplantation for fulminant hepatic failure. Histological examination of the explanted liver revealed panacinar necrosis with hemorrhage, consistent with toxic injury. She progressively improved after the transplant and was discharged on corticosteroids and immunosuppressive agents after a three-week hospitalization. She continues to do well one year after transplantation.

DISCUSSION

Mycetismus and *Amanita* Species

Amanita species are most poisonous

Mycetismus, or mushroom poisoning, is a rare and sometimes fatal cause of acute liver failure, and as such is a true medical emergency. There are over 5000 species of mushrooms, but fewer than 100 are poisonous to humans. Unfortunately, even trained mycologists can be fooled and may accidentally consume these potentially lethal fungi. The *Amanita* family of mushrooms accounts for the majority of mushroom poisonings worldwide, with *Amanita phalloides* thought to be responsible for more than 90% of all fatal cases. This mushroom has been referred to as the "death cap," "death angel," or "destroying angel," and as little as 0.1 mg/kg of amatoxin can be lethal for adults. There are approximately 5–15 mg of amatoxin per dried *Amanita* cap, therefore as few as two caps can result in death.

Amanita poisoning is on the rise in the United States

Fatalities are more common in Western Europe, where amateur mushroom gathering is popular, but the increasing popularity in the United States has been associated with a rise in the incidence of poisoning. *Amanita* species are frequently found in the Pacific Northwest, and have also been reported to grow in the Blue Ridge Mountains, Pennsylvania, New Jersey, New York, and Ohio. The mushrooms tend to grow in oak woodlands in the fall or rainy season. Poisonous members of the *Amanita* genus are not distinctive in appearance and can be easily confused with edible members of this family. Ensuring that wild mushrooms are not eaten unless identified as nonpoisonous by a competent mycologist can largely prevent accidental ingestion of toxic *Amanita* species. Field guides are not sufficient to differentiate toxic from nontoxic species.

Toxins and Targets

Two types of toxins

Two distinct groups of toxins cause the clinical features associated with ingestion of *Amanita* mushrooms—the *phallotoxins* and the *amatoxins*. The toxins are highly concentrated in the ring, gills, and cap, and to a lesser extent in the stalk as well. The phallotoxins, including the cyclical heptapeptide, phalloidin (Figure 31.1), are credited with the gastrointestinal symptoms associated with *Amanita* mushroom poisoning. Although there is some doubt as to the mechanism of toxicity, they are known to interfere with actin polymerization.

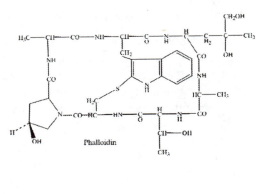

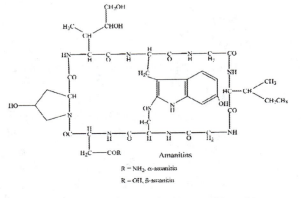

FIGURE 31.1 Chemical structures of phalloidin, α-amanitin, and β-amanitin.

Amanitins inhibit RNA polymerase II

There is little doubt as to the toxicity of the amatoxins (α-, β-, γ-, ε-amanitin), a group of cyclic octapeptides that are not destroyed by cooking, drying, or digestive enzymes. They, especially α- and β-amanitin (see Figure 31.1), are responsible for most of the morbidity and the mortality associated with mushroom poisoning. The amanitins bind noncompetitively to the largest subunit of RNA polymerase II, encoded by the *RPB1* gene, and interfere with pre-messenger RNA (pre-mRNA) synthesis. Crystallographic data have shown that α-amanitin binds the "bridge helix" that spans the nucleic acid binding cleft of the RNA polymerase II holoenzyme. These, together with biochemical and genetic data, indicate that the amanitins inhibit the translocation of the DNA and RNA through the holoenzyme, thereby inhibiting incorporation of nucleotides into the growing RNA chain. RNA polymerase III is also affected by amanitins; however, the effect is transient and the clinical consequences are relatively benign.

Thus, unlike cellular transcription factors and repressors, which regulate transcription by binding the DNA (or DNA-bound factors) and either recruiting or inhibiting the binding of other factors (see Chapter 31 of *Biochemistry* 5e), amanitins bind directly to the polymerase near the active site and inhibit the enzymatic activity.

Toxic effects are the result of lack of protein synthesis

The toxic effects of the amanitins are thought to result from the arrest of translation due to a lack of mRNA transcripts. In the absence of translation, proteins vital to cell survival are not be synthesized and the cell soon dies. The liver and kidneys are particularly affected, as exemplified by this patient. The amanitins undergo *enterohepatic circulation,* in which metabolites (such as bilirubin and the bile salts) are conjugated in the liver and secreted into the intestine as bile; they are subsequently deconjugated in the intestine by bacterial enzymes, absorbed through the intestinal mucosa, and ultimately returned to the liver through the circulation.

Amanitins thus become concentrated in the liver, increasing the risk of severe liver damage. The kidneys rapidly clear amanitins; however, the direct toxic effect on the tubular epithelium often results in renal failure. Fatalities are usually the result of a combination of hepatic and renal failure. In fatal cases or those requiring transplantation, pathological examination of the liver reveals massive centrilobular necrosis; vacuolar degeneration; and a positive acid-phosphatase reaction, a reflection of increased catabolic activity of liver cells. The kidneys often show signs of acute tubular necrosis and hyaline casts in the tubules.

Mushroom Poisoning—A Rapidly Progressive Intoxication

There are four clinical stages

The clinical syndrome associated with *Amanita phalloides* poisoning is characterized by four stages. The first stage, or the incubation stage, is a period of quiescence that lasts for 8–12 hours after ingestion. The second stage is characterized by gastrointestinal symptoms, including crampy abdominal pain, profuse watery diarrhea, and vomiting; it may take as little as 6 or as long as 16 hours for the patient to become symptomatic. Ideally, treatment is begun at this stage, but this only occurs if the clinician has a high index of suspicion and elicits a history of mushroom ingestion. The third stage occurs 24–48 hours after ingestion, and is characterized by rapidly progressive clinical and biochemical evidence of liver damage.

The terminal stage of poisoning occurs 4–7 days after ingestion. The patient develops oliguric (no urine output) renal failure, worsening liver failure, rising ammonia levels, hepatic coma and convulsions, hemorrhage, respiratory failure, and ultimately death. Factors that are associated with fatal mushroom poisoning are the size of the dose, the latency of symptom onset, severity of liver dysfunction, and age, with a very high mortality in children less than 10 years old.

This patient clearly passed through all four clinical stages. Her gastrointestinal symptoms began 10 hours after mushroom consumption, and after a brief respite, she began to show signs of liver and kidney disease, confirmed by the laboratory evaluation. She suffered from a progressive *coagulopathy* (many clotting factors are synthesized in the liver), and the serum levels of liver enzymes—AST, ALT and alkaline phosphatase—released upon necrosis of hepatocytes, were elevated. The elevated bilirubin level is a further sign of liver disease, as this metabolite of hemoglobin is normally eliminated by the liver via the bile. She also exhibited increasing signs of *hepatic encephalopathy,* a consequence of severe liver dysfunction. Her renal function was also impaired, as her BUN and serum creatinine levels were above normal. Fortunately, a graft was available, and she received a life-saving liver transplant and returned to health.

Treatment

Treatment focuses on support and removal of toxins

There is no antidote for *Amanita phalloides* poisoning. Treatment focuses on support and removal of as much of the toxin as possible. The mainstays of therapy include vigorous intravenous fluid replacement and correction of the electrolyte disturbances that result from the gastrointestinal symptoms. Treatment of the consequences of liver dysfunction, in particular, correction of coagulopathy with vitamin K and fresh frozen plasma (FFP), is essential to reduce the risk of hemorrhage.

Removal of the toxins is most effective if instituted early (within 30 minutes of ingestion); however, most patients, like this one, do not present until the gastrointestinal symptoms occur, typically no earlier than six hours after ingestion. Gastric lavage and administration of repeated doses of activated charcoal may remove any unabsorbed mushrooms from the gut and interrupt the enterohepatic circulation of the toxins. The network of pores in activated charcoal can adsorb as much as 100–1000 mg of toxin per gram of charcoal. Forced diuresis will also speed the clearance of the toxins from the circulation.

Adjunctive therapies

A number of pharmacologic therapies have been proposed for treating *Amanita* poisoning; however, the rarity of the condition makes for a lack of well-designed clinical studies. Nevertheless, several therapies have been reported to be beneficial. Administration of intravenous *penicillin G* is often recommended because of its ability to displace amanitin from plasma protein sites, thus promoting its renal excretion. Penicillin G may also antagonize uptake and penetration of toxins into the hepatocyte. Cimetidine, a cytochrome P450 inhibitor, may inhibit the conversion of mushroom toxins to more toxic forms, and the antioxidant properties of ascorbic acid may be beneficial. Silibinin (from the milk thistle *silybum marianum*), has putative hepatoprotective effects by inhibiting the penetration of amanitin into hepatocytes and competing with amatoxins for transmembrane transport. Silibilin is widely used in Europe; however, its availability is limited in the United States. Early hemoperfusion (within 24 hours of exposure) over a charcoal filter is often carried out, particularly if the patient has ingested a potentially lethal dose (>50 g). The only definitive treatment may be liver transplantation once fulminant liver failure occurs; this is always considered in patients with progressive elevation in liver enzymes and derangement of clotting factors.

QUESTIONS

1. Treatment of mammalian tissue culture cells with 100 nM α-amanitin and analysis of RNA levels three hours later revealed a profound reduction in mRNA synthesis, while having little effect on rRNA and tRNA levels. Can you explain this? How would the outcome have differed had the α-amanitin concentration been 10μM?

2. Although not directly inhibited by amanitins, a decrease in RNA polymerase I-dependent transcription is eventually observed in amanitin-treated cells. How might this be explained?

3. What is the cause of the blood clotting abnormality in patients with liver failure?

4. What accounts for the hypoglycemia associated with cases of *Amanita* mushroom poisoning?

5. Would you expect that high levels of ribonucleotides would overcome the inhibition caused by α-amanitin binding?

6. Phosphorylation of the C-terminal domain (CTD) of RPB1 has been found to be an important mediator of the activity of RNA polymerase II. How does the CTD modulate gene transcription, and do the amanitins modulate transcription by targeting this domain?

7. Do cellular transcriptional repressors generally work via mechanisms similar to the amanitins?

8. Rifampicin and actinomycin D are two antibiotic inhibitors of gene transcription, and each acts via a distinct mechanism (see Section 28.1.9 of *Biochemistry* 5e). Which is more similar to the inhibitory mechanism employed by α-amanitin?

FURTHER READING

1. Vetter, J. Toxins of *Amanita Phalloides*. *Toxicon* (1998) 36:13–24.

2. MMWR. *Amanita Phalloides* mushroom poisoning—Northern California, January (1997) 46 (22):489–492.

3. Köppel, C. Clinical symptomatology and management of mushroom poisoning. *Toxicon* 31:1513–1540.

4. Klein, A. S., et al. *Amanita* poisoning: treatment and the role of liver transplantation. *Am. J. Med.* (1989) 86:187–193.

5. Courey, A. J., and Jia, S. Transcriptional repression: the long and the short of it. *Gen. & Dev.* (2001) 15(21):2786–2796.

6. Bushnell, D. A., Cramer, P., and Kornberg, R. D. Structural basis of transcription: Amanitin-RNA polymerase II cocrystal at 2.8 Å resolution. *PNAS* (2002) 99(3):1218–1222.

For further information, see the following web sites:

American Association of Poison Control Centers: www.aapcc.org/

Food and Drug Administration, Center for Food Safety and Applied Nutrition, mushroom toxins: http://vm.cfsan.fda.gov/~mow/chap40.html

CHAPTER 32

Sensory Systems
Pain: A Sensory System Without an Organ

CASE HISTORY

A 60-year-old, right-handed woman was evaluated in the neurologic clinic because of severe pain and tingling in her hands. She had been in fair health until three months earlier, when she had the sudden onset of paroxysms of intense burning pain in both hands, worse in the right hand, with a constant, less severe pain with numbness also worse on the right side. She also experienced weakness of both hands. The paroxysmal pain varied in intensity but was generally more severe when she was trying to sleep. She was unable to sleep for more than a few hours even with her hands elevated. For the past six months she noticed difficulty with any task that required her to grip an object such as opening a door or jar, turning on a faucet, even brushing her hair. Because of pain and weakness other routine tasks such as writing, cooking, and driving a car became impossible. *Nonsteroidal anti-inflammatory drugs (NSAIDs)* such as ibuprofen, naprosyn, or indomethacin failed to provide satisfactory relief from the pain. Mild narcotic analgesics, such as acetaminophen with codeine, did not provide pain relief.

How do the NSAIDs exert their effects? What is their target?

Do the narcotic analgesics function in the same way?

The patient had anorexia, constipation, and a weight loss of 23 pounds during the six months before evaluation in the neurologic clinic. She was known to have type II diabetes mellitus and hypertension. There was no prior history of heart disease. She did not smoke and had no significant history of occupational exposures or repetitive motion disease.

The Neurologic Exam of Her Extremities

Neurologic examination revealed limited motion of the metacarpophalangeal and interphalangeal joints, without warmth, tenderness, or redness. She could not form a fully clenched fist with either hand. There was mild atrophy of the thenar eminences (muscle bulges at the base of the thumbs) of both hands but no significant atrophy of the proximal arm or hand muscles. Percussion over the ventral aspect of the wrist elicited pain and tingling in the hand over the first through third fingers (Figure 32.1). There was weakness in palmar abduction of the thumbs, with normal strength in the other fingers. Sensation of light touch and pinprick and two-point discrimination were markedly impaired over the thumb, second, and third digits of both hands. Sensation was normal in the remainder of the hands and arms. There was reduced sensation to light touch and two-point discrimination in the lower extremities bilaterally to the mid calf. The deep tendon reflexes were normal and symmetric in the arms and legs; the plantar responses were flexor.

> *What is two-point discrimination and how is it measured?*
>
> *What are the deep tendon reflexes and plantar responses?*

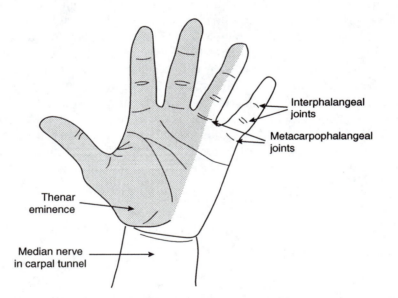

FIGURE 32.1 Anatomy of the hand and Tinel's sign. Pain and tingling in the thumb, first, second, and half of the third finger (shaded) elicited by light tapping of the inner wrist indicate a positive Tinel's sign. A positive Tinel's sign is associated with carpal tunnel syndrome. The metacarpophalangeal joints lie between the fingers and the hand, and the interphalangeal joints are the joints in the fingers. The thenar eminences are the bulges at the base of the thumb and contain several muscles, including the abductor pollicis brevis muscles, which exhibited abnormal patterns of activity in the EMG performed on this patient.

Laboratory and Radiographic Findings

Her medicines included insulin injected subcutaneously twice daily; lisinopril, 20 mg daily; aspirin, 81 mg daily; and acetaminophen, 650 mg/oxycodone 5 mg, every 6 hours as needed for pain.

Laboratory studies revealed moderate anemia with a packed red blood cell volume of 29% and a normal white blood cell count and differential count. The electrolyte levels were normal. Serum chemistries were remarkable for an elevated total calcium (12.3 mg/dl), uric acid (11.8 mg/dl), creatinine (2.4 mg/dl), and blood urea nitrogen (28 mg/dl). The erythrocyte sedimentation was 140 mm/hr (Westergren method <30 mm/hr). Urinalysis revealed proteinuria. Protein electrophoresis of the serum revealed an IgG monoclonal protein spike with a reduction in other immunoglobulins (Figure 32.2). Analysis of urinary proteins revealed an excess of immunoglobulin light chains (Bence Jones proteins). A radiographic survey of the skeleton revealed lytic lesions (loss of bone density) in the right femoral head, right ilium, and twelfth thoracic vertebral body.

How is the erythrocyte sedimentation rate measured and what does it reveal?

What is the basic structure of an immunoglobulin?

Where is most of the calcium in the body found?

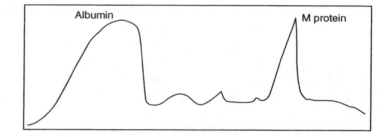

FIGURE 32.2 Representation of the monoclonal protein (M protein) spike on serum electrophoresis readout. The smaller peaks represent other immunoglobulins present in the serum.

EMG and Nerve Conduction Findings

Electromyography (EMG) and nerve conduction studies revealed abnormalities in the median nerve distribution bilaterally. The right median nerve motor conduction study revealed a distal latency of 14 msec (time from stimulation of the median nerve at the wrist to contraction of the abductor pollicus brevis [muscle at the base of the thumb] normal being less than 5 msec). The evoked motor response with stimulation of the median nerve above the wrist was 0.5 mV (normal 5 mV). The left median motor distal latency was prolonged at 10 msec, with a corresponding amplitude of 0.4 mV. The conduction velocity was mildly reduced at 41 and 43 meters per second in the right and left forearms respectively. The median-nerve action potentials from the hands were not measurable. EMGs of both abductor pollicis brevis muscles disclosed abundant spontaneous activity and a marked reduction of voluntary activity (consistent with denervation bilaterally).

How is EMG carried out, and what does it reveal?

How are nerve conduction studies carried out, and what do they reveal?

Diagnosis and Treatment

A diagnosis of Stage IIIB myeloma was made consistent with a high malignant plasma cell mass in the setting of abnormal renal function. The patient was hydrated with intravenous fluids and treated with high-dose corticosteroids. Allopurinol was used to treat the elevation in uric acid. The patient was then treated with a regimen of the alkylating agent, melphelan, and prednisone. She was also treated with the biphosphonate, pamidronate, to reduce bone resorption. Surgical decompression of the median nerve by carpal tunnel release dramatically improved the hand pain and weakness.

How is myeloma staged?

How do melphelan and prednisone exert their effects?

What is carpal tunnel syndrome?

DISCUSSION

The Neurobiology of Pain

It will probably come as no surprise that pain is the most common reason that people seek medical treatment. Unlike other sensory systems such as vision or olfaction (see Chapter 32 of *Biochemistry* 5e), pain does not have a dedicated sense organ. Rather, *nocipetors,* the peripheral sensory neurons that sense pain, are found throughout the body in the skin, bones, joints, muscles and internal organs. Nociceptors only sense strong stimuli, and transmit signals from the site of injury through the spinal cord to the brain, where pain is perceived. The molecular details of how nociceptors sense pain are now being uncovered, with several families of ligand- and voltage-gated ion channels now known to be involved. For example, the cation channel VR1 (see Section 32.5.1 of *Biochemistry* 5e) opens in response to noxious heat and acidic conditions, and mice lacking this protein are impaired in the sensation of these painful stimuli. In addition, the threshold for activation of VR1 is lowered in the presence of high levels of protons, which illustrates an important feature in the neurobiology of pain: the inflammation and acidification that occur at the site of injury *sensitize* nociceptors, rendering them more easily activated, such that even light touch can trigger a neuronal impulse and feel painful.

Two Main Classes of Drugs Are Used to Treat Pain

Nonsteroidal anti-inflammatory drugs

The medical community relies upon two main classes of drugs for treating pain: the NSAIDs (examples are aspirin, ibuprofen, acetaminophen) and the opioids (such as morphine, methadone, and fentanyl) (Figure 32.3). The NSAIDs inhibit *cyclooxygenases,* enzymes involved in the synthesis of prostaglandins, which are the chemicals released by damaged cells that cause pain and inflammation at the site of injury. Most NSAIDs act on the peripheral nervous system, at the site of the injury itself, and thus inhibit the sensitization of nociceptors. Although these drugs are widely used, they are of limited effectiveness in the treatment of intense pain, such as that experienced by this patient.

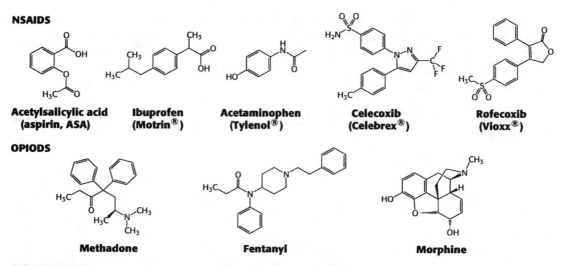

NSAIDS

Acetylsalicylic acid
(aspirin, ASA)

Ibuprofen
(Motrin®)

Acetaminophen
(Tylenol®)

Celecoxib
(Celebrex®)

Rofecoxib
(Vioxx®)

OPIODS

Methadone

Fentanyl

Morphine

FIGURE 32.3 Structures of some commonly used NSAIDs and opioids.

Opioids

The opioids are a class of chemicals structurally related to morphine, and are the most potent analgesics used clinically. They act on *opioid receptors,* G-protein-coupled (seven transmembrane helix or 7 TM) receptors in brain and spinal cord neurons, reducing neuronal excitability and the transmission of pain signals. The opioids thus do not act peripherally, but rather, on the central nervous system (brain + spinal cord). Side-effects, such as nausea and respiratory depression, together with the possibility for dependence, can preclude their use in the treatment of chronic pain. This patient had tried a mild opioid, codeine, together with acetaminophen, but even these did not provide relief. Once she was diagnosed with carpal tunnel syndrome secondary to myeloma, she underwent a surgical procedure, carpal tunnel release (see below), to decompress the median nerve. This alleviated the pain and should lead to improved sensation in her hands.

Myeloma

This patient is suffering from *myeloma,* a primary malignancy of the bone marrow characterized by uncontrolled growth of plasma cells. Plasma cells are the antibody-producing cells of the immune system and are thus critical to fighting infection. In a normal individual, plasma cells make up approximately 5% of bone marrow cells, but in those with myeloma this fraction rises to upward of 10%, and sometimes reaches 90%. Although myeloma is caused by a defect in an immune cell, its effect is far-reaching, causing bone and kidney damage, as well as the neurological symptoms that led this woman to seek treatment. Myeloma is sometimes referred to as *multiple myeloma,* because tumors often occur at more than one site.

Etiology and Incidence

The cause of myeloma is unknown. Exposure to environmental agents is thought to play a role, although this particular patient does not appear to have a history of occupational exposure. It is more common in those over the age of 65, and incidence is approximately 50% higher in men than women. Incidence varies from country to country, with higher rates reported in Western industrialized countries. Although incidence is reportedly climbing, it remains a rare disease, ranging from <1 per 100,000 in China to 4–5 per 100,000 in the United States. Although rare, the disease is serious, with a median survival time of three years from the time of diagnosis.

Pathophysiology

Myeloma causes bone damage

Bone damage is the most troubling complication of myeloma. Malignant plasma cells release growth factors and cytokines that signal osteoclasts to resorb bone, which can cause widespread osteoporosis throughout the skeletal system. This weakens the skeleton and renders it more susceptible to fractures. In addition, *lytic lesions,* localized areas of severe bone destruction where tumors have formed, are characteristic of the disease. They are most common in the spine and pelvis, as well as the skull and the long bones of the limbs, all areas where the bone marrow is especially active. The lesions observed on the radiograph of this patient are in the right femoral head (top of the thigh bone), the right ilium (a pelvic bone), and in the vertebral column and are thus localized at typical sites for the disease. Lytic lesions can cause the bone to fracture or even collapse, which can cause the sudden onset of intense pain or even paralysis when it occurs in the vertebral column. The dissolution of bone, with its high calcium content, leads to hypercalcemia in many patients, including this one, which causes fatigue, thirst, constipation, and nausea.

Anemia and renal dysfunction are common

The uncontrolled growth of plasma cells impinges upon the growth and development of other cells in the bone marrow, including erythrocytes, white blood cells, and platelets. This explains the patient's anemia, and although she does not seem to have suffered from susceptibility to infection or unusual bleeding, these are common symptoms as well.

Damage to the kidneys is a result of the large amount of immunoglobulin produced by the plasma cells. The immunoglobulin produced by these cells is often mutated, causing the chains to dissociate from one another. The light chains are small enough to pass into the urine, and due to their extremely high concentration, place a huge burden on the tubules of the kidney. In addition, they can precipitate in the tubules and damage the organ. Contributing to the problem is the hypercalcaemia that occurs in many patients, as calcium salts can also precipitate in the kidney, causing further damage. The high levels of uric acid, creatinine, and urea nitrogen in the patient's serum were indicative of serious kidney damage. This, together with the high malignant plasma cell mass as indicated by the presence of multiple lytic lesions, classified her disease as advanced, at stage IIIB.

Effects on the nervous system

The most common presenting feature of myeloma is the onset of pain of varying degrees. Although neurologic symptoms are usually a consequence of bone damage, the pain and numbness in the extremities experienced by this patient are most likely due to amyloidosis. The high levels of circulating immunoglobulin light chains can lead to their deposition in an insoluble form in bodily organs and tissues. When this occurs in the extremities nerves may become compressed, leading to the neurologic symptoms experienced by this patient. The neurologic exam uncovered reduced sensation in her lower extremities, and her upper extremities were still more affected, with extreme pain and loss of sensation. Her deep tendon reflexes and plantar responses were normal, suggesting that her symptoms were the result of peripheral neuropathy rather than upper motor neuron (CNS) disease. The positive Tinel's sign was a strong indicator of carpal tunnel syndrome, which is caused by compression of the median nerve in the carpal tunnel in the wrist (see Figure 32.4).

Carpal tunnel syndrome is caused by compression of the median nerve

Although carpal tunnel syndrome is most commonly caused by frequent repetitive motion of the hands or wrists, in patients with myeloma it is usually due to immunoglobulin light chain deposition in the carpal tunnel, causing compression of the median nerve. The median nerve enervates the palmar side of the thumb, first, second, and half of the third finger, and like most peripheral nerves, is a mixed nerve, consisting of both sensory and motor axons. Although these discrete axons are intermingled in the carpal tunnel, they branch out in the hand, with the motor neurons leading to the muscles at the base of the thumb, and the sensory neurons branching into the thumb and first through third fingers. Thus, damage to the median nerve in the carpal tunnel would be expected to affect both motor and sensory functions in the hand.

Indeed, both her symptoms and the neurologic exam revealed motor and sensory impairment. The weakness in her thumb, which made it difficult for her to grasp an object, was evidence of motor impairment, as was the atrophy observed in her thenar eminences. Sensory nerve damage was evident as well—she complained of severe hand pain and had a positive Tinel's sign (see Figure 32.1). The electrodiagnostic tests further supported damage to the median nerves. Both the velocity and amplitude of impulses traveling through them were reduced, and the abductor pollicis brevis muscles (which are enervated by the median nerves) exhibited abnormal patterns of electrical activity, consistent with denervation.

Diagnosis

Diagnosis of myeloma is multifactorial. Approximately 70% of patients present with pain of varying degrees. The presence of a monoclonal protein "spike" on a serum protein electrophoresis readout, which represents the antibody produced by the clonal plasma cell population, is a strong indicator of disease, but only in combination with other features such as evidence of bone and renal damage. Neurological symptoms are often reported, and hypercalcemia is also an indicator, occuring in approximately 30% of newly diagnosed patients.

Treatment

Treatment of the malignancy

Myeloma is not usually treated in its early stages, when evidence of excess monoclonal protein indicates disease but symptoms have not yet appeared. No treatment has yet been shown to slow the onset of illness at this stage. Most patients are not diagnosed until symptoms occur and the disease has reached an advanced stage, and in these cases it is most commonly treated with a combination of the alkylating agent melphalan and an anti-inflammatory, prednisone, which was the combination prescribed here. These are systemically administered, and approximately 60% of patients respond to this dual drug treatment.

Treatment of the hand pain

This patient's hand pain can be effectively alleviated by a surgical procedure called carpal tunnel release. The "sides" and "floor" of the tunnel are made up of the wrist bones, while the "ceiling" (the palmar side) consists of the transverse carpal ligament (Figure 32.4). The procedure is usually done with an arthroscope and consists of cutting the ligament, thus relieving pressure upon the nerve and allowing it to heal.

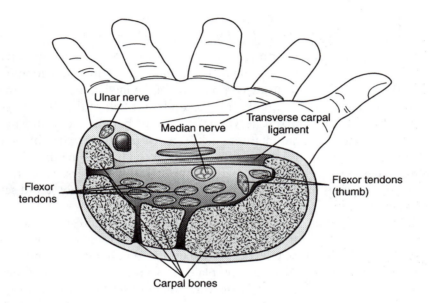

FIGURE 32.4 The carpal tunnel.

QUESTIONS

1. Myeloma cells are key components in the generation of monoclonal antibody-producing *hybridomas*. Can you briefly describe how hybridomas are produced?

2. Why do immunoglobulins not pass into the urine while Bence-Jones proteins (isolated immunoglobulin light chains) do?

3. There are two distinct types of neurons that transmit pain signals to the brain, the Aδ and C neurons. What are the differences between these, both anatomically and functionally?

4. Why are NSAIDS counterindicated in those with kidney dysfunction?

5. There exists a disorder characterized by a congenital inability to feel pain and anhidrosis, which is referred to as *CIPA (congenital insensitivity to pain)* or *HSAN IV (hereditary sensory and autonomic neuropathy IV)*. What is known about the cause of this condition?

6. Codeine and heroin are both structurally related to morphine, but codeine has about 15% the activity of morphine whereas heroin is 2–3 times more potent (Figure 32.5). What is the basis for these differences in activity?

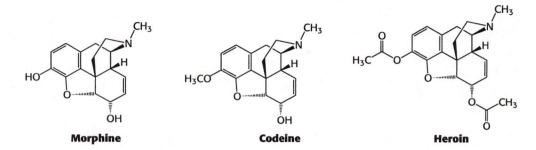

FIGURE 32.5 Structures of morphine, codeine, and heroin.

7. What is the basis of the difference between the first generation cyclooxygenase inhibitors such as aspirin, ibuprofen, and indomethacin, and the newer ones, such as celecoxib?

8. Naloxone is a commonly used drug to treat narcotic overdose. Can you explain the basis for its action?

9. Describe the "gate control theory" of pain.

FURTHER READING

1. Caterina, M. J. and Julius, D. Sense and specificity: A molecular identity for nociceptors. *Current Opinion in Neurobiology* (1999) 9:525–530.

2. Loeser, J. D. and Melzack, R. Pain: An overview. *Lancet* (1999) 353:1607–1609.

3. Melzack, R. and Wall, P. Pain mechanisms: A new theory. *Science* (1965) 150:971–979. (Original article describing the gate control theory.)

4. Tominaga, M., Caterina, M. J., et al. The cloned capsaicin receptor integrates multiple pain-producing stimuli. *Neuron* (1998) 21:531–543.

5. Kyle, R. A. Update on the treatment of multiple myeloma. *Oncologist* (2001) 6(2):119–124.

6. Goldschmidt, H., Lannert, H., Bommer, J., and Ho, A. D. Multiple myeloma and renal failure. *Nephrol. Dial. Transplant* (2000) 15:301–304.

For further information, see the following web sites:

American Pain Society: www.ampainsoc.org/

The International Association for the Study of Pain: www.iasp-pain.org

National Institute of Neurological Disorders and Stroke: www.ninds.nih.gov/health_and_medical/pubs/chronic_pain_htr.htm

National Institutes of Health Osteoporosis and Related Bone Disease— National Resource Center: www.osteo.org/myeloma.html

International Myeloma Foundation: www.myeloma.org.uk

CHAPTER 33

The Immune System
Failing Kidney in Systemic Lupus Erythematosus

CASE HISTORY

Patient History

A 38-year-old woman was referred to the nephrology clinic for worsening kidney function. She has carried a diagnosis of systemic lupus erythematosus (SLE) for 20 years. At age 18 she was diagnosed with SLE on the basis of a malar rash (over the bridge of the nose), nonerosive arthritis of the knees and ankles, proteinuria, mild leukopenia, and high titer positive antinuclear antibodies. Initially, her disease was mild and managed with nonsteroidal anti-inflammatory drugs, primarily indomethacin and ibuprofen, on an as-needed basis.

What are the diagnostic criteria for SLE?

What are antinuclear antibodies and what is their significance in SLE?

She has a healthy 13-year-old son, and 10 years prior to admission she had a miscarriage, complicated by a postpartum pulmonary embolus requiring treatment with warfarin for six months. She periodically has had pain and swelling in her joints as well as recurrent rashes and photosensitivity, but denied any symptoms of chest pain or shortness of breath, with the exception of the time after her second pregnancy. She has no history of seizures, psychosis, severe headaches, or depressed mood. She did not recall ever having had oral ulcers, and there was no history of Raynaud's phenomenon. There was no history of autoimmune disease in her parents or either of her two younger sisters. Both parents are alive and in their late 60s, and her father's hypertension is well controlled with medicines. Her mother has no significant medical illnesses.

What is the significance of a pulmonary embolus in a patient with SLE?

What is Raynaud's phenomenon?

Previous Hospitalization for Pericardial Serositis

She was hospitalized two years ago for nausea, vomiting, and pleuritic chest pain. She had a positive antinuclear antibody test (ANA test) with a titer of 1:480 in a speckled pattern, and her total hemolytic complement level (CH_{50}) was dramatically reduced. She underwent an echocardiogram, which revealed a small pericardial effusion with no compromise in filling of the heart. Urinalysis revealed 3+ protein with 1–2 white blood cells and 20–30 red blood cells (RBC) per high power field. A 24-hour urine collection revealed 3.2 gm of protein.

Laboratory evaluation revealed mild anemia with a hematocrit of 35%; leukopenia, with a white blood cell count (WBC) of 3200 per microliter; and a normal platelet count. Serum chemistries revealed a blood urea nitrogen of 32 mg/dl, creatinine of 1.6 mg/dl, and normal serum electrolytes and liver function tests. She underwent a renal biopsy that revealed focal proliferative glomerulonephritis with immunoglobulin and complement deposition in the mesangium and glomerular basement membrane.

What is the significance of the chest pain and pericardial effusion?

Why was a renal biopsy performed? What is the significance of the findings?

What is an ANA test, and what is the significance of the pattern of staining?

What does a low total hemolytic complement level signify?

She Was Given Immunosuppressive Medications to Control Her Condition

On the basis of the urine specimen, renal biopsy, and serositis involving the pericardium and abdomen, she was started on an intravenous corticosteroid pulse of 1 gm of methylprenisolone followed by a slow tapering dose of oral prednisone, together with azathioprine, 2 mg/kg daily. After discharge from the hospital she was also treated with hydroxychloroquine, 400 mg daily, and omeprazole, 20 mg daily. She did well for about 18 months after hospital discharge, with a reduction in the dose of steroids to 10 mg of prednisone every other day, and she was tapered off azathioprine. About six months ago her blood urea nitrogen and creatinine began to rise, and she started to spill larger amounts of protein in the urine. The azathioprine was restarted, and the prednisone was increased until she was ultimately taking 60 mg daily. The rise in urea nitrogen and creatinine slowed, and the amount of protein in the urine fell; consequently, the dose of prednisone was again tapered.

What is azathioprine, and how does it work in SLE?

What is serositis?

What is the danger of stopping treatment with corticosteroids (e.g., prednisone) too quickly?

Physical Examination

At the time of presentation to the renal clinic she appeared well. Her temperature was 37°C, the pulse was 68 per minute, and the respirations were 18 per minute. The blood pressure was 144/86 mmHg. She had an erythematous rash on her face, most prominently on the bridge of her nose. Her extremities were cool, but there was no cyanosis (blue discoloration).

The lungs and heart sounds were normal; specifically, there were no pericardial or pleural friction rubs. Abdominal examination was remarkable for tenderness to deep palpation in the epigastrium without rebound. The peripheral pulses were full and symmetric without bruits. There was no lymphadenopathy (swelling of the lymph nodes). There was mild swelling with effusions in both knees, but without redness or warmth.

What is a pericardial friction rub?

Laboratory Evaluation and Treatment

A urinalysis was 2+ positive for protein, and the sediment contained 4–6 red cells and 1–2 white cells per high-power field, with occasional granular casts. Complete blood count revealed a mild anemia and leukopenia, as previously described. The erythrocyte sedimentation rate was 28 mm per hour, with a normal mean corpuscular volume of the RBCs. The serum chemistries revealed essentially no change in the blood urea nitrogen (43 mg/dl) and creatinine (1.9 mg/dl). Treatment with prednisone and azathioprine was continued with plans to taper the dose of steroids as long as there was no deterioration in her renal function.

What does the erythrocyte sedimentation rate measure? Was it abnormal in this case?

DISCUSSION

General Features of Autoimmune Diseases

Autoimmune diseases result from the breakdown of self-tolerance

The immune system has evolved to maximize reactivity toward foreign antigens, while minimizing reactivity to self-antigens. Thus, the watchdogs of the human body are usually rapidly alerted to the presence of a foreign invader, yet rarely turn on the cells they are meant to protect. Although it was once thought that negative selection of lymphocytes (see Section 34.6 of *Biochemistry* 5e) early in the development of a fetus and child accounted for self-tolerance, it later became clear that small numbers of autoantibodies (antibodies to self-antigens) remain into adulthood. Although autoantibodies generally have low affinity for self-antigens, it became clear that other mechanisms must exist to keep their reactivity in check.

Other mechanisms thought to be at work in limiting autoantibody reactivity include lymphocyte anergy (when binding occurs but response is not elicited), and regulatory mechanisms, such as the production of anti-idiotypic antibodies. Anti-idiotypic antibodies bind the antigen-binding site of another antibody, thus interfering with antigen binding. Sequestration of self-antigens in tissues not easily accessed by the immune system, such as the brain or the cornea, may also contribute to self-tolerance. Breakdown of any one of these mechanisms—or more likely, a combination thereof—may lead to the development of an autoimmune disease. An important distinction must be made between *autoimmunity* and *autoimmune disease*. Even healthy individuals possess antibodies to self-antigens; indeed, these serve a number of physiologically important roles, such as the removal of antigen–antibody complexes. Only in rare instances does the development of autoantibodies have pathogenic consequences that lead to disease.

Several possible mechanisms for breakdown of self-tolerance

A number of mechanisms have been invoked to explain the breakdown of self-tolerance. Exogenous factors may play a role, such as infection with a pathogen bearing an antigen similar to a self-antigen. The pathogen-derived antigen may be part of a larger protein that acts as an adjuvant, rendering the self-antigen more immunogenic. Intense stimulation of the immune system by a pathogen may also cause nonspecific (non-antigen-dependent) activation of immune cells, some of which may react with self-antigens.

Other possible mechanisms include exposure of sequestered antigens as a result of tissue damage following an injury, or changes in protein structure as a result of drug treatment or normal physiological processes, such as inflammation or cellular senescence. Indeed, certain drugs, such as hydralazine and procainamide, have been known to induce lupus. Genetic factors may also predispose one to autoimmune diseases, with the most frequent association having been made with particular alleles of the major histocompatibility locus (MHC). Some genetic factors appear to confer susceptibility to a particular autoimmune disease; however, the increased occurrence of different autoimmune diseases within some families suggests that some factors may predispose to autoimmune disease in general.

Etiology and Pathogenesis of SLE

Pathogenesis results from autoantibodies with multiple specificities

SLE is an autoimmune disease that affects multiple tissues and organs. It is thus a prototypical *systemic* autoimmune disease, as opposed to *organ-specific* autoimmune diseases, such as type I diabetes and multiple sclerosis. Although the etiology of SLE is unknown, the generation of IgG autoantibodies to nuclear components such as DNA, RNA, histones, and RNA/protein complexes such as U1 snRNP and the RNA-binding proteins, Ro and La, plays a prominent role in pathogenesis. In particular, antibodies directed against double-stranded DNA (dsDNA) feature prominently in the disease. Damage to tissues is thought to be largely the result of a *type-III hypersensitivity response,* in which antibody–antigen complexes are deposited in blood vessels and tissues, causing activation of the complement cascade (see below) and thereby stimulating acute inflammation and producing tissue injury.

In addition, autoantibodies directed against cell-surface molecules found on neuronal and blood cells are also associated with the disease, resulting in problems such as anemia and central nervous system disturbances. Autoantibodies to phospholipids, or to phospholipids complexed with serum proteins such as prothrombin, are also associated with the disease and may lead to arterial and venous thrombotic events.

Genetic factors

Familial clustering and the high concordance rate in monozygotic twins highlight the importance of genetics in SLE. However, inheritance patterns do not follow simple Mendelian genetics, and even in inbred mouse models more than a dozen genes appear to mediate susceptibility to, and manifestations of, SLE. Homozygous deficiencies in some of the early components of the *complement cascade* (C1q, C2, and C4), which work in concert with immune cells to destroy foreign intruders, are the most common genetic factors associated with SLE, although the underlying mechanism is still unclear. Multiple genetic factors combined with environmental influences are likely to mediate susceptibility to SLE in most patients.

Incidence

The incidence of SLE is between 50 and 100 per 100,000 worldwide. It appears to be particularly common in China, Southeast Asia, and the Caribbean. The disease is far more common in women, accounting for 90% of cases, and it usually strikes during the childbearing years, suggesting that hormonal factors influence disease onset.

Clinical Manifestations

Clinical manifestations are highly variable

SLE is a disease of young women that affects multiple organ systems, prominently the skin, joints, central nervous system, and the kidneys. The latter two are the leading causes of mortality. The severity of the disease may vary greatly, ranging from mild to debilitating, and the clinical manifestations are also highly variable and appear to be related to the type and amount of antibody produced. For example, high titers of dsDNA antibodies are associated with nephritis, while antibodies to Ro are associated with cutaneous manifestations of disease. In most cases the disease is characterized by intermittent flare-ups within relatively symptom-free periods. The disease may have a sudden onset with a high fever, or may begin gradually with a feeling of malaise that may last for years. Fortunately, the disease can usually be medically controlled, and long-term survival is the rule.

Articular and cutaneous manifestations

These are among the most common manifestations of the disease, with approximately 90% of patients suffering from joint pain and 50% exhibiting sensitivity to sunlight. Pain and swelling of the joints is common, and may precede other symptoms by many years; however, it is nondestructive in most cases. The malar ("butterfly") rash found across the bridge of the nose seen in this patient is common, and rashes on the neck, chest, hands, and elbows are also typical. Alopecia and oral ulcers also may occur, although this patient does not appear to have experienced these symptoms.

Renal manifestations

Renal dysfunction is a common and serious consequence of SLE, accounting for significant mortality and morbidity associated with the disease. Patients with mild nephritis can maintain adequate kidney function; however, when kidney damage is more advanced it may lead to kidney failure if left untreated. Distinguishing between the two is consequential as it influences treatment decisions. The high doses of immunosuppressive medications used to treat the condition have adverse effects that are best avoided unless absolutely necessary. The kidney is therefore often biopsied in patients with evidence of kidney dysfunction.

The abnormal levels of blood urea nitrogen and creatinine, together with the proteinuria, hematuria, and granular casts, revealed significant renal dysfunction in this patient. Although the renal biopsy revealed *focal,* rather than *diffuse,* proliferative glomerulonephritis, there was immunoglobulin and complement deposition in the mesangium and the glomerular basement membrane, indicating a poorer prognosis. She was thus aggressively treated with immunosuppressive and anti-inflammatory medications.

Serositis

Inflammation of the serous membranes surrounding the heart and lungs, which may cause chest pain, is also frequently encountered in those with the disease. This patient had suffered from pericarditis, a result of inflammation of the pericardium, which surrounds the heart. Fortunately, this rarely leads to cardiac abnormalities and is rarely life-threatening.

Hematologic manifestations

The anemia and leukopenia observed in this patient are also typical of those with SLE. Although autoantibodies to nuclear antigens are prominent in the disease, autoantibodies directed to cell-surface molecules found on erythrocytes and lymphocytes are also found in those with SLE, and are thought to be the principal cause of the hematologic manifestations.

Vascular complications

Autoantibodies to phospholipids, or to proteins associated with them, such as prothrombin or β2-glycoprotein I, may increase the risk of thrombosis and have serious consequences. Indeed, this patient suffered a pulmonary embolus and a miscarriage 10 years ago, both typical vascular complications of SLE. Miscarriages and stillbirths in SLE patients are usually the result of thrombosis in a placental blood vessel. Low-dose anticoagulant therapy, typically coumadin or heparin, is administered to reduce the risk of further embolism.

Other manifestations

Other manifestations include nervous system, gastrointestinal, and ocular disturbances. Nervous system disturbances, in particular, are common and may be serious. Fortunately, this patient, with no history of headaches, depression, psychosis, or seizures, does not appear to have nervous system involvement.

Diagnosis

The diagnosis of SLE may be made on the basis of symptoms, but the presence of characteristic antibodies confirms it. An antinuclear antibody (ANA) test is positive in close to 100% of patients with SLE; however, other autoimmune diseases and viral infections may also yield a positive result. Antibodies to dsDNA are more specific for the disease, as are antibodies to Sm, a small nuclear RNA-binding protein. A more comprehensive analysis of the antibody profile is often carried out, as this may predict clinical manifestations. The various autoantibodies produce distinct patterns of nuclear staining. For example, antihistone antibodies produce diffuse staining, while a speckled pattern is thought to reflect antibodies to small nuclear ribonucleoprotein particles (snRNPs). Evaluation of the complement system is also often carried out, as depression of the complement level is associated with active disease. In some cases, reduction in hemolytic complement may precede overt clinical manifestations and may be useful in sorting out manifestations, such as early kidney disease, that may be relatively silent.

Treatment

There is no cure for SLE, and thus the focus of medical therapy is the administration of anti-inflammatory and immunosuppressive medications to minimize clinical manifestations. Mild disease may be treated with nonsteroidal anti-inflammatories, such as the indomethacin and ibuprofen this patient was using early in the course of her disease. More advanced disease is typically treated with a combination of glucocorticoids, which have an immunosuppressive effect, and cytotoxic agents such as azathioprine. High doses of steroids have adverse side-effects, and combination with a cytotoxic agent allows for reduction of the dose without loss of effectiveness. Higher doses are typically administered during active disease and gradually tapered as symptoms improve; however, flare-ups may then recur.

QUESTIONS

1. Genes involved in the regulation of apoptosis, such as CD95 (Fas) and CD95L (Fas ligand), have been associated with autoimmune disease in humans. What might explain this?

2. Analysis of the dsDNA antibodies found in those with lupus revealed a subset with a preponderance of arginine residues in the complementarity-determining regions (CDRs). Does this suggest anything to you regarding the type of interaction involved in binding between this subset of antibodies and the dsDNA antigen?

3. Antibodies to multiple epitopes of cellular complexes, such as snRNPs, are found in those with SLE. Given this fact, do you think SLE is likely to be triggered by an exogenous antigen such as that borne by an infecting pathogen?

4. It has been found that in patients with SLE and in animal models of the disease autoantibodies produced by B cells undergo *affinity maturation,* a process in which autoantibody affinity for dsDNA increases with time. Describe how the phenomenon of affinity maturation occurs.

5. Although autoantibodies to dsDNA feature prominently in SLE, CD4+ helper T cells specific for dsDNA do not appear to be involved. Given the dependence of antibody production by B cells on helper T cells, how might this be explained?

6. A SLE patient under your care has a genetic deficiency in the early complement component, C1q, which is strongly associated with the disease. However, she is puzzled because she is well informed of her family medical history, and there is apparently no history of SLE or other autoimmune diseases. How might you explain the apparent anomaly to her?

7. SLE has been called "the great pretender." Why might the disease be referred to in this way?

FURTHER READING

1. Doyle, H. A., Yan, J., Liang, B., and Mamula, M. J. Lupus autoantigens: Their origins, forms, and presentation. *Immunol. Res.* (2001) 24(2):131–147.

2. Davidson, A., and Diamond, B. Autoimmune Diseases. *N. Engl. J. Med.* (2001) 345(5):340–350.

3. Gordon, C., and Salmon, M. Update on systemic lupus erythematosus: Autoantibodies and apoptosis. *Clin. Med.* (2001) 1(1):10–14.

For further information, see the following web sites:

Lupus Foundation of America: http://www.lupus.org/

Medline plus Health Information: http://www.nlm.nih.gov/medlineplus/lupus.html

National Institute of Arthritis and Musculoskeletal and Skin Diseases, SLE: www.niams.nih.gov/hi/topics/lupus/slehandout/

National Institute of Allergy and Infectious Diseases, autoimmune disease: www.niaid.nih.gov/publications/autoimmune/autoimmune.htm

CHAPTER 34

Molecular Motors
A Dangerously Heavy Heart

CASE HISTORY

Patient Suffered a Cardiac Arrest at Age 13

A 20-year-old man returns to the clinic for routine follow-up. Now well, he had experienced a normal childhood until age 13 when he suffered a cardiac arrest while tossing a football with a friend. The friend's mother, who was a registered nurse, promptly initiated CPR (cardiopulmonary resuscitation) while her son summoned an ambulance. Upon arrival of emergency medical services (EMS) the patient was noted to be in ventricular fibrillation (a rapid, chaotic, and quickly lethal heart rhythm). They promptly defibrillated his heart and after defibrillation the heart rate was 110 beats per minute (bpm) and blood pressure was 160/100 mmHg. He remained unconscious and an endotracheal tube was inserted to protect his airway. He was transported to a local hospital emergency department, and *en route* he began to awaken and breathe spontaneously. On arrival in the emergency department he was intubated but agitated, and an arterial blood gas on 100% inspired oxygen revealed a pH of 7.38, pCO_2 (partial pressure CO_2) of 32 mmHg, and pO_2 of 650 mmHg. The endotracheal tube was removed, and he maintained excellent blood oxygenation on low-flow oxygen inspired through nasal cannulae.

What is ventricular fibrillation, and what are the causes and treatment?

He had no recollection of the events prior to the cardiac arrest but according to his friend he did not complain of chest pain, shortness of breath, palpitations, dizziness, or weakness prior to his collapse. He was not exerting himself heavily at the time of his collapse and the episode was sudden and associated with seizure-like activity, although he was not incontinent of stool or urine.

Is the presence of seizure-like activity helpful in this case?

No Family History of Heart Disease

He had no prior history of syncope or collapse, he denied the use of alcohol, took no medicines regularly, and was not taking any at the time of the arrest. There was no family history of significant heart disease; specifically, there was no history of premature sudden death, syncope, cardiac arrhythmia, or skeletal or cardiac muscle disease. Both parents were alive and well without heart disease in their late 30s, and all four grandparents were alive and in their 70s. His maternal grandfather has hypertension, and his paternal grandmother has hypertension and adult-onset diabetes mellitus.

Can you name some of the heritable causes of unexplained sudden death or heart disease that may predispose to sudden death in the young?

Physical Examination

On physical examination after the endotracheal tube was removed he was afebrile, with a blood pressure of 110/66 mmHg, a heart rate of 72 bpm and regular, and a normal respiratory rate. He was thin but well developed. He had no cyanosis or rashes. The head, eyes, ears, nose, and throat examination was normal. His neck was supple without jugular venous distension or thyromegaly. His chest was clear and resonant. The heart examination was remarkable for a dynamic apical impulse that was laterally displaced from the midclavicular line; he had an S4 gallop sound and a systolic ejection murmur. His bowel sounds were active, and the liver and spleen were not enlarged. The peripheral pulses were symmetrical in the arms and legs. His neurological examination was remarkable for normal muscle bulk, tone, and strength; normal sensory examination; and there were no signs of cerebellar dysfunction. The reflexes were normal with normal Babinski reflex. He exhibited mild short-term memory deficit.

What is cyanosis and what is the significance of this finding in someone who is suspected of having heart disease?

What does the heart exam suggest, and what test would you recommend to confirm your suspicion?

What is the significance of the symmetrical pulses in the arms and the legs?

Laboratory Evaluation and Imaging Studies

The laboratory evaluation was remarkable for normal blood chemistries and liver-function tests. Urinalysis was unremarkable and a toxicology screen revealed no substances of abuse. Blood count was normal. Thyroid function tests were unremarkable.

His electrocardiogram (ECG; Figure 34.1) demonstrated increased QRS voltage over the septal and anterior precordial (V1-V4) leads with abnormalities of the ST segment and T waves. The chest radiograph revealed mild cardiac enlargement without other abnormalities. An echocardiogram was performed and demonstrated left-ventricular hypertrophy that was global, but particularly pronounced in the septum with a septal wall thickness of 1.9 cm (up to 1.0 cm is normal) (Figure 34.2). The left ventricle was hyperdynamic with a supernormal ejection fraction of 85%. The left atrium was mildly enlarged with a diameter of 4.6 cm (up to 4.0 is normal), but there was no evidence for abnormalities of the right ventricle. A Doppler flow study revealed turbulent flow in the left ventricular outflow tract and systolic anterior motion (SAM) of the mitral valve (the valve between the left atrium and left ventricle).

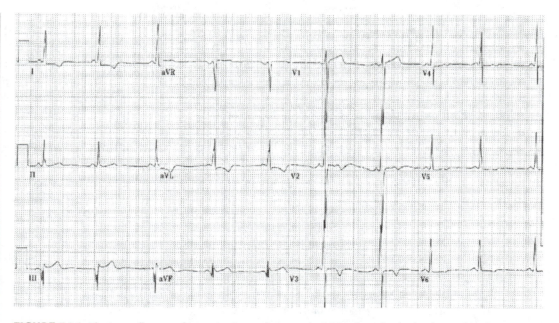

FIGURE 34.1 Electrocardiogram demonstrating an increase in QRS voltage in the leads are that closest to the interventricular septum (V1-V3) consistent with IHSS.

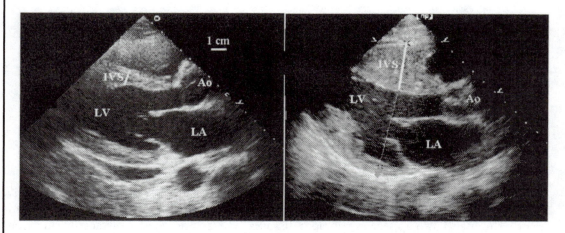

FIGURE 34.2 Echocardiographic view of a heart with idiopathic hypertrophic subaortic stenosis (IHSS). A left parasternal long-axis view is shown, the anterior chest wall is on top of both panels, this view cuts the left ventricle along its long axis. The IHSS heart (left) has a massively thickened interventricular septum (IVS) of approximately 4 cm; normal thickness is ~1 cm (right). Ao, aorta; LA, left atrium; LV, left ventricle.

What do the findings on the ECG suggest?

Is this confirmed by the echocardiogram?

What does the pattern of ventricular hypertrophy on the echocardiogram reveal?

What does SAM signify?

Diagnosis and Treatment

He underwent cardiac catheterization and coronary angiography that revealed normal coronary arteries without evidence for an occlusive myocardial muscle bridge. Pressure recordings revealed no evidence for ventricular outflow obstruction. The diagnosis of hypertrophic obstructive cardiomyopathy (also known as idiopathic hypertrophic subaortic stenosis, IHSS) was made. In light of the life-threatening nature of his presentation he was taken back to the catheterization laboratory for an electrophysiological study. During the study ventricular fibrillation was easily induced with programmed stimulation of the ventricle. He subsequently had an internal defibrillator implanted. He was treated with verapamil SR, 240 mg daily, and atenolol, 50 mg daily, in the hopes of preventing progression of the cardiac hypertrophy.

What is a muscle bridge and what is its significance?

What is IHSS, and how common a cause of cardiomyopathy is this entity?

What is the molecular basis of IHSS, that is, what proteins are abnormal?

Why was a cardiac catheterization with coronary angiography performed?

What is an electrophysiological study? What information does it provide?

An implantable defibrillator (ICD) was placed. What does this device do?

What are verapamil and atenolol, and how might they prevent progression of cardiac hypertrophy?

DISCUSSION

Hypertrophic Cardiomyopathy—Molecular Genetics

Disease may be sporadic or inherited in an autosomal dominant manner

Hypertrophic cardiomyopathies (HCM) are primary disorders of the myocardium characterized by myocyte fiber disarray, fibrosis, and an increase in muscle mass. A subset of syndromes characterized by hypertrophy that is especially prominent in the interventricular septum is variably referred to as hypertrophic obstructive cardiomyopathy (HOCM), IHSS, or asymmetric septal hypertrophy (ASH). These disorders are inherited in an autosomal dominant fashion and also appear as sporadic diseases. There is no gender or race predilection, and genetic-linkage analyses have revealed that HCM is heterogeneous, with a number of mutations in different genes producing similar phenotypes. In the United States unrecognized HCM may be the most common cause of sudden death in young athletes, with an estimated incidence of 1/500.

Mutations in sarcomeric genes are the culprits

More than 100 mutations in 10 different sarcomeric genes have been identified in patients with HCM. Mutations in five sarcomeric genes—β-myosin heavy chain (*MYH7*), cardiac troponin T (*TNNT2*), cardiac myosin binding protein-C (MyBP-C, *MYBPC3*), α-tropomyosin (*TMP1*), and cardiac troponin I (*TNNI3*)—account for over three-quarters of all cases of HCM. Other loci encoding sarcomeric genes more rarely associated with HCM include myosin essential light chain (*MYL3*), myosin regulatory light chain (*MYL2*), cardiac α-actin (*ACTC*), titin (*TTN*), and α-myosin heavy chain (*MYH6*). Mutations in nonsarcomeric genes, such as the potassium channel *KCNQ4* and the γ-subunit of protein kinase A *(PRKAG2)*, have also been associated with HCM, as well as mutations in mitochondrial DNA.

Extent of hypertrophy does not always correlate with prognosis

A number of different mutations have been identified in each of the sarcomeric disease genes, with variable clinical phenotypes. The important phenotypic features are the degree and onset of hypertrophy of the myocardium, which is generally ascertained by echocardiography, and lethality or life expectancy. For example, over 40 mutations have been identified in the β-myosin heavy chain gene, accounting for over 30% of all cases of HCM. The majority of these are missense mutations in the head or head-rod junction (see Figure 34.4 of *Biochemistry* 5e). In general, β-myosin heavy-chain mutations are characterized by significant ventricular hypertrophy but quite variable survival. Mutations in cardiac troponin T are the second most frequent cause of HCM, and are characterized by modest ventricular hypertrophy with a disproportionately poor prognosis.

MyBP-C mutations are also a frequent cause of HCM. Phosphorylation of this protein by cAMP-dependent and calcium/calmodulin-dependent protein kinases modulate cardiac contraction.

Clinical Manifestations

Disease may have innocuous or devastating onset

The diagnosis of HCM is made in the presence of hypertrophy of the left ventricle in the absence of an underlying cause such as high blood pressure, or valvular or vascular disease imposing a pressure or volume load on the heart. The initial symptoms may be nonspecific and innocuous, such as a mild reduction in exercise tolerance or shortness of breath, or may be quite devastating, such as heart failure or sudden death. Still others with HCM may be discovered incidentally by electrocardiography or echocardiography. HCM is often caused by a de novo mutation; therefore, the family history may not be informative, as in this case.

Symptoms

In general, the heart in HCM has preserved contractile function; indeed, the ejection fraction (the percentage of the total volume of the left ventricle ejected with each contraction) is normal or supernormal. In contrast, relaxation and therefore filling of the left side of the heart is impaired, and this, in combination with the obstruction to the ejection of blood exhibited by about 25% of patients, leads to pulmonary vascular congestion and dyspnea. Other symptoms commonly experienced by patients with HCM are chest pain, palpitations, fatigue, and, occasionally, syncope. Patients with HCM often experience cardiac arrhythmias such as atrial fibrillation (rapid and chaotic activation of the atria) and ventricular tachycardia.

Identification of high-risk patients is a major challenge

Most patients have minimal or no symptoms and are identified during screening of relatives of a patient with HCM. This patient highlights a major clinical concern in the management of HCM, which is identifying those patients who are at greatest risk for sudden and unexpected death. There are no features of HCM that will identify patients at high risk with absolute certainty. However, specific mutations, a malignant family history, early age of onset of symptoms, a history of syncope, the presence of malignant ventricular arrhythmias (spontaneous or induced in the cardiac electrophysiology laboratory), and exercise-induced hypotension have all been observed in patients with the worst prognosis. There is a correlation between the magnitude of hypertrophy and symptoms but this relationship is far from absolute.

Diagnosis

In the absence of symptoms of heart failure, the physical examination may be normal or only subtly abnormal in patients with HCM. Often, a ventricular filling sound or gallop is heard, as is a murmur. It is important to distinguish HCM from aortic valve disease. The laboratory evaluation, particularly the electrocardiogram and imaging studies, are the key to the diagnosis. The electrocardiogram is often abnormal, and invariably so in patients with symptoms. Echocardiography is arguably the single most important study in identifying patients with HCM. The cardinal feature is left-ventricular hypertrophy but the echocardiogram provides substantially more information regarding the heart structure and function.

Treatment

The management of patients with HCM is directed at alleviation of symptoms and prevention of sudden death. The former is generally achieved with medications that improve the diastolic performance of the heart, improving relaxation. In patients with septal hypertrophy, surgical resection of part of the septum or controlled infarction of the septum by alcohol infusion into a branch of a coronary artery has helped to control symptoms in some patients. The prevention of sudden death is more problematic and generally involves the avoidance of competitive athletics, and for those at highest risk, the implantation of a cardioverter-defibrillator, as was done in this patient. This is a pacemaker-like device that can shock the heart when an excessively fast heart rate is detected.

Pathogenesis

The sliding of myosin along actin filaments underlies contraction

The sarcomere is the elemental unit of contraction in the heart, with well-defined light and electron microscopic structures (see Figure 34.13 and Section 34.2 of *Biochemistry* 5e). The intertwined strands of thick (myosin heavy chain, myosin binding-protein C, and light chains) and thin (actin, troponin complex, and tropomyosin) filaments that constitute the sarcomere are revealed by a view in cross-section. The energy-dependent sliding of myosin along actin filaments is the molecular motor that underlies contraction of the heart.

Contraction is mediated by Ca^{2+}

Depolarization of the muscle cell initiates contraction by opening calcium-selective channels in the cell membrane, producing a small rise in intracellular Ca^{2+}, which signals a much larger release of Ca^{2+} from the sarcoplasmic reticulum (a process known as *Ca^{2+}-induced Ca^{2+} release*). Ca^{2+} binds to the troponin complex and relieves troponin I-mediated interference with actin–myosin interactions, such that myosin-ADP (S1 fragment) binds to actin. Exchange of ATP for ADP causes myosin to dissociate from actin; the hydrolysis of ATP to ADP and inorganic phosphate (P_i), which remain bound to myosin, then causes reassociation of actin and myosin. The release of P_i leads to a conformational change in the lever arm, displacing the S1 fragment domain along the actin filament (the power stroke). Myosin-ADP binding to actin will occur as long as sufficient levels of Ca^{2+} are present in the myoplasm. A number of processes are occurring simultaneously with contraction that reduce the levels of intracellular Ca^{2+} and restore the heart cell to its resting state.

Haploinsufficiency is probably not the cause of HCM

The detailed mechanisms of pathogenesis of HCM are not well understood. The proteins that constitute the sarcomeres are present in highly regulated stoichiometries. As a consequence, a number of disease mechanisms have been suggested for HCM. Mutation in a single allele is sufficient to cause disease (autosomal dominant transmission); thus haploinsufficiency in such a highly regulated macromolecular structure could disrupt sarcomere function, as could dominant negative effects of mutant proteins. In this regard, the nature of the mutations in the most frequent types of HCM and transgenic animal models have been invaluable in our understanding of the mechanisms of the structural and physiologic derangements in HCM. Mouse models of HCM suggest that in most cases haploinsufficiency is not the underlying cause of hypertrophy or altered sarcomere function. However, null mutations in some proteins may disrupt sarcomere structure and function purely on the basis of gene dosage.

Gain-of-function mutations may be to blame in some cases

A related question is whether the disease-causing mutations lead to a loss or gain of sarcomeric function; the latter has the appeal of the clinical correlate of hyperdynamic contraction in patients with HCM. Indeed, at least one mutation, β-MHC R403Q, produced enhanced sarcomere function and enhanced cardiac contraction when expressed in mouse heart. Such a gain of function may have a number of liabilities, including heterogeneity of contractile performance and increased energy consumption with suboptimal efficiency. It is currently not known whether other mutations in either β-MHC or other sarcomeric protein exhibit similar gains in function.

Defective regulation of Ca^{2+} concentration may contribute to pathogenesis

Another theme that has emerged from the study of animal models of both genetic and acquired myocardial hypertrophy are abnormalities in the regulation of intracellular Ca^{2+} concentration. The precise change in intracellular Ca^{2+} homeostasis is difficult to predict and may vary depending upon the particular type of HCM. There are a number of mechanisms by which defective regulation of intracellular Ca^{2+} concentrations could contribute to the HCM phenotype, including Ca^{2+}-induced initiation of cell hypertrophy, activation of programmed cell death (apoptosis) pathways, activation of Ca^{2+}-dependent enzymes that could induce cell damage, and alterations in electrical signaling.

QUESTIONS

1. What are the mechanisms by which mutations in patients with HCM may produce dysfunctional sarcomeres and cardiac hypertrophy?

2. A number of mutations of the β-MHC gene in the heavy meromyosin tryptic fragment have been described in HCM. What functional defects might be produced in myosin by these mutations?

3. Genetically altered animal models are important systems for studying the effects of mutations that cause HCM. Why are such models important in our understanding of these genetic diseases?

4. A subset of patients with HCM are at high risk of dying suddenly from lethal irregularities of heart rhythm. What is the link between muscle contraction and disturbed electrical activity of the heart?

5. Alterations in the way the heart cell handles Ca^{2+} is a prominent feature of many forms of structural heart diseases. How might changes in intracellular Ca^{2+} contribute to the phenotype in HCM?

6. What features of the molecular genetics of HCM suggest that these disorders occurred relatively recently in human evolution?

FURTHER READING

1. Seidman, J. G., and Seidman, C. The genetic basis for cardiomyopathy: from mutation identification to mechanistic paradigms. *Cell* (2001) 104:557–567.

2. Marian, A. J., and Roberts, R. The molecular genetic basis for hypertrophic cardiomyopathy. *J. Mol. Cell. Cardiol.* (2001) 33: 655–670.

3. Towbin, J. A., and Bowles, N. E. The failing heart. *Nature* (2002) 415:227–233.

4. Spirito, P., et al. The Management of Hypertrophic Cardiomyopathy. *New Engl. J. Med.* (1997) 336:775–785.

5. Maron, B. J. Hypertrophic Cardiomyopathy:A systematic review. *JAMA* (2002) 287:1308–1320.

For further information, see the following web sites:

The Hypertrophic Cardiomyopathy Association: http://www.hcma-heart.com/

OMIM Familial Hypertrophic Cardiomyopathy:
http://www.ncbi.nlm.nih.gov/entrez/dispomim.cgi?id=192600

CHAPTER 1

1. In ex vivo gene transfer cells are removed from the patient and transfected in vitro (in a test tube) and returned to the patient. It has the advantage of higher efficiencies of gene transfer; however it is a costly procedure and is not suitable for all cell types. For example, ex vivo gene transfer has been used successfully on lymphocytic cells that are relatively easy to remove and return to the body, while it is much more difficult to imagine how such a procedure might be used to treat cystic fibrosis! In vivo gene transfer is the direct administration of the gene transfer vector to the patient. Although the efficiency of gene transfer is lower than what can be attained using ex vivo methods, it can be carried out on a variety of cell types, and is a far less costly procedure.

2. Use of viral capsid or envelope proteins, or of cell-type-specific promoters. Many viruses selectively infect particular cell types through "lock and key"-type interactions between their capsid or envelope proteins and cell surface proteins on their target cells. For example, HIV infects a subset of T lymphocytes through an interaction between its envelope glycoprotein (gp120) and a molecule called CD4 on the surface of some T cells. These viral proteins, when engineered into viral vectors used for gene therapy, have the potential to target transgenes to specific cell types. Alternatively, cell-type-specific promoters may be used to achieve the same end. Some promoters (the region of a gene that regulates transcription, the first step in the generation of a protein from a gene) are only active in certain cell types; for example, the myosin light chain-1 promoter is only active in muscle cells. By engineering this type of promoter to regulate the transcription of a transgene, even vectors that transfect many cell types can be used to selectively express the gene in a specific cell type.

3. Although CF is usually caused by the ΔF508 mutation, hundreds of other mutations can cause the disease. Of these, only a handful have been identified, and current tests detect only approximately 25 disease-causing mutations in the CFTR gene. This explains why these tests are less than 100% effective.

4. The CFTR gene is a transmembrane protein manufactured in the endoplasmic reticulum (ER), from where it is transported through the Golgi apparatus to the cell surface. The ΔF508 mutation does not appear to affect the synthesis or function of the protein but rather its transport to the cell surface. Thus, most of the protein never reaches its destination, but remains "stuck" in the ER.

5. Obstructions in the gastrointestinal tract cause fluid and air to back up behind them, and the air-fluid level is where the fluid ends and the air begins. These are visible on an abdominal radiogram, and although one or two are normal, more than that can indicate an impending obstruction. Note that in the case discussed here, there was an obstruction in the ileum yet there was a lack of air-fluid levels. This is a common feature of meconium ileus and has been attributed to the thick meconium found in the gastrointestinal tract of newborns with cystic fibrosis.

6. Somatic-cell gene therapy is the transfer of a gene to any cell type besides the germ cells (egg and sperm), and thus the transfer of CFTR to epithelial cells in the airways of those with cystic fibrosis is an example of somatic-cell gene therapy. Germ-line gene therapy refers to the transfer of a gene to an egg or sperm. An organism derived from such an altered cell will express the transgene in every cell within its body, including the germ cells. The transgene will thus be transmitted to subsequent generations, and germ-line gene therapy is therefore a topic of intense bioethical debate and is banned in most countries.

7. The CFTR ΔF508 mutation is autosomal recessive, and each member of the couple bears a single copy of the mutation (they are *carriers*). Two faulty copies of the gene are necessary for disease, as a single normal copy is sufficient for normal mucus production. Thus, their next child (and any subsequent children) will have a 25% chance of having cystic fibrosis (Figure 1.2).

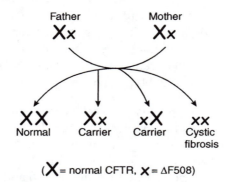

FIGURE 1.2 Genetic cross.

8. CFTR is a chloride (Cl^-) channel of the ATP-binding cassette (ABC) family of proteins. It is regulated by ATP and by phosphorylation of specific amino acids in its sequence. It consists of 1480 amino acids, and has 12 transmembrane segments that form the pore through which the ions flow, two cytoplasmic nucleotide binding domains, and a cytoplasmic regulatory domain. The regulatory domain blocks the opening of the pore until it is phosphorylated by protein kinase A, causing a conformational change that exposes the ATP-binding sites of NBD1 (nucleotide binding domain 1) and NBD2. Upon binding of ATP to NBD1 and NBD2 the pore opens and allows efflux of Cl^-, and NBD2 subsequently plays a role in closure of the pore. The number of sites (between 1 and 4) phosphorylated on the regulatory domain correlate with how long the channel remains open.

CHAPTER 2

1. The earliest life forms, although simple by today's standards, consisted of the same components that make up today's more complex organisms, including lipids, proteins, carbohydrates and nucleic acids. The generation of these components, and their collaboration in the formation of an early "cell" required many, many cycles of variation and reproduction. Drug resistance in bacteria, on the other hand, can occur via a single genetic mutation during a single round of replication, or instantaneously through the acquisition of a mobile genetic element. This explains why drug resistance appears so rapidly upon introduction of a new drug. Resistance to streptomycin, the first drug used to combat tuberculosis, was reported within a few months of its introduction onto the market.

2. In general, resistance occurs upon sub-optimal use of antimicrobials, such that bacterial growth is only partially inhibited. Under these conditions bacteria that already possess low intrinsic resistance can survive and multiply, and through mutation or acquisition of a mobile genetic element, become increasingly resistant. One of the most important factors contributing to this is poor adherence to drug regimens, caused either by irregular drug supply or poor patient compliance. Patients often stop taking medication once symptoms improve, especially in poor countries. The prescribing of antimicrobials for viral infections also contributes to the problem. The use of antimicrobials in animal feed and in agriculture is also thought to contribute to the problem, as is the addition of antibacterials to home cleaning products.

3. Most bacteria divide rapidly, with doubling times on the order of an hour or less. Mycobacteria, on the other hand, divide relatively slowly, with doubling times of 12–24 hours. It thus typically takes between 2–4 weeks, and sometimes up to two months for a culture to yield evidence of *M. tuberculosis*. This explains why the culture initially only yielded the usual respiratory tract flora, and only later turned up the disease-causing mycobacteria.

4. A number of factors point to his having been infected prior to entry into the United States. Most importantly, except for the immunocompromised (which he does not appear to be; he was HIV negative and his blood count was normal), most people who get TB disease get it from reactivation of a previous infection. Secondly, his profession as a physician and his country of origin are both risk factors for TB infection. Since entering the United States, he has not worked in a clinical setting, and is thus less likely to have come into contact with tuberculous patients. Finally, his previous vaccination was probably taken into account when the tuberculin skin test was administered. The reaction to PPD diminishes with time since vaccination, and thus only a strong reaction would likely have been interpreted as positive.

5. Although the duration of treatment for tuberculosis is between 6–12 months, many patients cease taking their medication once symptoms improve, after a few weeks. This can lead to the development of drug-resistant strains, which are a significant public health hazard. Infection with multidrug-resistant tuberculosis is very expensive to treat and can be fatal, thus justifying the extreme measures.

6. The market for tuberculosis drugs is huge: approximately one third of the world's population is infected with the bacterium, and the number of cases proceeding to clinical illness and resistant to first-line drugs is likely to increase with the ongoing AIDS epidemic.

Thus, a lower profit margin would be counterbalanced by the size of the market. A second point is that high cost of the medicine might lead to its improper use, with patients only taking what they can afford rather than the recommended course of treatment. This, of course, would lead to drug resistance, which would limit its effectiveness, and sales may begin to drop in favor of other drugs.

7. For a drug to be effective it must target an essential process. Thus, for example, the drugs rifampicin and isoniazid target bacterial transcription and mycolic acid synthesis, respectively, both essential to survival. A bacterium must possess an efficient transcriptional apparatus if it is to make the RNAs and proteins it needs, and cell-wall mycolic acids are essential if it is to resist the inhospitable environment supplied by the host's immune response. Mutation of enzymes involved in these processes thus may come at a cost of reduced efficiency, which the organism must somehow cope with.

8. Modification of the drug's target, rerouting of cellular pathways to avoid that inhibited by the drug, decreasing cellular permeability, inactivation of the drug, increased efflux of the drug.

9. *Pneumocystis carinii* pneumonia (PCP) shares many symptoms with tuberculosis, including weight loss, cough, progressive dyspnea, and a low-grade fever. Like tuberculosis, it infects the lung, and can lie dormant for decades, resurfacing and causing pneumonia in those who become immunocompromised. PCP is the most common opportunistic infection amongst those with AIDS, developing in 60–85% of patients. He was probably tested because he was suspected of being immunocompromised or of having AIDS.

10. The culturing and drug-susceptibility testing of clinical isolates of M. *tuberculosis* is extremely slow due to the slow growth properties of the bacterium. While awaiting the results, standard treatment with first-line drugs is usually initiated, with little information from the onset as to whether or not they will prove effective. A rapid assay would thus be invaluable and would greatly assist in the making of treatment decisions. Because most of the mutations that confer resistance to rifampicin and isoniazid, the most potent drugs against the disease, have been mapped to specific nucleotides within a few genes, a nucleic acid-based assay such as PCR could readily be applied to the detection of resistance in strains isolated from patients.

11. Availability of anti-tuberculous drugs and access to a health care provider are not universal. In addition, application of DOT to the highly mobile refugee and homeless populations is difficult if not impossible.

12. Resistance to the first-line drugs in M. *tuberculosis* occurs through genomic mutations. Resistance to rifampicin, streptomycin, and ethambutol result from genomic mutations that modify the drug's target, whereas isoniazid and pyrazinamide resistance result from mutations in enzymes that convert the drugs into their active forms within the cell, or from mutations that cause overexpression of the target enzyme.

13. Although they do so via different mechanisms, both ethambutol and isoniazid inhibit the accumulation of mycolic acids in the cell wall, weakening it and rendering it more permeable. Thus, drugs such as rifampicin and streptomycin can more easily enter the cell and gain access to their targets when administered together with ethambutol, which explains the synergistic effect. Because ethambutol acts by inhibiting mycolic acid deposition, no further advantage is gained by inhibiting their synthesis with isoniazid, which explains the lack of synergy in this case.

CHAPTER 3

1. The incidence of a disease refers to the number of new cases of a disease in a given period of time, whereas the prevalence refers to the total number of people afflicted with the disease within a time period. Thus, for example, the prevalence of a disease such as type II diabetes is much higher than its incidence because patients typically survive long after the disease has been diagnosed. For a rapidly fatal disease such as CJD, incidence and prevalence are roughly equal because virtually all cases of the disease are new cases because most patients succumb soon after diagnosis.

2. The most unusual feature of prion diseases compared to other infectious diseases is the chemical nature of the infectious agent. Most infectious agents rely upon genes encoded in their DNA or RNA genomes to multiply and propagate themselves. In contrast, prions do not contain nucleic acids and are entirely proteinaceous. Their infectivity is thus not enciphered in genes, but rather in the tertiary structure adopted by the protein.

3. Many cell-surface proteins are glycosylated, including PrP. PrP is glycosylated at two asparagine residues, at positions 181 and 197 of the amino acid chain (see Section 11.3 of *Biochemistry* 5e for more details on the formation of glycoproteins).

4. Amyloid plaques or fibers form upon the extracellular deposition and self-association of insoluble polypeptides, resulting in cell death and disease. Examples of amyloid diseases are the prion diseases, type II diabetes mellitus, and Alzheimer disease. Although the proteins involved in each disease share little sequence or structural similarity, the fibrils share a common structure, as determined from X-ray diffraction patterns and electron micrographs. They are unbranched, 70–120 Å in diameter, and of variable length. The fibril has a characteristic *cross-β repeat structure* in which the β-sheets are arranged parallel to the axis of the fibril, while the strands that make up the β-sheets are oriented perpendicular to the axis. The β-sheets twist around the axis of the fibril to form repeating units of fixed length. The amyloid *fibers* consist of several fibrils coiled around one another.

5. The active forms of many proteins, such as the digestive enzymes, are generated by cleavage of a larger precursor polypeptide. Cleavage of PrP also occurs, but in this case the cleavage events are involved in targeting the protein to the appropriate subcellular location, the outer leaflet of the cell membrane. The N-terminal *signal sequence,* which targets the protein to the endoplasmic reticulum (from where it will be transported to the cell membrane), is cleaved upon entry into this organelle. The C-terminus is also cleaved in a process involving the addition of the GPI anchor, which serves to attach PrP in the cell membrane. Forty-five amino acids are ultimately removed, resulting in a protein 208 amino acids in length.

6. In the early 1980s, a 27–30-kd protein specific for tissues infected with transmissible spongiform encephalopathies (e.g., CJD) was identified by its resistance to digestion by proteinase. This protein was named prion protein (PrP), a term derived from *proteinase-resistant protein*, and was found to exist in both normal and diseased tissues. PrP molecules from both sources have the same amino acid sequence and bind to PrP antisera. However, the isoform of PrP from normal brains, called PrP^{sen} or PrP^C, is soluble in non-denaturing detergent and is sensitive to proteinase K. In contrast, the isoform of PrP from brains infected with transmissible spongiform encephalopathies, called PrP^{res} or PrP^{Sc}, is insoluble, and proteinase K removes only part of the polypeptide, converting the

protein into 20–30-kd truncated forms (depending on the species and subtype of PrP^{Sc}). This form can aggregate to form amyloid. The function of PrP^C and the mechanism by which PrP^{Sc} (or the removal of PrP^C) causes disease are unknown. The chief difference between the PrP^C and PrP^{Sc} isoforms is conformational, with PrP^C having more α-helical content and much less β-sheet content than PrP^{Sc}.

7. Prion diseases are the result of a major conformational transition in PrP, and in going through this transition the protein is likely to go through one or more unfolded intermediates. Chaperones, in binding to "sticky" hydrophobic surfaces that can become exposed during unfolding, could potentially break the chain reaction that leads to aggregates of PrP^{Sc}, and could thus prevent disease. In fact, the chaperone Hsp104 has been shown to do just this in yeast. Overexpression of Hsp104 cured cells of the yeast equivalent of prion disease. Regulation of chaperone levels thus holds much promise in the development of treatments to control or cure mammalian prion diseases.

8. FFI is associated with a mutation that changes amino acid 178 from aspartic acid to asparagine, together with the presence of methionine at amino acid 129. The thalamus, where the brain's "sleep center" is found, is selectively destroyed in this disease.

CHAPTER 4

1. The aromatic amino acids—phenylalanine, tryptophan, and tyrosine—have large side chains that would not fit into the sterically restricted spaces of the triple helix structure of type I collagen. They are not found in the helical regions of the chains and mutations that result in the substitution of glycine for an aromatic amino acid generally cause type II OI.

2. Ascorbic acid (vitamin C) is necessary to keep the enzyme prolyl hydroxylase in the active form. Prolyl hydroxylase hydroxylates a number of the proline residues in type I collagen to hydroxyproline. Hydrogen bonds involving the hydroxyproline residues stabilize the helix, and in their absence the triple helix is unstable and tends to fall apart. The abnormal polypeptides accumulate in the endoplasmic reticulum and are gradually degraded. Patients feel weak and have joint pain, stiff and swollen limbs, and are susceptible to bleeding (due to a shortage of collagen in the blood vessel walls). Bleeding gums and skin lesions are typical. The disease was once common among sailors on long sea voyages, and was referred to as "sea scurvy." When it was discovered that citrus fruits could combat the illness, concentrated forms of lemons and limes were brought aboard ships destined for long voyages.

3. No. Although most mild mutations are caused by decreased expression of one of the collagen chains, which would be detected by SDS-PAGE, many of the more serious mutations are caused by point mutations that do not significantly affect the mobility of the polypeptide, under either reducing or non-reducing conditions. In the cases discussed here, however, the cysteine substitution resulted in the formation of a novel species that was detectable by non-reducing SDS-PAGE. Partial gene deletions, insertions, and duplications may be detected if the mutation causes a significant change in size. RNA splicing mutations, yielding aberrantly spliced products, would also be detected by this technique. Defects in postranslational modifications may also be detected if the modification significantly alters the size of the peptide.

4. CNBr cleaves polypeptides on the carboxy side of methionine residues, and thus may be used to cleave the $\alpha1(I)$ and $\alpha2(I)$ chains into peptide fragments of predictable sizes when the protein sequence is known, as in this case. Cleavage followed by non-reducing SDS-PAGE would result in the substitution of one of the predicted fragments with a higher molecular weight species, representing the disulfide-linked peptide. The "missing" predicted fragment will contain the substitution. Because the *COL1A1* and *COL1A2* genes are large, this will greatly expedite the identification of the specific mutation by DNA sequencing, because only a portion of the gene will need to be analyzed.

5. It would probably be less. Only half of the collagen molecules would be affected rather than 75% because there is only one $\alpha2(I)$ chain per molecule; in addition, each molecule would contain only one defective $\alpha2(I)$ chain. Remember that 25% of molecules in the patient described here will have two defective $\alpha1(I)$ molecules, increasing the chance that these will be nonfunctional. Indeed, a study has shown that an identical glycine-to-serine substitution causes a far more serious phenotype in $\alpha1(I)$ than in $\alpha2(I)$.

6. The formation of the type I collagen triple helix occurs in the endoplasmic reticulum, where the prolyl and lysyl hydroxylases are found. Disturbance of helix formation will result in the chains remaining unfolded for greater lengths of time. The hydroxylases thus have a longer time in which to hydroxylate their substrates, leading to overhydroxylation. The degree of overhydroxylation in these mutants roughly correlates with phenotypic severity.

7. Mutations that simply reduce gene expression are generally the least severe because residual expression leads to the formation of functionally normal collagen fibrils. In contrast, mutations that affect protein structure—in particular, the folding of the triple helix—are more detrimental because they lead to the formation of abnormal collagen molecules. Among these types of mutations, the most serious usually affect the *COL1A1* gene rather than the *COL1A2* gene (see question 5), and mutations closer to the C-terminus, where helix formation initiates, usually lead to more serious disease. Finally, the nature of the substituted amino acid is important, with the larger, bulkier residues generally causing the more serious types of OI.

CHAPTER 5

1. Acute leukemia is characterized by the accumulation of immature blast cells that are unable to carry out their function. Acute leukemias usually progress rapidly, leading to disease and death, unless treated promptly. Chronic leukemias are characterized by an accumulation of somewhat more differentiated cells, which can largely still carry out their functions. The blast cell burden increases more gradually in the chronic leukemias, with a more gradual onset of illness.

2. The different types of blood cells (erythrocytes, leukocytes, and platelets) are formed from a common precursor cell known as a *stem cell.* The stem cell will divide and form several types of immature *blast cells,* which will then go on to differentiate into the three main types of blood cells. This process of blood cell development is continuously occurring in the bone marrow and at several other sites in the body, and hence the presence of a limited number of blast cells in a bone marrow aspirate is normal.

3. Malignant cells are characterized by loss of growth control and rapid proliferation. The goal of treatment is to inhibit proliferation, and irradiation and medications such as methotrexate achieve this by inhibiting DNA replication; a cell will not usually divide until its DNA has been replicated. Most cells within the body do not replicate rapidly if at all, and are thus relatively unaffected by such treatments. However a few cell types do continuously proliferate, and this accounts for some of the adverse effects associated with them. For example, the continuously dividing cells of the gastrointestinal tract and immune system are affected and cause the nausea and susceptibility to infection common in patients undergoing chemotherapy.

4. A host-derived tRNA molecule serves as primer by annealing a portion of its 3′ end to a complementary sequence at the 3′ end of the retroviral genome. In the case of HTLV-1, the particular tRNA used is that specific for the amino acid proline, tRNA^Pro.

5. Yes. Reverse transcription of the RNA genome yields a double stranded DNA template, which integrates into host cell chromosomes. From this point the flow of genetic information travels in the conventional forward direction. The proviral DNA is transcribed into RNA, which serves as a template for protein synthesis, and as viral genomic RNA.

6. The high error rate of reverse transcriptase accounts for the hypervariability found in retroviruses. The high error rate generates mutations in the genome, which result in alterations in viral proteins. Many mutations have detrimental effects, and produce viral particles that are unable to replicate. However, some mutations will alter protein structure without significantly affecting function, and this type of mutation will allow the virus to replicate normally while evading the host immune response. Incidentally, this hypervariability also contributes to the rapid development of resistance to antiretroviral medications, and to the difficulty in developing effective vaccines.

7. The LTRs play a critical role at several points in the viral life cycle: they play a role in reverse transcription, as well as integration of proviral DNA into host cell chromosomes and in the subsequent transcription of viral genes. The LTR contains the *primer-binding site (PBS)* for the tRNA molecule that primes reverse transcription of the RNA genome. In addition, the viral integrase directs the integration of the proviral DNA into host cell DNA via the LTRs; and finally, the cis-acting sequences necessary to catalyze transcription of viral genes are found in the LTRs.

CHAPTER 6

1. In most cases, the validity of microarray analysis rests upon the concept that cellular mRNA levels reflect the corresponding protein levels. This is generally a fair assumption: although some gene-regulatory mechanisms function posttranscriptionally, gene transcription is the major point of regulation for most genes.

2. The Southern blot. In the Southern blot, DNA is digested with restriction enzymes and the resulting fragments are separated on a gel. The fragments are then blotted onto a matrix such as nitrocellulose, and probed with a radioactive DNA fragment complementary to the gene of interest. The intensity of the radioactivity of the band corresponding to the gene can then be used as a measure of gene copy number. Because expansion occurs at the level of the DNA in these tumors, both Northern (measures RNA levels) and Western blotting (measures protein levels) can be used diagnostically as a measure of Her-2 expression. However, neither of these techniques would have identified the underlying mechanism behind increased expression.

3. Western blotting. Microarray analysis is most often used to analyze the expression patterns of multiple genes by comparing *mRNA* levels in cells treated in different ways, whereas Western blotting is a technique that makes use of antibodies to detect *proteins*. *PCR* is usually used to generate the oligonucleotide "dots" (that represent each gene to be analyzed) on the slide. *cDNA* is generated from each mRNA sample to be tested, and is labeled with a fluorescent dye *(probe labeling)*. Subsequently the probes are *hybridized* to the slide to obtain the expression data.

4. In contrast to diseases such as cystic fibrosis and Huntington's disease (see Chapters 1 and 27 of *Biochemistry* 5e), which involve a single gene, the risk of developing breast cancer and the malignancy of the tumor are likely to involve many genetic factors. Indeed, microarray analyses have already shown that the gene expression profiles of tumors from different patients differ. Thus, many different gene expression patterns can apparently lead to breast cancer. Herein lies the strength of microarray analysis: previous to the advent of this technology only one or a few genes could be analyzed at a time, whereas now thousands of genes can be analyzed simultaneously. Eventually, the analysis of data accumulated from many tumors will allow physicians to make meaningful predictions as to risk and prognosis. There may nevertheless be a role for gene expression profiling in monogenic diseases, particularly those with variable clinical expression, where the role of other so-called modifier genes may be crucial to the expression of the disease phenotype.

5. Tamoxifen acts through the estrogen receptor (ER), and the molecular analysis of the tumor revealed that her tumor was ER-negative. Thus, the drug would have no beneficial effect in this particular patient.

6. Both chemotherapy and hormonal therapy are administered systemically and thus affect the whole body. However, chemotherapeutic agents act fairly nonspecifically on any cell that is dividing, thus accounting for the hair loss, nausea, and immunosuppression associated with these drugs. In contrast, hormonal therapies only affect cells that express the appropriate hormone receptor, and thus the side-effects are more limited.

7. Three mutations (two in BRCA-1 and one in BRCA-2) account for the majority of cases of breast cancer associated with these genes, and thus genetic testing typically only tests for these specific mutations (185delAG and 5382insC in *BRCA-1,* and 6174delT in *BRCA-2*). First, the regions in which the mutations lie are amplified from a sample of the test subject's DNA using PCR. The fragments generated by PCR are then spotted in duplicate onto a membrane and each spot is hybridized to a radioactively labeled probe: one corresponding to the wild-type sequence and the other to the mutant sequence. The spot that emits radioactivity indicates whether the wild-type or mutant sequence is present in the individual.

CHAPTER 7

1. No, this does not necessarily indicate a genetic component. Remember that families usually share a common environment, with similar diets and lifestyles, and one of these shared environmental factors might just as easily confer susceptibility to a disorder. For many common ailments such as migraine, susceptibility appears to be determined by a combination of environmental and genetic factors.

2. Variations in the drug target, transport, or metabolism could all account for differing responses to a particular drug between individuals.

3. A completely penetrant genetic trait affects virtually all those who possess it, while an incompletely penetrant trait only affects a fraction of those who possess it.

4. Many commonly used drugs, including timolol, are metabolized by the hepatic enzyme cytochrome P450 2D6 (CYP2D6) as a step in the elimination of the drug. However, there exist a number of genetic polymorphisms in CYP2D6, which are reflected in the wide range of functional expression of the enzyme found throughout the general population. Differing responses between individuals to a number of drugs can partially be accounted for by CYP2D6 polymorphisms. β-adrenergic blockers bind adrenergic receptors, reducing heart rate, as well as preventing vasoconstriction (hence their utility in treating migraine). This individual developed a profound slowing of heart rate—an exaggerated response—which could be explained by prolonged bioactivity of the drug. This patient may thus possess an allele of CYP2D6 that results in impaired function of the enzyme. Incidentally, a number of commonly used antidepressants are also metabolized by CYP2D6, leading some to speculate that some apparently intentional drug overdoses among mentally ill patients may have been accidental.

5. Caffeine inhibits the enzyme phosphodiesterase, which degrades cyclic 3′,5′ adenosine monophosphate (cAMP). The elevated cAMP level that results upon caffeine consumption has many effects, including constriction of cerebral blood vessels. This effect can be useful in counteracting the vasodilation that occurs during the painful phase of the migraine headache.

6. No. Remember that human cells are diploid, and thus have two copies of each gene. You have only sequenced one copy, and although it contains a polymorphism, it may not be the one associated with migraine. You will need to sequence several more clones before you can be certain that both gene copies from this individual are represented, and compare the sequences with those obtained from other family members before you can unambiguously identify the polymorphism associated with migraine in this family.

CHAPTER 8

1. Each viral particle bears two copies of the viral genome, and hence the concentration of viral particles would have been approximately 15,500/ml.

2. As discussed in Chapter 2 of *Biochemistry* 5e, subtherapeutic drug doses provide an ideal setting for the development of drug resistance. Under these conditions a pathogen can continue to replicate and mutations that confer resistance will soon result in a predominantly resistant population. Thus, allowing for this possibility, the drug regimen was promptly changed, ensuring immediate effectiveness and a decrease in viral load.

3. Thymidine triphosphate is a substrate for the cellular DNA polymerases and hence the thymidine analog, zidovudine, would be expected to inhibit it as well. Indeed, in vitro studies have shown the drug to bind and inhibit β and γ polymerases in particular, and this may account for some of the toxic effects associated with zidovudine. In contrast, thymidine triphosphate is not a substrate for cellular RNA polymerases (remember that uridine triphosphate is used in its place in RNA), and hence these drugs would not be expected to inhibit these enzymes to a significant extent.

4. Yes. By acting as a competitive inhibitor of thymidylate kinase, zidovudine monophosphate ultimately acts to reduce the amount of thymidine triphosphate produced by the cell. The lower level of thymidine triphosphate allows zidovudine triphosphate to compete more effectively for reverse transcriptase, which ultimately increases the potency of the drug.

5. No. Both of these drugs are thymidine analogs and would compete with one another to inhibit reverse transcriptase. Thus, the drug combination would show little or no added effectiveness over one of the drugs alone.

6. Upon administration, valacyclovir is rapidly converted to acyclovir by intestinal and hepatic enzymes, acyclovir is then converted to acyclovir monophosphate by the viral thymidine kinase (Figure 8.5). Cellular enzymes then convert acyclovir monophosphate into acyclovir triphosphate, the active form of the drug. Acyclovir triphosphate, a deoxyguanosine analog, competitively inhibits the viral DNA polymerase and inhibits DNA replication by acting as a chain terminator. It thus acts via the same basic mechanism as the antiretroviral nucleoside analogs. Acyclovir is highly selective for infected cells because it serves as a poor substrate for the cellular thymidine kinase. Valacyclovir is more readily absorbed by the intestinal tract than acyclovir: after conversion the bioavailability is 54%, which is 3–5 times higher than oral acyclovir.

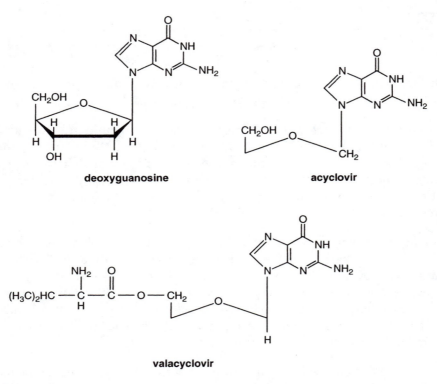

FIGURE 8.5

7. There are several possible explanations. First, the K_i is measured under purified conditions, and thus one can assume that the concentration of drug used is the concentration "seen" by the enzyme. In contrast, in tissue culture cells, the concentration of drug added to a petri dish is typically far greater than the concentration attained within the cell. The drug may not be efficiently taken up by the cell, and degradative enzymes may render some fraction of it inactive. In addition, remember that stavudine must be converted into the triphosphate form to serve as a reverse transcriptase inhibitor, and the efficiency of this conversion will not be 100%. Finally, the K_i represents the concentration of drug necessary for 50% inhibition of the enzyme, while the IC_{50} represents the concentration of drug required to inhibit viral replication by 50%; and although the activity of reverse transcriptase is certainly related to the rate of viral replication, they may not necessarily be identical.

CHAPTER 9

1. Within the nephron, carbonic anhydrase is especially active in the lumen of the proximal tubule (where the urine begins to be collected), where it catalyzes the dehydration of H_2CO_3 to carbon dioxide and water, a critical step in the reabsorption of HCO_3^- into the bloodstream. In the absence of this reaction HCO_3^- is not reabsorbed but rather excreted, together with sodium, water, and potassium. Thus, acetazolamide enhances renal excretion of HCO_3^-, inducing a metabolic acidosis that counteracts the alkalosis induced by hyperventilation (respiratory alkalosis).

2. Solubility in water
   Solubility in lipid
   Kd for CAII
   Kd for CAIV (Although less active that CAII in the ciliary body, CAIV is also considered a target for therapy.)

3. A benzene ring with a sulfonamido group and a primary amine para to the sulfur side chain are essential for antibacterial activity. These sulfonamides are structurally related to para-amino benzoic acid (PABA) (see Figure 9.4, part B), with which they interfere by competitive inhibition of the enzyme, dihydropteroate synthase, which is involved in the biosynthesis of folic acid. Folic acid, in turn, is used in the biosynthesis of DNA and amino acids. The sulfa drugs only inhibit the growth of microorganisms (and thus are referred to as *bacteriostatic* rather than *bacteriocidal*) that depend on their own biosynthetic pathways to obtain folic acid. Humans are not affected because we absorb folic acid from dietary sources.

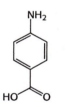

FIGURE 9.5 Para-amino benzoic acid (PABA).

4. Carbonic anhydrase is ubiquitous in the human body and thus systemic inhibitors have multiple effects. In the kidney, carbonic anhydrase is involved in the reabsorption of electrolytes. In the absence of enzyme activity, electrolytes are excreted, together with water, explaining the systemic diuretic effect.

   Carbonic anhydrase also plays a role in the transport of CO_2 from the tissues to the lungs. The CO_2 generated by actively metabolizing tissues is converted to HCO_3^- by carbonic anhydrase, which is present at high levels in red blood cells. The HCO_3^- is then transported through the bloodstream to the lungs where carbonic anhydrase reconverts it to CO_2 and it is released into the alveoli.

   Finally, carbonic anhydrase contributes to the formation of H^+ ions by the gastric mucosa of the stomach. An acidic environment is necessary to denature proteins, activate digestive enzymes, and to kill any bacteria present in ingested food.

5. The bitter taste has been attributed to the drug-laden fluid draining into the oropharynx and stimulating the taste buds located there. Additionally, inhibition of carbonic anhydrase will lead to the accumulation of HCO_3^- in the oropharynx, which also contributes to the bitter taste.

6. When a compound is formulated into eyedrops, a number of other inactive ingredients are included for such purposes as improving the solubility of the compound, maintaining a reasonably physiological pH, or as a preservative. Benzalkonium chloride is a preservative commonly added to eyedrops, including dorzolamide preparations, and can accumulate in soft contact lenses. For this reason, the wearing of soft contact lenses is somewhat contraindicated with the use of dorzolamide.

7. Acute closed angle glaucoma is caused by an anatomical defect and so cannot be treated with medication. Treatment for acute closed angle glaucoma is a surgical procedure called an iridectomy. By cutting an opening through the iris an iridectomy allows the passage of fluid and a return to normal IOP.

8. Yes. Dorzolamide acts by inhibiting fluid flow into the eye, while latanoprost stimulates outflow, and thus, an additive effect would be expected. In addition, latanoprost has the interesting side effect of increasing iris pigmentation in those with light colored eyes!

9. Carbonic anhydrase II is expressed early in the differentiation of bone resorbing cells or osteoclasts. CA II is essential for differentiation of progenitor cells into osteoclasts and modulates the intracellular pH of mature osteoclasts allowing them to function optimally.

CHAPTER 10

1. Although used at lower dosages in treating sickle-cell disease than in treating cancer, it is not yet known how a drug that inhibits cell proliferation will affect the growth and development of children. Thus, hydroxyurea is rarely used in children.

2. (a) A drug that increases the affinity of hemoglobin for oxygen would tend to keep hemoglobin in the R form, which does not form fibers.

   (b) A drug that inhibits the adhesion molecules on the surface of sickled cells might reduce sticking to the walls of blood vessels (certain adhesion molecules are expressed at unusually high levels in HbS erythrocytes).

   (c) A drug that targets the intermolecular interactions between hemoglobin molecules—either within a fiber or between fibers. (Choosing this route will be a challenge given the very large amounts of hemoglobin in erythrocytes. Large amounts of drug may be necessary, increasing the chance of toxicity.)

   (d) A drug that increases the expression of fetal hemoglobin would inhibit fiber formation.

   (e) A drug that reduces red cell dehydration would reduce the intracellular hemoglobin concentration and might minimize fiber formation.

   (f) A drug that reduces α-globin expression (lowered α-globin expression [α-thalassemia] is associated with reduced severity of symptoms in those with sickle-cell disease, perhaps due to the lower levels of hemoglobin associated with α-thalassemia).

3. The R form of hemoglobin does not form fibers, which are what ultimately result in limiting the progression of the infection in heterozygous individuals. Thus, if the parasite were to have induced the R form, the prevalence of the HbS mutation would probably never have become so nearly so wide.

4. The mutation in HbS facilitates a hydrophobic interaction between β-globin chains that occurs when the molecule is in the T form (deoxygenated). Supplemental oxygen will increase occupancy and favor the R form, preventing β-globin chain interaction and sickling, with consequent vaso-occlusion.

5. Damage to the lungs caused by repeated vaso-occlusive events results in poor oxygenation of the blood. The heart then tries to compensate by pumping more blood, which results in enlargement. However, hypoxemia and vaso-occlusive events in the lungs need not be present in sickle-cell patients with enlarged hearts (cardiomegaly). Anemia is also a cause of cardiac enlargement because it imposes a chronic volume load on the heart. The heart must pump more blood to deliver the same amount of oxygen to the tissues.

6. Most of the pathophysiology of the disease is caused by sickling of red blood cells, with the resultant blockage of a blood vessel. The fewer the circulating red blood cells, the less likely a blockage is to occur; should one occur, the more likely it is to dissolve before it becomes a more severe obstruction leading to a painful crisis.

7. It is an advantage. Inhibition of erythropoiesis in a patient experiencing ACS will inhibit the production of the patient's HbS erythrocytes. It will thus keep the percentage of circulating HbS erythrocytes low for a longer period of time, thus improving the prognosis.

8. The patient expresses both HbA and HbS and so has "sickle-cell trait" rather than "sickle-cell disease." The substitution of a negatively charged residue (glutamic acid) with an uncharged residue (valine) in HbS causes the protein to migrate more slowly than wild type toward the positive pole in an electric field, thus explaining the pattern depicted in the figure.

9. The disease is inherited in an autosomal recessive manner, and, thus, with both parents carrying a copy of the disease gene, each of their children has a 25% chance of having the disease. Prenatal testing of a fetus and pre-implantation genetic testing of an embryo can be done to test for the disease. In the United States, those of African descent are most likely to be carriers of the disease, with 8% carrying the HbS allele.

10. In the United States sickle-cell disease is most common among those of African descent, while the blood used for transfusion is derived from individuals of all backgrounds, especially those of European ancestry. It is thus possible for the recipient of a blood transfusion to elicit an immune reaction to minor antigens present on donor red blood cells. This can result in a deadly reaction to subsequent transfusions unless donor and recipient are carefully matched.

11. 2,3-BPG shifts the Hb-oxygen dissociation curve to the right, decreasing the oxygen affinity of hemoglobin (an effect similar to that of low pH). The consequence of this shift is that at any given oxygen tension hemoglobin is less saturated (has lower affinity) and more easily releases bound oxygen in the tissues. Acute and chronic hypoxemia, such as that associated with high altitude and heart or lung disease, is associated with higher 2,3-BPG levels, enhancing tissue delivery of oxygen. See Section 10.2.3 of *Biochemistry* 5e.

12. The spleen of patients with sickle-cell disease is hypofunctional, possibly the result of recurrent vaso-occlusion and infarction (destruction due to the absence of blood flow). The spleen is a central location for clearance of microorganisms, particularly encapsulated bacteria. Antibiotic therapy, in part, compensates for reduced function of the spleen in patients with sickle-cell disease.

CHAPTER 11

1. Tight binding is likely achieved through multiple, simultaneous HA-sialic acid interactions. Thus, cooperation is thought to be involved in effecting a physiologically significant interaction.

2. Antibiotics did not become widely used until after World War II, and thus the lack of antibiotic drugs was a major factor in the enormous number of fatalities that occurred as a result of influenza in 1918–1919. The subsequent pandemic, in 1957–1958, and others after it were far less deadly, thanks in part to the availability of antibiotics, as well as to the development of vaccines.

3. A number of bacterial pathogens, such as *P. aeruginosa*, are known to bind the mucins via their carbohydrate moieties. Once bound, they are removed from the airways by muco-ciliary clearance. This and other host defenses are very effective, keeping the lower respiratory tract sterile. However, pathogen-mucin binding is a double-edged sword; in addition to the secreted forms, a number of mucins are also found embedded in the cellular membranes of the underlying epithelials cells. They thus may enhance bacterial colonization of the airways if an organism is able to traverse the mucus layer to the underlying cells.

4. Aquatic birds are evolutionarily quite divergent from humans, whereas we are somewhat more closely related to swine (the evidence lies in the DNA). Transmission of an aquatic virus to a new host will result in the *adaptation,* through mutation and reassortment, of the strain to its particular host. Once adapted to a mammalian host, the virus is then more easily transmitted and propagated in humans.

5. The pandemic was caused by an H1N1 strain, which had not been circulating in the population since 1957, when it was supplanted by an H2N2 strain in the 1957–1958 pandemic. Thus, those born after 1957 were immunologically naïve to this variant and were particularly susceptible to infection.

6. The same virus isolate can result in the evolution of differing variants when grown in different cell types. This reflects the adaptation to a particular host cell, and results from mutation and selection of the "fittest" viruses. Thus, virus grown in human (or mammalian) cells would be more likely to reflect the initial virus isolate than virus grown in other cell types. However, in practice, human isolates can effectively infect chicken eggs, indicating that the type of sialic acid linkage is not a major factor in the species barrier between human and avian influenza viruses.

7. Hemagglutinin was so named because of its ability to agglutinate red blood cells. The multiple HA molecules found on the surface of each viral particle allows for the binding of multiple erythrocytes, each of which, in turn, can bind other viral particles. This results in extensive virion–erythrocyte networks that appear as red blood cell clumps.

CHAPTER 12

1. TSD results from a mutation in the *HEX A* gene (encodes the α subunit), while Sandhoff's disease results from a mutation in the *HEX B* gene (encodes the β subunit). Consequently, Tay-Sachs patients completely lack the Hex A isozyme, yet retain Hex B, whereas those with Sandhoff's disease lack both isozymes. The most important physiological effect of loss-of-function mutations in *either* the *HEX A* or *HEX B* genes is the loss of Hex A activity (loss of Hex B activity is relatively inconsequential) and the accumulation of ganglioside G_{M2}, and thus TSD and SD have almost identical clinical features. However, SD is commonly associated with organomegaly due to lipid deposition in the organs, and organomegaly is typically absent in TSD.

2. Total Hex activity is derived from three isozymes, Hex A (αβ heterodimer), Hex B (ββ homodimer), and Hex S (αα homodimer); however, Hex S is unstable and is a very minor contributor. Patients with TSD lack Hex A and Hex S, but retain Hex B. In the group of patients observed in the study the Hex B activity was increased somewhat, compensating for the lack of the other isozymes, and thus masking the deficiency. In addition, the artificial substrate used in the assay was effectively hydrolyzed by *either* Hex A or Hex B, unlike ganglioside G_{M2}, which is preferentially hydrolyzed by Hex A.

3. No. Glycolipids are not found in the outer leaflet of the lysosomal membrane. Glycolipids are found only on the inner leaflet of intracellular membrane-bound compartments because glycosylating enzymes are found only within the endoplasmic reticulum (ER) and the Golgi apparatus, from which the lysosome derives. Because transverse diffusion (between leaflets) is very slow, especially for a charged molecule such as ganglioside G_{M2}, it will remain in the inner leaflet.

4. In vivo, G_{M2} activator protein "extracts" the lipophilic ganglioside G_{M2} from the membrane, bringing it into the hydrophilic interior of the lysosome. The activator–ganglioside G_{M2} complex then specifically binds Hex A and the enzymatic reaction proceeds with the conversion of ganglioside G_{M2} into ganglioside G_{M3}. The activator then reinserts the product into the lysosomal membrane, and thus serves as a sort of shuttle between the membrane and the interior of the lysosome. Detergents, with their polar heads and lipophilic tails (see Chapter 12 opening illustration, *Biochemistry* 5e), can solubilize membranes in aqueous medium and render lipids accessible to water-soluble enzymes such as Hex A. Thus, in vitro, the use of detergents renders ganglioside G_{M2} accessible to Hex A without the need for activator protein.

5. Given their vast number (nearly 100), it is not feasible to screen all potential carriers of TSD for all the known disease-causing mutations by DNA testing. Instead, an enzymatic assay is first carried out and, if a deficiency is identified, further DNA-based tests are performed to determine the nature and severity of the mutation. Hex A and Hex B activities are distinguished by their differing heat stabilities. Hex B has a greater heat stability than Hex A, and thus measurement of Hex activity from a serum or leukocyte sample before and after heat inactivation of Hex A may be used to determine Hex A activity.

6. Enzymatic screening assays commonly utilize the artificial substrate, 4-methylumbelliferyl-glucosaminide (4MUG) in assessing Hex A activity. A number of mutations in the *HEX A* gene exist that cause a reduced catalytic activity of Hex A for its natural substrate,

ganglioside G_{M2}, yet the catalytic activity using 4MUG is unaffected. Individuals with these types of mutations are said to be of the *B1 variant phenotype,* and would incorrectly be assigned the *noncarrier* status following screening. Conversely, other mutations in *HEX A* cause reduced catalytic activity of Hex A using the 4MUG substrate, while ganglioside G_{M2} is effectively hydrolyzed. These patients, said to be *Hex A pseudodeficient,* would incorrectly be assigned the *carrier* status. Finally, the sensitivity of the enzymatic screening test is not sufficient to distinguish between mutations causing the different forms of TSD. The difference in Hex A activity between carriers of the adult and juvenile forms and carriers of the infantile form is not large, and thus DNA-based assays are used to determine the severity of the mutation.

7. Yes. A congenital defect is a defect that has been present since birth, even if it is not manifested until later. A congenital defect may be caused by a genetic factor (like TSD) or by an environmental factor. Examples of the latter are rubella infection or heavy intake of alcohol during pregnancy, either of which may cause mental and physical abnormalities in an infant.

CHAPTER 13

1. KvLQT1 encodes the α-subunit of the potassium channel, and four α-subunits must come together to form a functional channel. Patient #1, with one normal copy of the gene and a complete lack of expression of the mutant copy, experiences a 50% reduction in potassium channel function compared to a normal individual. However, in patient #2, the mutant copy is expressed but defective, and due to the multimeric nature of the channel, has the potential to interfere with the function of the normal copies. Thus, due to a dominant negative effect, patient #2 may be more severely affected by the disease than patient #1.

2. Functional KvLQT1/MinK potassium channels are necessary for hearing to develop (perhaps related to the production of endolymph), and thus a complete loss of function or lack of expression of either of these genes will cause loss of hearing. Thus, single, normal copies of KvLQT1 and MinK will suffice for hearing to develop, but a complete loss of one or the other will cause deafness.

3. Because it is often a genetic disease, ECGs of family members may be helpful in diagnosing the syndrome. In addition, repeating the ECG may be helpful, especially if carried out under varying conditions such as an exercise ECG, or an ambulatory ECG. Finally, a hearing test may be useful as deafness is a characteristic of Jervell-Lange-Nielsen syndrome, a rare form of LQTS. However, by childhood or adolescence the hearing deficit is usually apparent without the need for formal testing.

4. Gap junctions between heart myocytes allow the passage of ions, thus allowing these specialized cells to act in unison in response to an action potential.

5. Abnormalities in the electrical patterns of the heart, as determined by ECG, are the most telling diagnostic features of LQTS, and as the heart stops upon death, an ECG would be useless. Genetic testing, however, may reveal a mutation indicative of the syndrome, but only in approximately 50% of cases. Unfortunately, LQTS is underdiagnosed, and hence the cause of death when it occurs often goes unexplained.

6. Without a detailed structure of voltage-gated potassium channels, the effect of the mutation on channel function cannot be known with certainty. However, the potassium channel pore is formed in part by transmembrane helices S5 and S6 (see Section 13.5.5 of *Biochemistry* 5e) and hence any changes to these regions may affect ion flow through the channel. In particular, the substitution of the small alanine residue for the bulkier valine or glutamic acid residues might obstruct the pore and slow the passage of potassium ions through it. It is possible that residues in S5 and S6 interact with other transmembrane segments, such as S4 (the "voltage sensor"), which are important in mediating channel opening in response to a change in membrane voltage. Inhibition or a change in the function of the voltage sensing S4 may impair channel opening. Either would be consistent with reduced K^+ flux through the channel, delayed repolarization, and prolongation of the QT interval associated with the syndrome.

CHAPTER 14

1. No. Type 1 diabetes mellitus typically has a juvenile onset; however, it has been estimated that between 5% and 10% of those diagnosed with diabetes after age 30 have type 1.

2. The first step in ketone body formation involves the condensation of two acetyl-CoA molecules to form acetoacetyl-CoA via the action of thiolase (Figure 14.1). A third acetyl group is added via HMG-CoA synthase to form HMG-CoA, from which *acetoacetate* is formed by the removal of acetyl-CoA by HMG-CoA lyase. Reduction of acetoacetate yields D-*3-hydroxybutyrate,* and spontaneous decarboxylation yields acetone. Note that many of the recurring motifs of metabolic reactions are represented in ketone body formation: acetyl-CoA serves as an activated donor of acetyl groups in the first two steps, while NADH serves as an activated carrier of electrons in the formation of D-3-hydroxybutyrate from acetoacetate. In addition, of the six main types of metabolic reactions (see Table 14.3 of *Biochemistry* 5e), three are represented: oxidation–reduction, group transfer, and addition/removal of functional groups.

   At physiological pH acetoacetate and D-3-hydroxybutyrate are acids, and hence the acidity of the blood increases. The kidneys are able to excrete some into the urine, but they may become overwhelmed; when the ketone body level in the blood reaches a certain point, the patient is said to have ketoacidosis, a dangerous condition that can lead to coma and death. Prior to the discovery of insulin virtually all type 1 diabetics died of ketoacidosis.

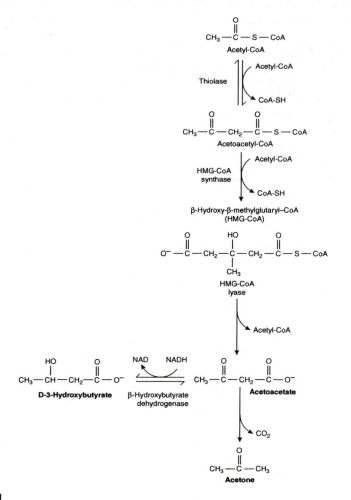

FIGURE 14.1

3. The liver is one of the key sites for the metabolism of carbohydrates. It responds to elevated glucose and insulin levels by stimulating the synthesis of glycogen and fats, long-term energy storage materials, concomitantly reducing the blood-sugar level. Thus, in healthy individuals the pancreas and liver act in concert to rapidly reduce blood-glucose levels following a meal, while in diabetics this process may be somewhat delayed, resulting in longer exposure to the damaging effects of hyperglycemia.

4. Coumarin is an *anticoagulant,* and it acts to prevent thrombus formation. This is in contrast to a class of drugs called the *thrombolytics,* which can actually act to dissolve a thrombus once it has formed. Coumarin interferes with the function of vitamin K in γ-carboxylating coagulation factors such as factors I, II, VII, and IX. These factors are then inactive and unable to participate in the process of blood coagulation. Aspirin inhibits thrombosis via a different mechanism. It acetylates and thus irreversibly inhibits cyclooxygenases, thereby inhibiting prostaglandin synthesis. By inhibiting the synthesis of certain platelet prostaglandins involved in platelet aggregation, aspirin inhibits thrombosis.

5. Creatine kinase is found in tissues that have high and fluctuating energy needs, such as skeletal muscle, the brain, spermatozoa, and the myocardium. Creatine kinase can regenerate ATP from ADP using creatine phosphate as a phosphoryl donor, and thus the enzyme is important in the rapid generation of usable energy (see Section 14.1.5 of *Biochemistry* 5e). The enzyme is also released upon skeletal muscle damage and in certain diseases such as muscular dystrophy, and thus a more accurate indicator of MI is the level of the myocardial isozyme (MB) rather than total enzyme activity.

6. Hemoglobin A1c is a glycosylated form of hemoglobin that is formed via non-enzymatic mechanisms when blood-glucose levels are high. It is used as a measure of long-term glucose control in diabetic patients. It reflects the degree of glucose control during the previous 2–3 months because the life span of erythrocytes (the carriers of hemoglobin) is approximately 120 days.

CHAPTER 15

1. Many neurotransmitters act through G-protein-coupled receptors to activate second messengers or to mediate the activities of ion channels. Indeed, neuronal function depends upon the precisely regulated activity of a number of signal transduction pathways. One might predict that aberrant regulation of one or more of these pathways would lead to neurological disturbances. Indeed, a number of signal transduction molecules, including G-proteins, kinases, and transcription factors, have been implicated in bipolar disorder.

2. Lithium is an uncompetitive inhibitor of IMPase, meaning that it binds only the enzyme-substrate complex (not the enzyme alone). An uncompetitive inhibitor has a greater fractional inhibition at higher substrate concentrations, as would be the case in patients with bipolar disorder according to this particular hypothesis. Increased signaling through the pathway in these patients would result in higher IMP levels (and higher IMP-IMPase levels) and a greater fractional inhibition by lithium. The drug would have a lesser effect in normal individuals with lower IMP levels.

3. Lithium is a simple metal ion and appears to inhibit its various known targets by displacing a divalent metal ion (such as Mg^{2+}) necessary for enzyme activity. Li^+ and Mg^{2+} have similar ionic radii and thus can bind to the same sites on macromolecules. Many unrelated enzymes rely upon divalent metal ions for their activity, which is a contributing factor in the diversity of enzymes inhibited by lithium. Li^+ may also substitute for Na^+ and influence excitability in the brain, muscle, and heart.

4. $G\alpha_s$ stimulates adenylate cyclase (AC), and elevation of AC activity has indeed been observed in postmortem brain samples from patients with bipolar disorder. The cyclic AMP (cAMP) generated by AC in turn stimulates the activity of protein kinase A (PKA), and elevation of PKA activity has also been observed in these samples. Thus, increased activity of three components of the PKA pathway has been associated with the disorder, suggesting a role for this pathway in the pathogenesis of the disorder. In support of this, some studies have shown that lithium treatment normalizes the activities of these enzymes in patients with bipolar disorder.

5. Inositol is a polar molecule and thus does not readily cross the lipophilic blood–brain barrier. The brain thus relies heavily on synthesis and recycling of inositol for the generation of phosphoinositides.

6. Tamoxifen has been found to be a selective inhibitor of protein kinase C (PKC), and would thus act to dampen signaling through the phosphoinositide pathway (like lithium). Stimulation of phospholipase C produces IP_3 and DAG, which subsequently activate PKC (see Section 15.2.2 of *Biochemistry* 5e). While larger clinical trials are needed to confirm the preliminary promising results, these data further support the importance of the phosphoinositide pathway in bipolar disorder.

7. It may take some time before modulation of the activities of upstream effectors in a signal transduction pathway are translated into a cellular response, particularly when the response involves changes in gene expression patterns. This involves changes in transcription and translation patterns, and there will be a lag before the new steady state is reached. Indeed, changes in gene expression have been implicated in bipolar disorder; in particular, changes in the activity of a transcription factor called *cAMP-responsive element binding protein* (*CREB*) have been associated with this disorder and other mental illnesses.

CHAPTER 16

1. Sulphonic acid groups are polar and are *ionized* under physiological conditions; thus the drug is unable to cross the lipophilic blood–brain barrier.

2. Trypanosomes have a large repertoire of surface antigens (close to 1000 variants) and can switch among them to evade immune destruction. This renders the development of an effective vaccine extremely challenging, if not impossible.

3. The selective advantage of a system for rapidly metabolizing glucose would exist only if high concentrations of glucose were available, such as in the bloodstream of a vertebrate host. The evolution of glycosomes has been proposed to predate the evolution of vertebrates, suggesting that the organelle provided some other selective advantage.

4. Glycerol together with salicylhydroxamic acid result in the accumulation of glycerol-3-phosphate, leading to inhibition of the glycerol-3-phosphate dehydrogenase catalyzed conversion of dihydroxyacetone phosphate to glyerol-3-phosphate (see Figure 16.1). This reaction generates NAD^+, which is necessary for the reaction catalyzed by glyceraldehyde-3-phosphate dehydrogenase. Thus, NAD^+ is rapidly depleted, glycolysis comes to a halt, and the cell dies.

5. In *T. brucei,* phosphofructokinase is found in the glycosomes, where no net ATP production occurs as a result of the glycolytic pathway. Instead, net ATP production occurs in the *cytosol,* in the reaction catalyzed by pyruvate kinase. Thus, the mechanism whereby the enzyme activity could be modified according the cellular energy charge (i.e., the ATP/AMP ratio) appears to have been lost in the *T. brucei* enzyme.

6. Inhibition of an enzyme in the glycolytic pathway (or any other metabolic pathway) could result in accumulation of a metabolite to toxic levels, perhaps as a result of osmotic stress caused by the high metabolite concentration. Indeed, computer modeling has indicated that inhibition of pyruvate export from the parasite may lead to the accumulation of this metabolite to toxic levels.

7. No. Although the parasite depends upon oxidative metabolism for ATP during the insect phases, in the human phase the mitochondrion is poorly developed (it is sometimes referred to as a *pro-mitochondrion*) and the parasite depends exclusively on glycolysis to meet its energy needs. Thus, inhibitors of oxidative metabolism may be useful in treating tse tse flies, but not humans.

CHAPTER 17

1. It was the ANA line. A high ADH activity and low ALDH activity would result in high levels of acetaldehyde, a toxic compound that causes unpleasant symptoms. Differences in acetaldehyde metabolism would thus account for the differences in ethanol intake between these rat lines. Indeed, it has been shown that ethanol leads to higher serum acetaldehyde levels in the ANA rats compared to the AA rats. Differences in acetaldehyde metabolism have also been observed in humans. Some people have a mutation in the low K_M, mitochondrial ALDH and have adverse reactions to ethanol as a result of acetaldehyde toxicity.

2. The increased levels of lactate in the blood impair the ability of the kidneys to excrete uric acid, the end product of purine metabolism. As a result uric acid can build up in the blood and may be deposited in crystalline form in the joints, thus precipitating attacks of gout.

3. Ethylene glycol and methanol are converted to oxalate and formate, respectively, via the enzymatic activity of ADH. Both of these are toxic compounds, which can cause life-threatening complications, such as extreme metabolic acidosis. Ethylene glycol and methanol poisoning accounted for 3% and 2%, respectively, of deaths due to poisoning in the United States in 1998. Infusion of ethanol will compete with these toxins for the active site of ADH, and thus allow for their excretion before being converted into their toxic forms.

4. FAS is characterized by microcephaly and mental retardation, facial abnormalities, cardiac defects, and both pre- and postnatal growth retardation. Both alcohol and acetaldehyde can cross the placenta, and both are thought to contribute to the pathogenesis. The precise amount of alcohol consumed during pregnancy that will cause the syndrome is unknown, as is the period during which the fetus is most susceptible. However, chronic heavy drinkers and alcoholics, especially those who smoke, are at higher risk of giving birth to infants with FAS.

5. The "French paradox" refers to the relatively low incidence of ischemic heart disease in France, despite consumption of high levels of saturated fats. The paradox has been attributed to the consumption of red wine, which is also common in France. Elevated levels of high-density lipoprotein (HDL), inhibition of platelet aggregation and improved endothelial cell function has been associated with moderate consumption of red wine. These cardio-protective effects have been attributed to the flavenoids, antioxidant compounds that are abundant in red wine. However, a number of studies have shown beneficial effects from other types of alcoholic beverages, such as beer and spirits, indicating that ethanol itself may confer the protective effect.

6. The lactate:pyruvate ratio increases. In humans, pyruvate, the end product of glycolysis, may be converted into lactate or into acetyl CoA and enter the citric acid cycle. In addition, pyruvate may be converted into glucose via the gluconeogenic pathway. Conversion to lactate is favored because it is an NADH-consuming reaction and NADH is present at high concentrations as a result of ethanol metabolism. In contrast, conversion of pyruvate to acetyl CoA requires NAD^+, and high levels of NADH inhibit the *pyruvate dehydrogenase complex* that catalyzes this reaction. Thus,

the lactate level rises at the expense of the pyruvate level, resulting in acidosis and inhibition of gluconeogenesis.

7. Ethanol is a small molecule that can easily diffuse across the intestinal lining and into the bloodstream, and thus does not necessitate an active transport process. It is absorbed in all parts of the GI tract, from the mucosa in the mouth to the esophagus, the stomach, and the small and large intestines; however, it is most efficiently absorbed in the small intestine. Because absorption occurs by simple diffusion, the concentration gradient across the intestinal lining is an important factor in the rate of ethanol absorption. Thus, more dilute forms of alcohol or concurrent food intake reduce the ethanol concentration in the GI tract and reduce the rate of absorption.

CHAPTER 18

1. The cytoplasm of a fertilized egg is derived almost entirely from the oocyte, with the spermatozooite largely contributing nothing but its genetic material. Thus, diseases linked to mtDNA mutations are transmitted through the maternal line.

2. Mitochondria segregate randomly during cell division, which may largely explain the variation in clinical manifestations of diseases, such as KSS, caused by mtDNA mutations. In many diseases (including KSS), not all of the mitochondria carry the mutation, and thus only some fraction are defective—these diseases are called *heteroplasmic* diseases. Unequal segregation of defective mitochondria during oogenesis can result in differing clinical manifestations, even between siblings. Subsequently, unequal segregation during development may ultimately result in some tissues being more seriously affected by the defect than are others. The condition is more serious when cells with high-energy needs bear a large burden of defective mitochondria. The uneven partitioning of mitochondria between daughter cells during cell division, combined with other genetic and environmental factors, contributes to the range of symptoms observed among those with the disease.

3. The L/P ratio is often measured in patients suspected of having a mitochondrial disease because it can help differentiate between a defect in oxidative phosphorylation and a defect in pyruvate metabolism, such as pyruvate dehydrogenase deficiency. Pyruvate dehydrogenase deficiency would likely yield a *marked* increase in pyruvate levels, yielding L/P ~ 10–20, whereas defects in oxidative phosphorylation would give somewhat lower levels of pyruvate and higher L/P ratios (L/P > 20). In the latter case pyruvate could still enter the citric acid cycle, thus reducing its level to some extent.

4. Mitochondrial DNA polymerase (also called the γ-DNA polymerase) is fairly error-prone, with an error rate approximately ten times higher than its nuclear counterpart. In addition, the lack of recombination in mtDNA adds to the high mutation rate, as recombination repairs some errors in nuclear DNA. Finally, mitochondrial DNA is not packaged into protective chromatin-like nuclear DNA, and is thus more susceptible to the damaging effects of electromagnetic radiation and chemical mutagens. It is also particularly subject to the damaging effects of reactive oxygen species that are produced as a result of oxidative phosphorylation. Finally, like prokaryotic genomes, mtDNA possesses no introns, thus increasing the liklihood that a mutation will strike a functional sequence and cause a defect.

5. There are several reasons. First, biological samples usually contain more mtDNA than nDNA. Remember that there are only two copies of the nuclear genome per diploid cell, whereas there are hundreds of copies of the mitochondrial genome. In addition, the circular nature of the mitochondrial genome renders it less susceptible to exonucleases and more stable to heat and other environmental stresses. Additionally, the noncoding region of the mitochondrial genome is particularly polymorphic and thus differs from individual to individual, and it is transmitted, unchanged, for generations through the maternal line. It is thus particularly useful in identifying long-dead servicemen or victims of crimes: any maternal relative can serve as a reference sample for identification. Mary-Claire King has pioneered the use of mtDNA typing in identifying children (now teenagers) of those who "disappeared" during the military dictatorship in Argentina between 1975–1984, and returning them to their grandmothers.

6. Neurons maintain an ion gradient across the cellular membrane largely via the action of a pump called the Na^+-K^+ ATPase. This pump uses the energy from ATP to transport Na^+ out of the cell and K^+ into the cell, thus forming an electrochemical gradient across the membrane. Transmission of a nerve impulse involves the opening of channels and destruction of the gradient, which must subsequently be restored to bring the cell back to the resting state. The maintenance of the gradient largely accounts for the high-energy needs of neuronal cells.

7. The mitochondrial theory of aging was first proposed in the 1960s as an alternative to the notion that aging was caused by an accumulation of nuclear DNA mutations. Nuclear DNA is efficiently repaired, unlike mitochondrial DNA, which tends to accumulate mutations. In addition, mtDNA is particularly susceptible to oxidative damage because oxidative phosphorylation takes place in the mitochondria. In support of the mitochondrial theory of aging, it has been found that a number of features associated with aging are found in those with mitochondrial diseases, such as muscle weakness, loss of hearing and vision, and other neurologic manifestations, such as dementia and seizures. In addition, although the mechanisms have yet to be elucidated, mutations in mtDNA have been found to be associated with Alzheimer's, Parkinson's, and Huntington's disease patients.

8. In a mitochondrial disease a mother may transmit the disease to all of her children, both male and female. Conversely, because sperm contribute such a small fraction of the mitochondria to the zygote, a father does not transmit the disease to any of his children, nor is the disease expressed in any subsequent generations. For X-linked disorders, an affected father will transmit the mutant allele to his daughters, and thus paternal transmission occurs. In addition, X-linked disorders are more frequently manifested in males, whereas in most cases there is no predilection for either gender in mitochondrial diseases. Finally, although the inheritance of mitochondrial diseases is known to occur solely through the maternal line, the expression of the disease is unpredictable and does not follow the laws of Mendelian genetics as do X-linked disorders. The unusual inheritance pattern of certain mitochondrial diseases is what first led researchers to suspect mtDNA as the source of the mutant alleles.

CHAPTER 19

1. 7-dehydrocholesterol, the vitamin D_3 precursor, absorbs UV light in the range of 290–315 nm, and thus competes with melanin for light photons. Thus, one would expect that the rate of synthesis of vitamin D_3 would be greater in a light-skinned person than in a dark-skinned person. Indeed, children with dark complexions are at increased risk of rickets.

2. Although vitamin D may be obtained from dietary sources, most is derived from photolysis of 7-dehydrocholesterol in the skin. The smog produced by the proliferation of factories during the industrial revolution blocked out much of the UV light, thus blocking the generation of the vitamin. The incidence of rickets reached 75–98% among children in some northern cities during this time, and thus the claim that rickets was the first disease caused by environmental pollution may well be justified. Even today, the risk of rickets is greater among children living in inner city areas.

3. Humanity is thought to have originated in tropical climates in Africa, where sunlight is plentiful throughout the year. In these areas the ample sunshine would ensure adequate endogenous production of vitamin D_3, and an exogenous source would not have been necessary. Had humanity evolved in more northern climates, there may have been more intense evolutionary pressure for the extraction of vitamin D from the mother into human breast milk.

4. Vitamin D is technically neither a vitamin nor a hormone. Although vitamin D is derived from cholesterol and is related to the steroid hormones, it is not a classical hormone because it is secreted from the skin, rather than a discrete endocrine gland such as the thyroid gland. Technically, vitamin D is not a vitamin either because it can be made by the body and is thus not a necessary constituent of the diet. However, in certain parts of the world, particularly in the winter, vitamin D is effectively a vitamin because sunlight is insufficient for endogenous production.

5. Synthesis and secretion of parathyroid hormone (PTH) is stimulated by hypocalcemia. Thus, a typical laboratory finding in those with vitamin D deficiency is elevated PTH level. PTH acts to increase serum calcium levels via several mechanisms. First, PTH stimulates the activity of 25-hydroxyvitamin D-1α-hydroxylase, thus increasing the level of 1,25-dihydroxyvitamin D, the most active metabolite of vitamin D, thereby stimulating calcium absorption from the intestine. 1,25-dihydroxyvitamin D feeds back on PTH, inhibiting its synthesis. PTH can also act directly on bone and the kidneys, increasing bone resorption (and calcium release) and decreasing calcium loss in the urine.

6. The last step in the formation of 1,25-dihydroxyvitamin D, the most biologically active metabolite, occurs in the kidneys and thus patients with renal failure are at increased risk of rickets. These patients must be treated with oral phosphorus and 1,25-dihydroxyvitamin D, rather than vitamin D.

CHAPTER 20

1. The *G6PD* gene lies on the X chromosome, which explains this phenomenon. *X inactivation,* a process in which one of the X chromosomes in females is inactivated by packaging into chromatin, occurs early in embryonic development. However, the particular X chromosome (either paternally derived or maternally derived) inactivated in each cell is random, leading to distinct populations of cells with regard to genes present on the X chromosome. This accounts for the finding that heterozygous females express either wild-type or mutant *G6PD*—but not both—in any one cell.

2. Large deletions, frameshift mutations, and nonsense mutations are likely to result in complete abrogation of enzyme activity, as would mutation of the substrate binding site. The fact that these types of mutations have not been observed indicates that a total lack of G6PD activity is incompatible with life.

3. NADP$^+$-dependent malic enzyme. There are two versions of the enzyme—a mitochondrial isozyme and a cytosolic isozyme—encoded by the *ME3* and *ME1* genes, respectively. Both oxidatively decarboxylate malate, generating NADPH. Erythrocytes do not possess mitochondria and thus do not express *ME3*.

$$\text{malate} + \text{NADP}^+ \rightarrow \text{pyruvate} + \text{NADPH} + \text{CO}_2$$

4. Iron is a component of heme, which is necessary for the oxygen-binding property of hemoglobin. Iron deficiency leads to depletion of bodily iron stores, and subsequently to reduced production of hemoglobin, and a fall in the hematocrit. Eventually, the bone marrow (the site of red-blood-cell development) will compensate by stimulating the production of erythrocytes. However, these are smaller than normal, leading to the characteristic appearance of *microcytic* cells in a blood smear, a hallmark of iron-deficiency anemia. The underlying cause of anemia is thus distinct from the hemolytic anemias (such as G6PD deficiency), which are caused by premature destruction of erythrocytes rather than impaired production.

5. Any test based on G6PD enzymatic activity, such as the fluorescent spot test, may yield a false-negative result immediately following a hemolytic crisis. During a crisis old erythrocytes, with lower levels of G6PD activity, are preferentially lysed, and reticulocytes, with higher levels of activity, are beginning to replace them. Thus, a nonquantitative test such as the fluorescent spot test may come back negative, especially among those with the A− variant, which is associated with only moderately impaired enzyme activity.

6. Catalase, like glutathione peroxidase, depends upon NADPH for activity, and thus its activity plummets during a hemolytic crisis in G6PD deficient patients, when the NADPH level is very low. Under normal circumstances, however, catalase is active and contributes to the disposal of reactive peroxides in erythrocytes. The activity of catalase likely contributes to the episodic nature of hemolytic anemia in most G6PD-deficient patients by acting to keep cellular peroxide levels low under all but the most stressful conditions.

CHAPTER 21

1. The rational for creatine supplementation is that its phosphorylation by creatine kinase yields creatine phosphate, thus increasing the amount of this important cellular energy reserve.

$$\text{Creatine} + \text{ATP} \leftrightarrow \text{creatine phosphate} + \text{ADP} + \text{H}^+$$

   Upon exercise, within a few seconds cellular ATP is depleted and the major source of energy shifts to creatine phosphate. Subsequently, the cell's energy requirements are met by metabolic processes. Creatine phosphate thus serves as an important immediate source of energy during exercise. Pyridoxal phosphate (derived from vitamin B_6) is a cofactor for phosphorylase, and hence supplementing with vitamin B_6 may enhance the activity of any residual phosphorylase enzyme in McArdle disease patients. In addition, the majority of muscle cell vitamin B_6 is complexed with phosphorylase and thus supplementation increases vitamin B_6 availability in general.

2. Under conditions when oxygen is limited, such as when blood flow is cut off during the ischemic exercise test, glucose metabolism will ultimately yield lactate (see Figure 21.1). The pathway leading to the citric acid cycle requires oxygen, which would be limited under these conditions. In patients with McArdle disease an inability to break down glycogen results in a shortage of glucose, and ultimately in a failure of the lactate level to rise during exercise, as would occur in healthy individuals.

3. Insulin plays an important role in regulating glycogen metabolism (see Section 21.5.2 of *Biochemistry* 5e). By binding to receptors on the surface of cells, insulin stimulates glycogen synthesis via the activation of glycogen synthase and inhibits its breakdown via the inactivation of phosphorylase kinase. Insulin thus signals the "fed" state and stimulates the storage of glucose as glycogen. Overtreatment of diabetics with insulin can lead to excessive storage of glycogen, which may damage tissues and organs, especially the liver, a major glycogen repository in the body.

4. During exercise the enzyme *myoadenylate kinase* generates ATP from two molecules of ADP (see [1] below). The AMP produced is then converted to IMP and ammonia (NH_3) via the enzyme *myoadenylate deaminase* (see [2] below). Most of the ammonia generated is ultimately excreted in the urine; however, it is removed from the bloodstream with a half-life of between 20 and 30 minutes, and hence the rise that occurs upon exercise may be measured.

$$\text{ADP} + \text{ADP} \rightarrow \text{ATP} + \text{AMP} \tag{1}$$

$$\text{AMP} \rightarrow \text{IMP} + \text{NH}_3 \tag{2}$$

5. Glycogen storage disease type 0 results from a defect in glycogen *synthesis,* rather than its *breakdown,* and thus does not lead to the accumulation of glycogen in bodily tissues seen in other glycogen storage diseases. It is thus technically not a glycogen storage disease. However, defects in either the breakdown or synthesis of glycogen result in its unavailability as a source of energy, and thus they cause diseases with some similarities. A liver biopsy and a test for enzyme activity can distinguish type 0 from the other glycogen storage diseases.

6. No. Defects in branching enzyme cause *type IV glycogen storage disease,* which is usually fatal by age 5. The α-1,6 linkages increase the solubility of glycogen, as well as increasing its rate of synthesis and breakdown. However, the pathogenesis of this disease stems from the decreased solubility of glycogen. Deposition of insoluble glycogen aggregates damages bodily tissues, especially the liver, leading to cirrhosis and death from liver failure in most cases.

CHAPTER 22

1. Triacylglycerols are purely hydrophobic compounds and thus have the tendency to co-alesce into droplets to minimize contact with the aqueous medium. Hence the formation of large fat vesicles in the aqueous cytosol of hepatocytes of those with NAFLD. This property of triacylglycerols renders them poorly suited to the formation of biological membranes. In contrast, membrane lipids have both hydrophobic and hydrophilic moieties and adopt a *bilayer* configuration in which the hydrophobic portions are oriented inward, and the hydrophilic portions outward, toward the aqueous medium. The *amphipathic* nature of phospholipids and glycolipids is thus critical to their role as constituents of cellular membranes.

2. Glycerol is converted to D-glyceraldehyde-3-phosphate in the liver via the activities of glycerol kinase and glycerol phosphate dehydrogenase. D-glyceraldehyde-3-phosphate is an intermediate in the glycolytic pathway (it can also enter the gluconeogenic pathway). The two-carbon acetyl group from acetyl CoA is transferred to oxaloacetate in the citric acid cycle to form citrate. It is important that acetyl CoA can enter the citric acid cycle only in the presence of glucose, as this is the source of oxaloacetate. In its absence, acetyl CoA is instead converted into ketone bodies (see Section 22.3.5 of *Biochemistry* 5e). The NADH and $FADH_2$ produced are both electron donors in oxidative phosphorylation. NADH enters the chain at complex I, while FADH2 enters the chain at complex II.

3. Very low calorie diets or fasting would signal the "starved" state, which would trigger a number of physiological responses, including lipolysis. The hormone glucagon would simultaneously inhibit fatty acid synthesis through inhibition of acetyl-CoA carboxylase and stimulate fatty acid breakdown by induction of lipases. The rapid release of free fatty acids from adipocytes would result in their influx into the liver, where TAGs are formed. From this point, the pathogenesis would closely resemble the proposed pathogenic mechanism involved in NASH, which includes oxidative stress, lipid peroxidation, and inflammation. In addition, fasting has been associated with hepatic depletion of glutathione, a potent antioxidant, which may contribute to the pathogenesis.

4. Triacylglycerols are exported from hepatocytes in the form of VLDLs. Triacylglycerols are hydrophobic molecules and thus must be transported in the bloodstream via these specialized transport particles. A shortage of the lipoproteins (called *apo B-100* and *apo E*) would lead to the accumulation of triacylglycerols in hepatocytes and eventually to steatosis.

5. Troglitazone is an agonist of peroxisomal proliferator-activated receptor$_{gamma}$ (PPAR$_{gamma}$), which increases the transcription of insulin-responsive genes and acts to stimulate glucose uptake into cells. Troglitazone thus acts to increase insulin sensitivity in NASH patients, targeting what is thought to be the first step in the pathogenic mechanism. Troglitazone has been discontinued in the United States as a result of harmful side-effects. In particular (and ironically), a number of cases of liver damage were reported in patients taking the drug. The mechanism of liver damage appears to result from drug-induced inhibition of taurocholate export. This bile acid is important in maintaining bile flow, and in its absence susceptibility to intrahepatic cholestasis appears to be increased.

6. β-oxidation of fatty acids takes place in the mitochondria, thus mitochondrial injury would lead to impairment of this process. As a result, fatty acids would accumulate and

form TAGs. The development of NASH would be favored by the oxidative stress generated by defective mitochondrial function, the site of oxidative phosphorylation, and the shortage of ATP would contribute to cellular dysfunction and death. The harmful side-effects of perhexiline are accentuated in those with a particular allele of cytochrome P450 CYP2D6, a hepatic oxidase involved in the metabolism of the drug. The drug is metabolized more slowly in patients with the polymorphism, and it thus remains active longer.

7. Leptin (derived from the Greek *leptos,* meaning "slender" or "thin") is a peptide hormone released by adipocytes upon feeding. In binding to its receptor, it initiates neural pathways that act through the hypothalamus to ultimately decrease appetite and increase energy expenditure. Like many obese humans, *ob/ob* mice are insulin-resistant, and thus these mice have been a useful model for human obesity. Genetic mutations in the leptin gene and its receptor have been identified in humans and are associated with morbid, early-onset obesity. However, these mutations are rare, and most obese individuals have normal leptin and leptin-receptor genes. Nevertheless, most obese individuals have elevated leptin levels and thus may be said to exhibit "leptin resistance." Administration of recombinant leptin to individuals with genetic leptin deficiencies has proven very effective; however, the response in the more common cases of obesity is limited at best.

CHAPTER 23

1. Cryptogenic cirrhosis is cirrhosis with no identifiable cause, and approximately 10% of cases are of this type.

2. Metabolism of drugs is largely carried out by the liver and generally occurs via a two-phase process. In the first phase the drug is oxidized, reduced, or hydrolyzed; in the second phase it is coupled to a polar molecule for more ready elimination, either in bile or urine. These reactions are usually less efficient in those with liver disease, and thus drugs remain active for longer periods. This explains the heightened sensitivity of patients with liver damage to many drugs. The effect of liver dysfunction on each particular drug is unpredictable, and does not correlate well with the type of liver damage, liver-function tests, or severity of disease, and thus the correct doses must be determined empirically. Although most drugs are inactivated in the liver, some are *activated* by hepatic metabolic enzymes, and this explains the ineffectiveness of certain drugs in those with liver dysfunction.

3. Virtually all exocrine glands are affected by cystic fibrosis, including the liver, which secretes bile into the duodenum via the bile duct. The thick secretions characteristic of the disease can block the duct, causing chronic inflammation and, eventually, cirrhosis.

4. Humans do not possess a mechanism for metabolism and elimination of iron, and thus what is lost is through loss of skin cells and cells from mucosal surfaces, such as the gastrointestinal tract. Premenopausal women lose additional iron through menstruation, and pregnant and nursing mothers lose still more, in furnishing the needs of their fetuses and infants. Thus, iron stores typically do not rise as rapidly in women, thus lessening the damage to bodily organs that occurs in the disease.

5. Transferrin saturation represents the proportion of total transferrin iron-binding sites occupied by iron. It is equal to the serum-iron concentration/total iron-binding capacity. The total iron-binding capacity is measured by taking a blood sample and completely saturating all the sites. The TS in this patient is 169 μg/dl ÷ 295 μg/dl, multiplied by 100 = 57%.

6. Proteins are composed of amino acids, and excess amino acids cannot be stored and are thus metabolized. The first step in amino acid metabolism is the deamination step (Section 23.3.1 of *Biochemistry* 5e), which produces an ammonium ion, which normally then goes on to participate in the urea cycle within the liver. When liver function is impaired, the urea cycle does not proceed efficiently and ammonium ions can accumulate in the bloodstream. Although, the substance(s) that cause hepatic encephalopathy have not definitively been determined, much evidence indicates a role for ammonium ions.

7. Lactulose is a non-absorbable, synthetic disaccharide, consisting of one molecule each of galactose and fructose. Its beneficial effects are believed to be due to its lowering of blood-ammonium levels, which is thought to be an important factor in hepatic encephalopathy. Metabolism of lactulose into organic acids, especially lactic acid, by colonic bacteria lowers the pH of the contents of the colon, which favors the conversion of ammonia (NH_3) into the less easily absorbed ammonium ion (NH_4^+). In addition, through an effect of bacterial metabolism, lactulose may inhibit bacterial production of ammonia. Finally, lactulose serves as an osmotic laxative, thus decreasing absorption of peptides and amino acids.

8. The retention of water and Na⁺ is common in those with cirrhosis, although the under-lying causes are still a matter of debate. It appears that portal hypertension leads to the pooling of blood in the splanchnic vascular bed, which results in a lowering of the effective circulating blood volume. This triggers Na⁺ reabsorption by the kidneys and water retention. Contributing to the problem is the low level of plasma albumin (*hypoalbuminemia*), a blood protein produced by the liver, found in many cirrhotic patients. Hypoalbuminemia causes reduced plasma oncotic pressure, which favors the extrusion of water from the blood vessels into interstitial spaces. In cirrhotic patients, the excess fluid is initially primarily found in the peritoneal cavity, which lies in close proximity to the portal venous system, causing *ascites*. However, peripheral edema, which was observed in the patient discussed here, is often manifested in later stages of the disease.

9. No one test definitively indicates hepatic dysfunction and thus multiple tests are performed, with the accumulation of evidence leading to the diagnosis. In addition, the various tests assist the physician in determining the *cause* of liver dysfunction. ALT and AST are present in high concentrations in the liver, and damage to hepatic cells causes leakage of these enzymes into the bloodstream; however, these enzymes, especially AST, are not specific to the liver, and thus only suggest hepatocellular necrosis. Alkaline phosphatase is found in high concentrations in the biliary tract, and thus elevated levels suggest cholestasis (blockage of bile flow). Again, the enzyme is not specific to the biliary tract, but is also found in bone and placenta, and thus does not definitively implicate the biliary tract. Finally, bilirubin becomes conjugated in the liver before being secreted into the duodenum in bile, and thus the levels of total and conjugated bilirubin can differentiate between hepatic dysfunction and cholestasis. When total bilirubin is elevated while the conjugated bilirubin level is low, the likely cause is hepatic cell damage; conversely, cholestasis is likely when the level of conjugated bilirubin is high. Note that this patient had elevated levels of both conjugated and unconjugated bilirubin, indicating both hepatic cell damage and cholestasis.

CHAPTER 24

1. Vitamin B_6 is a necessary cofactor for the enzymatic production of cystathionine from homocysteine, whereas vitamin B_{12} is necessary for the activity of methionine synthase, which generates methionine from homocysteine. Vitamin supplementation will promote the activities of these enzymes, thereby diminishing the level of homocysteine and limiting its pro-atherogenic effects.

2. The pathway leading to the synthesis of cysteine is known as the transsulfuration pathway. It begins with the condensation of serine and homocysteine (via cystathionine β-synthase) to form cystathionine, followed by deamination and cleavage to α-ketobutyrate and cysteine (via cystathioninase). In this pathway the *sulfur* atom from homocysteine is *transferred* to the carbon skeleton of serine, hence the designation of *transsulfuration*. Homocysteine is methylated via the enzymatic activity of methionine synthase. Thus, methionine *recovers* the *methyl* group it lost earlier in the activated methyl cycle (see Figure 24.14 of *Biochemistry* 5e), and hence the designation of *remethylation* pathway.

3. Betaine donates one of its three methyl groups to homocysteine to generate methionine in a reaction catalyzed by *betaine:homocysteine methyltransferase*. Thus, betaine supplementation promotes the remethylation pathway (homocysteine → methionine) and causes a reduction in homocysteine levels. The mitochondria are particularly reliant upon betaine in the generation of methionine as the folate derivatives traverse the mitochondrial membranes very slowly.

$$\text{Homocysteine} + \text{betaine} \rightarrow \text{methionine} + \text{dimethylglycine}$$

4. The animal studies revealed an association betweeen elevated homocysteine levels and neural tube defects, while folic acid supplementation in pregnancy reduced the risk of these defects. Given the known role of N^5-methyltetrahydrofolate in the conversion of homycysteine to methionine, this suggests that folic acid may reduce the risk of neural tube defects by lowering homocysteine levels.

5. After methionine loading, a *marked* increase in plasma homocysteine level indicates a defect in homocysteine metabolism. Hyperhomocysteinemia is defined as two standard deviations over the mean level obtained from normal individuals. An oral methionine dose of 100 mg/kg body weight is given, and the plasma homocysteine level is measured prior to the challenge and between four and eight hours afterward. Methionine leads to homocysteine via the activated methyl cycle, and any defect in homocysteine metabolism, either through the cysteine or methionine biosynthetic pathways, would result in marked elevation of homocysteine.

6. An elevated BUN is an indicator of kidney disease, and it is significant in this case because renal disease has been associated with hyperhomocysteinemia. Although the exact mechanims have yet to be elucidated, this is thought to be due to impaired metabolism of plasma homocysteine by the kidney (a major route of homocysteine clearance) rather than decreased excretion through the urine (a very minor route of homocysteine clearance).

CHAPTER 25

1. Most people who live in malaria endemic areas have been infected by the parasite at one time or another (often repeatedly) and have acquired some degree of immunity. Visitors from regions where the disease has been eliminated are immunologically naïve and are thus far more susceptible. This explains why foreign troops and other visitors to malarious regions are often ravaged by the disease, whereas the local adult population is relatively unharmed.

2. *Plasmodia* do not possess the enzymes necessary for de novo biosynthesis of purines and thus rely upon the salvage pathway (Table 25.1). HGPRT is an important enzyme in this pathway, converting hypoxanthine to inosinate, a precursor of both guanylate and adenylate. Thus, suppression of HGPRT results in a shortage of these purine nucleotides, and inhibition of parasite RNA synthesis and DNA replication.

3. Humans cannot synthesize folate, but instead must obtain it from dietary sources. Thus, humans do not possess a version of dihydropteroate synthase, which generates dihydropteroic acid (a precursor of tetrahydrofolate) from *p*-aminobenzoic acid (PABA) and pteridine. Sulfonamides like sulfadoxine are structurally similar to PABA, thus competitively inhibiting the enzyme. Pyrimethamine, which is prescribed in conjunction with sulfadoxine, inhibits dihydrofolate reductase; thus the drugs conspire to inhibit the biosynthesis of folate and thus prevent the replication of the parasite.

4. Glucose-6-phosphate dehydrogenase participates in the pentose phosphate pathway (see Chapter 20), which generates much of the NADPH found in the cell. *P. falciparum* needs NADPH and other products of the pentose phosphate pathway for optimal growth, and thus does not thrive in cells deficient for glucose-6-phosphate dehydrogenase. The enzyme deficiency is most often found in areas where malaria is endemic, suggesting that the mutation persisted because it provided a selective advantage in surviving malaria. The decreased level of reduced glutathione that results from NADPH deficiency causes susceptibility to reactive oxygen species (ROS), especially in erythrocytes, which do not possess alternative pathways for NADPH production. Thus, any drug that has oxidative potential, including several antimalarials (such as primaquine, dapsone, and sulfamethoxazole), and other drugs such as salicylates, phenacetin, and nitrates, can cause hemolysis and thus be harmful in those with the enzyme deficiency.

5. Like viruses, the malaria parasite replicates inside cells and is thus somewhat sheltered from the host immune system. In addition, *P. falciparum*–infected erythrocytes adhere to the walls of small blood vessels, thus sequestering themselves well away from the spleen, one of the principal components of the host immune system. Another factor is that the parasite's surface antigens may be varied. The major variant surface antigen in P. *falciparum* is *Plasmodium falciparum* erythrocyte membrane protein-1 (PfEMP1), which is encoded by a family of 40–50 genes. Transcriptional switching among these allows for variation in the surface antigen expressed and evasion of the immune response.

6. Animals possess only the mevalonate pathway and not the DOXP pathway, whereas *Plasmodia* possess only the DOXP pathway. This explains why inhibition of DOXP reductoisomerase blocks replication of the parasite without harming the host. Herein lies another example of how differences in biosynthetic pathways between host and pathogen may be exploited in treating disease. Evidence of the presence of DOXP pathway enzymes and the lack of the mevalonate pathway enzymes was revealed by the malaria genome project, thus exemplifying how acquisition of genome sequence data can accelerate the pace of identification of therapeutic targets and the development of pharmaceuticals.

CHAPTER 26

1. Steroid hormones, including testosterone, are able to simply diffuse across the cell membrane thanks to their small size and hydrophobicity. Within the cell, they bind to receptor proteins. These receptors alter the levels of gene expression through binding to regulatory sequences in DNA. (See Section 31.3 of *Biochemistry* 5e.) This mode of signaling through a cellular receptor is in contrast to the majority of signaling pathways, which involve hydrophilic ligands. In these cases the ligand binds the extracellular portion of a transmembrane receptor, inducing a conformational change in the intracellular portion which mediates the cellular response.

2. Dexamethasone is a synthetic glucocorticoid (Figure 26.2), which inhibits secretion of ACTH from the anterior pituitary. ACTH acts upon the adrenal glands to stimulate steroidogenesis via activation of membrane-bound adenylate cyclase, thus increasing the cyclic AMP (cAMP) level. The increased cAMP stimulates protein kinases, which in turn stimulate enzymes in the steroid biosynthetic pathways. The mineralocorticoids and glucocorticoids, end products of these pathways, normally act to suppress pituitary ACTH secretion as part of a negative feedback loop. Suppression of testosterone levels into the normal range by dexamethasone suggests an adrenal source, while incomplete suppression suggests an ovarian source. However, *autonomous* secretion of androgens, meaning secretion that is no longer under the control of ACTH (such as in a neoplasm), will not be suppressed by dexamethasone, even if it is of adrenal origin.

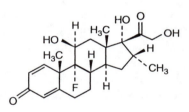

FIGURE 26.2

3. Hirsutism refers to the growth of hair in androgen-dependent areas of the body, areas where boys typically begin to grow hair at puberty, while hypertrichosis refers to excessive androgen-independent growth of hair. Distinguishing between the two may be accomplished by physical examination. Hair growth in hypertrichosis is not predominantly in androgen-dependent areas, but is found diffusely over the trunk, hands and face and is usually the vellus, rather than terminal, type of hair. Hypertrichosis may be caused by a genetic trait (familial hypertrichosis) or may be induced by medications or metabolic disorders. Distinction between hirsutism and hypertrichosis is consequential because while a variety of treatments are available to those with hirsutism, mechanical hair removal (or discontinuation of the offending drug) is the only means of treatment for hypertrichosis.

4. Finasteride (Figure 26.3) is a competitive inhibitor of type II 5α-reductase, the enzyme that converts testosterone into the more active metabolite, DHT. Although type I 5α-reductase predominates in the scalp, evidence suggests that type II is the isozyme involved in androgenic alopecia. The drug has been shown to reduce the level of DHT at the top of the scalp, and to stimulate hair growth there. This may be due to local reduction in the DHT level or, alternatively, to reduced serum DHT levels, which occur upon administration of the drug. The scalp is highly vascularized, and thus may be affected by a systemic reduction in DHT level.

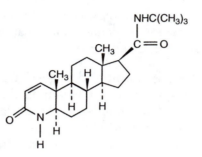

FIGURE 26.3 Chemical structure of finasteride.

5. 17,20-lyase and 17α-hydroxylase make up a bifunctional enzyme encoded by the P450-C17 gene, and thus deletion of the gene would result in accumulation of pregnenolone and progesterone.

6. StAR is a transport protein, which carries cholesterol from the outer through the inner membrane of the mitochondria, the rate-limiting step in steroid biosynthesis. StAR is essential to steroidogenesis. Individuals with mutations in the StAR gene have *lipoid congenital adrenal hyperplasia,* a condition characterized by dramatically impaired steroidogenesis and a lack of virilization in males. The mechanism of cholesterol transport remains to be elucidated; however the crystal structure of a StAR homolog suggests the existence of a hydrophobic tunnel in the structure, indicating that StAR may serve as a molecular shuttle.

CHAPTER 27

1. Symptoms do not usually appear until the fourth or fifth decade of life, after many victims of the disease have already had children. However, more recently, with the advent of unambiguous diagnosis brought about by DNA testing, the transmission of the disease may be curtailed and the incidence may begin to diminish. A case in point, the son of the patient described here was evaluated by DNA testing subsequent to his father's diagnosis, before he had had children. It is thus likely that transmission to one's grandchildren may soon be curtailed, if not to one's children.

2. Postzygotic changes in the number of repeats would account for variability in disease severity between monozygotic twins, and contraction of repeats would account for the disappearance of disease from a branch of an affected family. Although there is a strong bias toward repeat expansion, contraction does occur occasionally. Finally, a sporadic case could arise as a result of a mutation that increases the number of repeats to over the threshold level.

3. It indicates that the *HD* mutation is a toxic gain-of-function mutation, and does not result from loss-of-function of the huntingtin protein. If the *HD* mutation were a loss-of-function mutation, it would be expected to have the same or similar phenotypic effects as the deletion. The dominant inheritance pattern of the disease is further evidence for a toxic gain-of-function mechanism of pathogenesis.

4. No. Although in many cases mutations have deleterious effects, mutations can also have beneficial effects that provide a survival advantage. For example, the mutation that causes sickle-cell anemia provides a survival advantage for malaria. Mutation is what allows us to evolve in response to ever-changing environmental challenges. A perfect DNA replication apparatus would thus likely be selected *against,* as it would be an impediment to variation, a critical feature of Darwin's theory of natural selection.

5. The first patient, with 35 repeats, has the greater number of repeats; however, there are no other CAG repeats in close proximity to the stretch of 35 repeats. The second patient, although she possesses fewer repeats (33), has a stretch of 8 repeats separated from the first 33 by a single codon. Neither patient is likely to develop the disease as they are both under the threshold number of 36 repeats; however, patient #2 probably has a higher risk, with the risk being magnified in her children and grandchildren. In patient #2, a single point mutation (CTG→CAG), or strand slippage through the CTG codon, will result in an expanded tract of CAG repeats within the range associated with the development of the disease.

6. It is possible that the patient does not have HD. There are other causes of chorea and dementia; however, most of these can be distinguished from HD by their clinical features and laboratory tests or imaging studies. Another possibility is that the patient is a mosaic, meaning that some cells possess the mutation and others do not, and the cells tested by the laboratory (usually blood cells) were of the subpopulation that did not possess the mutation. A mosaic comes about from a mutation that occurs after fertilization has taken place.

CHAPTER 28

1. Splicing. The site of transcriptional initiation may vary by a few bases without significantly affecting the final gene product because all mRNAs possess 5' untranslated regions (5' UTRs) that are not part of the open reading frame (ORF). In fact, in many cases transcriptional initiation is not precise, which results in slight variations in the 5' end of the mRNA for a given gene. Similarly, the precise site of transcriptional termination is not crucial because the 3' end of the mRNA is ultimately cleaved prior to addition of the poly(A) tail. Precision in splicing, however, is critical. Should the site of cleavage differ by as little as one base, the reading frame will shift, resulting in a totally different amino acid sequence.

2. First, in many (~30%) cases the disease-causing mutation has arisen sporadically, in patients with no family history of disease. Like the inherited mutations, sporadic mutations tend to be private, that is, different mutations occur in each family or affected individual, and a vast number (over 130) of different mutations has been described. In addition, despite the already large number of known mutations, it is likely that still more remain to be discovered.

3. Mutant fibrillin-1 may be sufficiently similar to the wild type protein to be incorporated into microfibrils along with normal versions of the protein; however, this may create a macromolecular structure that is defective. Microfibrils with mutant fibrillin-1 may be structurally normal but in some way targeted for degradation by the mutation, resulting in destruction of both normal and mutant protein. In addition, this may elicit an inflammatory response, as is often observed in the media of blood vessels from Marfan's patients. Mutant fibrillin-1 may be produced by an insertion or deletion mutation that produces a frameshift and premature termination of the protein. If the truncated protein retains protein–protein interaction domains, it may bind to and sequester normal fibrillin-1 from nascent microfibrils.

4. This allows for further diversity of protein structure and function without increasing the number of genes. In addition, it provides an additional step on the path from gene to protein, which may be subject to regulation.

5. Donor splice site mutations will affect the binding of U1 and U5 snRNP. Acceptor splice site mutations will alter binding of U1snRNP. Mutations within the intron at the branch site will interfere with the binding of U2 snRNP (see Chapter 28 of *Biochemistry* 5e for details).

6. The mutation was in the splice 5' site (the donor site), creating a sequence that differed from the consensus. As a result, a cryptic site upstream of the normal site was used, which resulted in deletion of a portion of the exon (amino acids 1568 to 1582). The deletion resulted in the loss of 15 full codons and two partial codons (for glycine and serine, codons 1567 and 1583, respectively), with the new junction generating a codon for alanine. Thus a total of 16 amino acids were lost from the polypeptide, and one substitution was made.

CHAPTER 29

1. Puromycin is an aminonucleoside antibiotic that binds the A site of the ribosome, thus preventing the binding of the incoming aminoacyl-tRNA. In addition, it has an α-amino group that can form a peptide bond with the carboxyl group of the growing polypeptide chain, thus prematurely terminating chain synthesis. Puromycin binds prokaryotic and eukaryotic ribosomes with approximately equal affinity, which accounts for its toxicity in mammals. Studies in animals have shown that puromycin particularly affects the kidney, causing glomerular epithelial cell injury and impairment of renal function.

2. A bacteriostatic antibiotic prevents bacteria from dividing, whereas a bacteriocidal antibiotic can actually kill the organism. All things being equal, bacteriocidal antibiotics are preferable to bacteriostatic drugs; however, particular strain susceptibilities, drug resistance, and side effects also govern the choice of antibiotic. In many cases, bacteriostatic antibiotics are just as effective—halting bacterial reproduction until the body's own immune defenses can eradicate the offending organism. Most antibiotics that affect protein synthesis are bacteriostatic (except the aminoglycosides).

3. Although L4 and L22 are not close enough to interact with erythromycin, they are nevertheless found fairly close to the drug binding-site (within 10 angstroms). The most straightforward explanation for the genetic results would be that these proteins stabilize the structure of 23S rRNA in the region bound by the drug. Thus mutations in either one would *indirectly* result in resistance by perturbing the structure of this region.

4. AOM is more common in children than adults for several reasons. First, the immune system of young children is underdeveloped, which renders them prone to the upper respiratory tract infections that often precede AOM. In addition, the anatomy of the eustachian tube differs slightly in children: it is shorter and straighter than in adults, thus facilitating infection with organisms found in the nasopharynx. Finally, the adenoids, which are collections of lymphatic tissue found at the back of the nose, are particularly large in children and may obstruct the eustachian tubes, thus setting the stage for infection.

5. The mechanism of toxicity is likely to involve aminoglycoside binding to mitochondrial 12S rRNA and inhibition of translation. Indeed, in vitro experiments have shown that aminoglycoside binding affinity for the ribosomes with the polymorphism associated with susceptibility (G at position 1555) is high, compared to binding to the wild-type ribosomes. Diseases transmitted via mitochondrial DNA mutations often affect hearing, indicating a particular dependence of the hearing process on mitochondrial function, perhaps through an effect on the auditory nerve or cells of the cochlea. The high degree of similarity between mitochondrial and bacterial rRNAs likely reflects the prokaryotic origin of mitochondria and the selective effect on mitochondrial (as opposed to cytoplasmic) ribosomes.

6. Between 30% and 60% of *S. pneumoniae* strains, 55% of *H. influenzae* strains, and virtually 100% of *M. catarrhalis* strains are resistant to penicillin. The most common resistance mechanism is the production of β-lactamase, a serine peptidase that cleaves the

β-lactam ring of penicillin, thereby inactivating the drug. Drug inactivation is a common mechanism used by microorganisms to evade the toxic effects of antibiotics.

7. As opposed to the other antibiotic protein synthesis inhibitors, which target the ribosome, mupirocin targets a tRNA synthetase. Toxicity arises from a shortage of isoleucine-charged tRNA, which is necessary for protein synthesis. Selective binding to prokaryotic isoleucine tRNA synthetase accounts for its selective toxicity toward bacteria.

CHAPTER 30

1. Anorexia means "loss of appetite," but patients with the disorder do not lose their appetites, but instead willfully restrict food intake despite feeling hungry. Furthermore, they often become preoccupied with food, fantasizing about it, collecting recipes, and preparing elaborate meals for others. Perhaps a more appropriate name for the disorder would have been *autophagia,* which means self-consumption, because during prolonged starvation the metabolic consumption of bodily tissues occurs to maintain the survival of the most essential functions of the organism, in particular, the brain.

2. Early in the course of a fast proteins are utilized to form glucose (via gluconeogenesis), upon which the brain depends for fuel. However, this is detrimental as proteins carry out many important physiological functions, and the body does not store extra quantities to use as a potential source of energy. Thus, after a few days, the brain becomes adapted to the use of ketone bodies produced from fatty acid metabolism. The shift from glucose to ketone bodies as a primary source of energy for the brain spares the proteins at the expense of triacylglyerols, the dedicated energy storage molecules of the body.

3. Once glycogen stores have been depleted, triacylglycerols stored in adipocytes are mobilized as a source of fuel. In the first few days of fasting the glycerol backbone of triacylglycerols may be converted to dihydroxyacetone phosphate (see Figure 16.24 of *Biochemistry* 5e) and enter the gluconeogenic pathway to form glucose. However, most of the glucose is derived from proteins in the early days of a fast. After more prolonged fasting, fatty acid moieties of triacylglycerols are converted into the ketone bodies, which become a major source of fuel for the brain.

4. During fasting bodily phosphate stores become depleted; however, the serum phosphate levels are often in the normal range (although they were low in this patient) because phosphate is primarily stored inside cells, rather than in extracellular fluids. β-Oxidation of fatty acids, the primary source of energy during prolonged fasting, does not require a source of phosphate and thus it proceeds despite low phosphate levels. However, upon refeeding with glucose-based nutritional supplements, a source of phosphate is needed for metabolism via glycolysis, which utilizes ATP in the early steps. Phosphate is rapidly transferred into the cell (this is stimulated by insulin), and may lead to a precipitate drop in serum phosphate levels, or *hypophosphatemia.* Phosphate is needed for the proper functioning of cells, as many processes are regulated by reversible phosphorylation. In addition, phosphate, in the form of bisphosphoglycerate, is important in mediating the release of oxygen from hemoglobin (see Section 10.2.3 of *Biochemistry* 5e), and thus for ATP synthesis via oxidative phosphorylation. Hypophosphatemia can cause hemolytic anemia and muscle weakness, as well as life-threatening cardiopulmonary and neurological complications.

5. Glucagon stimulates glycogenolysis (by activating phosphorylase) and gluconeogenesis (by activating fructose bisphosphatase 2) in the liver, and lipolysis in adipose tissue (by activating lipases).

6. No. Although there is a bingeing and purging subtype of anorexia nervosa, the weight must be markedly decreased (<85% normal) to meet the diagnostic criteria for anorexia

nervosa. This patient has a BMI of 21.7, which is well within the normal range (roughly 18.5–25). This patient more likely has bulimia nervosa, a related illness characterized by bingeing/purging behavior but normal body weight in most cases.

7. Muscle cells do not contain the enzyme glucose-6-phosphatase, which converts glucose-6-phosphate (from glycogen catabolism) into glucose. Glucose-6-phosphate, unlike glucose, cannot traverse the cell membrane, and it thus enters the glycolytic pathway inside muscle cells. Thus, the energy stored in muscle glycogen is used exclusively by muscle cells, and may not be mobilized for use by other tissues. This dedicated muscle energy store may serve the organism well in situations that require intense bursts of activity, such as when the "fight or flight" response is elicited.

CHAPTER 31

1. α-amanitin binds RNA polymerase II with high affinity (K_d = 10 nM) and thus treatment of tissue culture cells with 100 nM α-amanitin would be sufficient to inhibit the enzyme. As RNA polymerase II synthesizes mRNA, inhibition of mRNA synthesis would be an expected outcome. RNA polymerase I, which synthesizes 18S, 5.8S, and 28S rRNA, is insensitive to α-amanitin, and RNA polymerase III, which synthesizes tRNA and 5S rRNA, is only sensitive to concentrations on the order of 1μM, and thus the synthesis of these RNAs would be unaffected. Conversely, had the cells been treated with 10μM α-amanitin, inhibition of both RNA polymerase II and RNA polymerase III would have occurred, and a reduction in mRNA, tRNA, and 5S rRNA would have been observed.

2. Although RNA polymerase I is insensitive to the amanitins, it is indirectly affected by their effects. The amanitins inhibit RNA polymerase II, which ultimately affects the synthesis of proteins, a number of which are necessary for RNA polymerase I activity. Thus, a shortage of RNA polymerase I transcription factors or polymerase components would result in decreased RNA polymerase I-dependent transcription. In addition, at higher concentrations amanitins inhibit RNA polymerase III, which would also result in the inhibition of protein synthesis due to a shortage of tRNAs and 5S rRNA.

3. All the coagulation factors except for factor VIII are produced in hepatocytes, and they have fairly short half-lives (6 hours–5 days), and thus liver dysfunction together with impaired protein synthesis inevitably leads to coagulopathy. The serum prothrombin time, which collectively measures the activities of factors II, V, VII, and X, is used as a measure of the integrity of part (intrinsic pathway) of the blood coagulation system. The prothrombin time of 24 seconds measured in this patient indicated severe hepatic dysfunction, as an elevation of five seconds or more is considered dangerous.

4. The liver plays an important role in regulating blood-glucose levels. It is a major bodily store of glycogen and is highly sensitive to the effects of insulin and glucagon, which trigger glycogen synthesis and breakdown, respectively. Thus, *Amanita* poisoning, with the acute hepatitis it may cause, may bring on the rapid onset of hypoglycemia and lead to hypoglycemic coma if left untreated.

5. No. Although the amanitins bind the active site of RNA polymerase II, they do not interfere with binding of ribonucleotides. Rather, they bind to the "bridge helix," inhibiting its movement and thereby inhibiting the translocation of the polymerase along the DNA template.

6. Many cellular signal transduction pathways ultimately exert their effects by modulating RNA transcription. The CTD is a target of a number of these pathways, largely via kinases that phosphorylate residues within it, thus modulating the activity of the polymerase. A *hypophosphorylated* CTD is associated with RNA polymerase II in the preinitiation complex that is formed at the promoter, whereas *hyperphosphorylated* CTDs are associated with elongating RNA polymerase II holoenzymes. The CTD has been found to be involved in a number of processes involved in gene transcription, such as promoter clearance, elongation, and pre-mRNA processing. Rather than directly altering the activity of the polymerase at the active site like the amanitins, the various phosphorylated states of the CTD are thought to exert their effects largely through the recruitment and

activation of other factors. For example, the mRNA capping enzyme has been shown to specifically interact with a phosphorylated form of the CTD.

7. No. Cellular transcriptional repressors usually bind cis-acting elements (i.e., DNA sequences) within genes and impede the binding or activity of RNA polymerases. In many cases (in eukaryotes) repressors act by altering the chromatin structure of the gene, often by recruiting histone deacetylases. Deacetylated histones are associated with reduced accessibility to the transcriptional machinery and with transcriptionally repressed genes. Alternatively, repressor binding to gene regulatory regions can directly prevent activator or polymerase binding, or obstruct the passage of the polymerase, a strategy commonly observed in prokaryotes (e.g., the *lac* repressor, see Section 31.1.3 of *Biochemistry* 5e). Thus, the mechanisms employed by cellular transcriptional repressors allow for significant flexibility in gene regulation. The complex interplay between transcriptional regulatory proteins can result in dramatically different expression patterns among cellular genes depending on environmental and developmental stimuli. In stark contrast, inhibition by the amanitins results in a complete block in RNA polymerase II transcription, which even if temporary would be deleterious, if not lethal, to the cell.

8. Rifampicin. Rifampicin binds to the bacterial RNA polymerase within the channel through which the DNA–RNA hybrid must pass, thus blocking it. α-amanitin acts via a similar mechanism by binding near the active site and preventing translocation. In contrast, actinomycin D binds to the DNA template, intercalating between neighboring base pairs, and preventing its use as a template for RNA synthesis.

CHAPTER 32

1. Myeloma cells grown in culture secrete a monoclonal antibody into the growth medium, however the antibody is rarely of any use for either medical research or clinical treatment. To generate a cell line that produces an antibody with a given, predetermined specificity, myeloma cells are fused with normal plasma cells from a mouse which has been immunized with the desired antigen. This procedure is called *somatic cell hybridization* and generates the *hybridomas* that produce the monoclonal antibody specific to the antigen of choice. Once fusion has occurred the hybridomas will randomly lose chromosomes, with the result that only a subset will produce an antibody to the chosen antigen (remember that the myeloma cell also produces an antibody, but not with the desired specificity). The hybridomas must thus be cloned and screened for production of an antibody with specificity to the preselected antigen (see Section 4.3.2 of *Biochemistry 5e*).

2. In order for a substance to enter the urine, it must pass through the glomerular capillary into the urinary space. This *filtration barrier* excludes most plasma proteins, excluding proteins any larger than approximately 70 kd. The entire immunoglobulin molecule (two heavy and two light chains) has a molecular weight of approximately 150 Kd, while Bence-Jones proteins weigh ~25 kd. The latter can thus traverse the barrier and collect in the urine.

3. Both Aδ and C neurons are small primary afferent neurons, however Aδ fibers are myelinated and have a more rapid speed of transmission (5–35 m/s). The C neurons are smaller (with diameters of 0.2–1.5 μm compared to 1–5 μm for Aδ neurons) and unmyelinated, with conduction velocities of 0.5–2 m/s. C neurons transmit pain from deep within the tissues and organs, which is often felt as a throbbing or burning pain, while the Aδ fibers transmit sharp, painful stimuli.

4. Renal prostaglandins play a role in maintaining renal blood flow under certain pathological conditions, including cirrhosis, congestive heart failure, and renal dysfunction. By acting as vasodilators they increase renal blood flow and glomerular filtration rate. NSAIDs, in inhibiting prostaglandin synthesis, would reduce renal blood flow, which may lead to renal failure.

5. CIPA is a developmental disorder characterized by a complete lack of small myelinated and unmyelinated nerve fibers, which sense pain and temperature. Although the sweat glands in the skin are normal, they are not enervated, and children affected with the disorder do not perspire and are thus extremely susceptible to febrile illnesses, often succumbing in early childhood. Most of those affected are mentally retarded, and self-mutilation is common. The disease is inherited in an autosomal recessive manner, and has recently been linked to the NTRK1 gene, which encodes the tyrosine kinase receptor for *nerve growth factor (NGF)*. This finding suggests an important role for NGF and NRTK1 in the development of nociceptive neurons as well as the establishment of thermoregulation via sweating.

6. Codeine does not act directly on opioid receptors but is converted to its active form, morphine, by the hepatic enzyme, cytochrome p450 2D6. This processing step accounts for the reduced activity of codeine. Interestingly, approximately 8% of Caucasians, 6% of African Americans, and 1% of Asians lack the enzyme and are insensitive to codeine.

Acetylation of the 3- and 6- hydroxyl groups of morphine yields 3,6 diacetylmorphine (heroin), with the increased potency being largely due to increased hydrophobicity, with more efficient and rapid penetration of the CNS. This accounts for the intense "rush" produced by heroin, which makes it extremely addictive.

7. The early NSAIDs were nonspecific cyclooxygenase inhibitors, meaning that they inhibited both isoforms of the enzyme, COX-1 and COX-2. COX-2 levels are low in most tissues, but its synthesis is stimulated in activated macrophages and in other cells at sites of inflammation, while COX-1 is expressed constitutively in many tissues and plays a role in a number of normal physiological processes. COX-2 is thus the primary target for pain relief, but the concurrent inhibition of COX-1 causes most of the side-effects associated with the use of NSAIDs, most notably gastrointestinal disturbances (COX-1 plays a role in maintaining the gastroduodenal mucosa). The newer NSAIDs, such as celecoxib (Celebrex) and rofecoxib (Vioxx), specifically bind and inhibit the activity of COX-2, with little effect on COX-1, thus providing pain-relief without the troubling gastrointestinal disturbances associated with the older drugs.

8. Naloxone (Narcan; see Figure 32.6) is a competitive antagonist of the opioid receptors, with especially high affinity for the μ subtype, through which the opioid drugs exert most of their effects. It thus competes with narcotics such as morphine and heroin for binding to receptors, and reverses their effects. Its close relative, naltrexone (Trexan), has a longer duration of action and is used to treat recovering addicts.

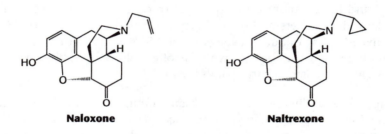

Naloxone **Naltrexone**

FIGURE 32.6 Structures of naloxone and naltrexone.

9. The gate control theory was devised in 1965 by Patrick Wall and Ronald Melzack to explain how nonpainful stimuli can reduce the sensation of pain, for example as in the easing of back pain by massage. They proposed a gating mechanism in the spinal cord, where nociceptors synapse with spinal cord neurons that transmit the signal to the brain, where pain is perceived (Figure 32.7). In the absence of any other stimulation, the "gate" (i.e., the synapse) would be open and pain would be felt. However, the sensation of touch (such as a massage) could ease pain via the action of sensory neurons that sense touch, which could close the "gate" (perhaps through the release of inhibitory neurotransmitters), such that the pain signal would be blocked and go no further than the spinal cord. In addition, signals originating in the brain are also thought to affect the transmission of pain signals. This would explain the effects of meditation, placebos, distraction, and other psychological factors in reducing the sensation of pain.

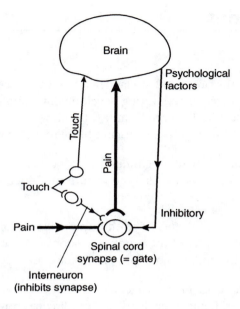

FIGURE 32.7 Model of gate control theory.

CHAPTER 33 SOLUTIONS

1. Apoptosis is critical to the process of clonal deletion—the process whereby immune cells reactive to self-antigens are deleted during development—and thus features prominently in self-tolerance. A defect in this process may thus allow the survival of immune cells reactive to self-antigens, thus increasing susceptibility to autoimmune diseases.

2. DNA is negatively charged, while arginine bears a positive charge at physiological pH. Thus, electrostatic interactions are likely to feature prominently in the binding of these autoantibodies to DNA.

3. No. The fact that autoantibodies exist to multiple epitopes of a complex suggests that the complex itself is the target of the immune response. A pathogen would be unlikely to possess a complex so similar to one present in the host as to share multiple epitopes.

4. Affinity maturation occurs as a result of somatic mutations that increase the affinity of an antibody for the antigen. Remember that B cell binding to antigen triggers a signal transduction pathway that leads to the proliferation of that particular B cell clone. As proliferation proceeds certain somatic mutations will result in increased affinity for the antigen, dsDNA in this case. These clones encoding higher-affinity antibodies will then begin to take over the population as they will more effectively compete for the antigen, and hence further stimulate their own proliferation. Affinity maturation is yet another example of how the principles of natural selection may be applied to the immune system.

5. A B cell expressing an antibody with a binding specificity for dsDNA may bind and internalize a fragment of DNA complexed to DNA-binding proteins such as histones. Histone antigens may then be presented by MHC class II molecules and recognized by helper T cells, which would ultimately result in the expansion of the B cell expressing an antibody with dsDNA binding specificity. Alternatively, peptides from the dsDNA autoantibody molecule *itself* could be presented by the B cell MHC class II molecules, recognized by helper T cells, ultimately enhancing the proliferation of the B cell clone. Studies in animal models of SLE have provided some evidence for both these possibilities.

6. A *homozygous* deficiency has been associated with SLE, and these deficiencies are rare; thus the disease would not be expected to appear frequently. She was unfortunate enough to inherit a mutation in the same complement component, C1q, from each of her parents and thus developed the disease.

7. SLE has many manifestations, from arthritic, cutaneous, gastrointestinal, ocular, cardiovascular, and obstetric problems to severe kidney and neurological disturbances. In addition, manifestations vary greatly in severity and differ between individuals. It can thus be difficult to diagnose, especially when disease is mild. Differential diagnoses include rheumatoid arthritis; dermatitis; various neurologic and psychiatric disorders; and certain hematologic, infectious, and connective tissue diseases. Even reliance on a positive ANA test may be misleading, as some fraction of patients are negative, especially in mild cases. To assist in diagnosis, the American Rheumatism Association has published a set of criteria: 4 of 11 different diagnostic criteria must be met for unambiguous diagnosis of SLE.

CHAPTER 34

1. Because sarcomeres have a highly regulated macromolecular structure, a reduction in the amount of a constituent protein, or haploinsufficiency, may alter sarcomere function. Mutant sarcomeric proteins may also function as a "poison peptide," producing a dominant negative disruption of sarcomere function. Alternatively, the mutant proteins may be functional but with altered properties that change the behavior of the sarcomere (and in some cases increase contractile performance!). Of note, how any of these putative effects on sarcomere function produce hypertrophy and myocyte disarray is not yet known, although compensatory hypertrophy as occurs in acquired heart disease has been suggested.

2. This portion of the myosin protein mediates actin binding and release, nucleotide binding and release, and lever arm movement associated with the power stroke of muscle contraction. Any or all of these functions could be disrupted by local or more global disruption of the structure of myosin.

3. Although HCM is a disease of the sarcomere, the main problem produced by mutations that cause the disease is disruption of the structure and function of the left ventricle. The absence of a cardiac cell model has impeded the mechanistic understanding of the basis of HCM. Intact animal models are a good alternative, providing the opportunity to study the behavior of the mutant protein in the context of a functioning heart. It has also been hypothesized that the hypertrophy in HCM is a compensatory response, and such responses may not be apparent even if a cellular model existed.

4. There are a number of mechanisms by which patients with HCM might die suddenly. Some are mechanical, such as obstruction of left ventricular output and drop in blood pressure and ischemia (lack of blood flow) due to the thickness of the left ventricle, with resultant mechanical failure. Most sudden deaths in patients with preserved left-ventricular function are likely to be the result of arrhythmias. Hypertrophic heart cells are not only remodeled mechanically but also electrically. The changes in the cellular action potential can predispose to potentially lethal arrhythmias. Changes in intracellular Ca^{2+} handling also influence the cellular electrophysiology of the myocyte and may further increase the risk of an abnormal and potentially lethal heart rhythm. Finally, ischemia not only alters the mechanical properties of the heart cell but also its electrical properties; ischemia even in otherwise normal hearts may be associated with ventricular fibrillation and sudden death.

5. Changes in the levels of intracellular Ca^{2+} may have a number of potentially untoward effects, including altering actin–myosin cross-bridge cycling and contractile performance, activation of signaling pathways that mediate hypertrophy and programmed cell death (apopotosis), and changing the electrophysiology of the cardiac myocyte.

6. The diversity of mutations in an individual gene and the number of genes mutated, the rarity of founder mutations, and the frequency of sporadic mutations all suggest an evolutionarily recent disorder.